Performance Characterization of Lubricants

The text discusses the fundamentals of lubrication science and technology linking the science concepts to engineering practices. It further explores the performance characterization of lubrication systems by utilizing sophisticated experiments and tests and motivates the readers to develop their conclusions and reach solutions based on modern tools and techniques.

This book:

- Presents the principles of surface and lubricant chemistry, and its implementation to devise engineering solutions for various application-based systems.
- Discusses viscosity index improvers, tribology of green lubricants, and biolubricants from non-edible oils.
- Highlights 2D nanomaterial lubricants, biogreases, hydrogel, and lubricants for extreme temperature and pressure conditions.
- Explains lubrication for electrical, biomedical, automobile, marine, turbine, and aerospace applications.
- Covers design considerations, formulations, and compositions of lubricants for high-temperature applications in diverse areas.
- Explores the simulation, computational, and empirical models to characterize, quantify, and mitigate the adverse effects of friction.

It is primarily written for senior undergraduate and graduate students and academic researchers in the fields of mechanical engineering, production engineering, industrial engineering, aerospace engineering, and manufacturing engineering.

Advances in Manufacturing, Design and Computational Intelligence Techniques

Series Editor – Ashwani Kumar – *Senior Lecturer, Mechanical Engineering, at Technical Education Department, Uttar Pradesh, Kanpur, India*

The book series editor is inviting edited, reference, and textbook proposal submissions in the book series. The main objective of this book series is to provide researchers with a platform to present state-of-the-art innovations, research related to advanced material applications, cutting-edge manufacturing techniques, innovative design, and computational intelligence methods used for solving nonlinear problems of engineering. The series includes a comprehensive range of topics and their application in engineering areas such as additive manufacturing, nanomanufacturing, micromachining, biodegradable composites, material synthesis and processing, energy materials, polymers and soft matter, nonlinear dynamics, dynamics of complex systems, MEMS, green and sustainable technologies, vibration control, AI in power station, analog–digital hybrid modulation, advancement in inverter technology, adaptive piezoelectric energy harvesting circuit, contactless energy transfer system, energy-efficient motors, bioinformatics, computer-aided inspection planning, hybrid electrical vehicle, autonomous vehicle, object identification, machine intelligence, deep learning, control–robotics–automation, knowledge-based simulation, biomedical imaging, image processing, and visualization. This book series compiled all aspects of manufacturing, design, and computational intelligence techniques from fundamental principles to current advanced concepts.

Hybrid Metal Additive Manufacturing: Technology and Applications
Edited by Parnika Shrivastava, Anil Dhanola, and Kishor Kumar Gajrani

Thermal Energy Systems: Design, Computational Techniques, and Applications
Edited by Ashwani Kumar, Varun Pratap Singh, Chandan Swaroop Meena, Nitesh Dutt

https://www.routledge.com/Advances-in-Manufacturing-Design-and-Computational-Intelligence-Techniques/book-series/CRCAIMDCIT

Performance Characterization of Lubricants

Edited by
Abhishek Kumar, Avinash Kumar,
and Ashwani Kumar

CRC Press
Taylor & Francis Group
Boca Raton London New York

CRC Press is an imprint of the
Taylor & Francis Group, an **informa** business

Front cover image: Andrey VP/Shutterstock

First edition published 2025
by CRC Press
2385 NW Executive Center Drive, Suite 320, Boca Raton FL 33431

and by CRC Press
4 Park Square, Milton Park, Abingdon, Oxon, OX14 4RN

CRC Press is an imprint of Taylor & Francis Group, LLC

ISBN: 978-1-032-65786-8 (hbk)
ISBN: 978-1-032-87930-7 (pbk)
ISBN: 978-1-003-53544-7 (ebk)

DOI: 10.1201/9781003535447

Typeset in Sabon
by Deanta Global Publishing Services, Chennai, India

Contents

Aim and scope

The proposed book titled *Performance Characterization of Lubricants* summarizes the past and current work in lubrication technology and its application and future scope to the tribological systems, i.e., systems dealing with wear, friction, and lubrication. This book gives a brief overview of the pivotal research work pertaining to chemical, mechanical, materials, biomedical, and surface science and engineering around multidisciplinary research in science and technology. The book has the potential to be valuable to a wide readership of researchers, academicians, professionals, and graduate students in engineering.

The purpose of lubricants is to reduce friction and cool surfaces in contact with one another, protecting them and the threat of damage and wear, and ever since the first bearings were invented, we can be sure that humans tried to make them operate more efficiently using lubricants. The earliest bearing dates back to the Copper Age (4500–3300 BCE) in Mesopotamia, which was used in the form of potters' wheels and wheeled carts. Egyptian murals dating to around 2000 BCE show liquids being poured in the path of sledges carrying statues, likely to ease their transportation, and olive oil-soaked timbers were used to move large stones in the construction of the pyramids. During the Greek and Roman eras, greases that combined lime and olive oil (calcium greases) were used to lubricate axles. With fats, oils, and primitive oily compounds in use throughout the world, the next true evolution in lubrication came from the father of invention, Leonardo da Vinci. Wishing to avoid friction, Da Vinci created self-oiling systems for axle ends and roller-bearing attachments. It is likely that he would have lubricated his inventions with tallow, suet, or either rape or poppyseed oils – all in use in Italy during his time. With the first industrial development of oil fields in the mid-19th century, petroleum-based mineral oils overtook natural oils – the new apex of lubrication technology. Today, lubricants are highly developed using complex chemical methods designed to push their potential to the absolute limit.

This book discusses the fundamentals of "Lubrication Science and Technology" linking the science concepts to engineering practices. It deals with the principles of "Surface and Lubricant Chemistry" and its implementation to devise engineering solutions for various application-based systems. This book explores the performance characterization of lubrication systems by utilizing sophisticated experiments and tests and motivates the readers to develop their conclusions and reach solutions based on modern tools and techniques. This book is not merely limited to the theoretical aspects of tribology and lubrication chemistry but also explores the simulation, computational, and empirical models to characterize, quantify, and mitigate the adverse effects of friction, namely wear, and compare the effectiveness of various lubricants using sophisticated tools and techniques. Thus, it incorporates a wider application-driven approach and summarizes the latest trends in lubricant research.

This book can be used as a textbook on Tribology, Lubricant Science and Technology, Materials Characterization, and Interface Science for an undergraduate engineering course or as a reference for postgraduate-level scholars. This book can also serve as a useful reference for academics; materials and biomedical researchers; materials, mechanical, and biomedical engineers; aerospace engineers; automotive engineers; and professionals in tribology, chemical, and electric vehicle-related industries. Moreover, each chapter in this book will have a literature review, research solution methodology, simulation, experimental setup, and result validation works. The chapters are well organized and easy to follow. The above helps to ensure the completeness of the book and to satisfy the needs of the potential audience in different areas of the world related to manufacturing, design, experimental, simulation, and computational techniques. Having high quality of research content, it will work as a reference book for researchers and scientists working on solving lubrication-related problems in science and engineering.

Editors
Abhishek Kumar, Avinash Kumar, and Ashwani Kumar

Preface

The proposed book titled *Performance Characterization of Lubricants* summarizes the past and current work in the area of lubrication with a special emphasis on its applications to design and manufacture various types of lubricants such as liquid, semi-solid, and solid lubricants; 2D materials; and their potential future scopes. This book discusses the fundamentals of lubrication chemistry and links it to various industrial applications such as electric vehicles, aerospace, and biomedical applications and the lubricant performance characterization attempts to solve the critical issues related to lubricant development by applying interface science and fundamental chemistry for suitable device development to work under extreme environments. It involves the pivotal research work of chemical, mechanical, materials, electrical, and biomedical engineering in multidisciplinary research related to science and technology. The book has the potential to be valuable to a wide readership of researchers, academicians, professionals, and graduate students in engineering.

This book is a valuable resource for anyone interested in lubricants, from students and researchers to engineers and industry professionals. It offers a well-structured journey, starting with a foundational understanding of lubrication chemistry (Chapter 1) before delving into the exciting realm of sustainable green lubricants (Chapter 2) and their environmental impact (Chapter 3). The book explores various types of lubricants, including ecofriendly options (Chapter 4), biolubricants derived from plant sources (Chapter 5), and those incorporating cutting-edge 2D materials (Chapters 6 and 7). Lubricants under electrified conditions (Chapter 8) and biogreases (Chapter 9) offer alternative solutions for specific applications. The innovative potential of hydrogels as lubricants is explored in Chapter 10.

Moving beyond lubricants themselves, the book delves into critical additives that enhance their performance. Chapter 11 examines recent advancements and challenges in viscosity index improvers, essential for maintaining optimal lubrication across varying temperatures. Similarly, Chapters 12 and 13 focus on formulating lubricants for extreme weather conditions and thermal stability, respectively. Understanding the behavior of lubricants under pressure is crucial, and Chapter 14 explores extreme pressure lubricants. Chapters 15–17 showcase how tribological studies can be used to evaluate the performance of specific lubricants and additives in contact.

The applicability of this book covers a wide range of industries such as chemical engineering, mechanical engineering, production sector, design engineering, research & development engineering, materials science and engineering, biomedical engineering, etc. and it will help to audience conducting research in these industries. Moreover, each chapter in this book will have a literature review, research solution methodology, simulation and/or experimental setup, and results validation works. The chapters will be well organized and easy to follow. The above help to ensure the completeness of the book and to satisfy the needs of the potential audience in different areas of the world related to manufacturing, design, and computational techniques. Having high quality of research content, it will work as a reference book for researchers and scientists working toward solving the issues related to lubricant technology, and the failure occurring due to improper lubrication of various interacting systems with special attention to the application-based systems in science and engineering.

Editors
Abhishek Kumar, Avinash Kumar, and Ashwani Kumar

About the editors

Abhishek Kumar is currently working as a Graduate Assistant (PhD scholar) in the J. Mike Walker '66 Department of Mechanical Engineering at Texas A&M University, College Station. He received his MS in Mechanical Engineering (2023) from the University of California Merced, USA, MTech in Design Engineering (2016) from Indian Institute of Technology Delhi, India in 2016, and Bachelor of Engineering degree (honors) in Mechanical Engineering (2014) from Rajiv Gandhi Proudyogiki Vishwavidyalaya, Bhopal. He has worked as Assistant Professor at Jabalpur Engineering College, Jabalpur (2018–2020), Galgotias University, Greater Noida (2017), and Parul University, Vadodara (2016–2017). He was a summer visiting researcher at the Indian Institute of Technology Kanpur (2018). He qualified for GATE-2014 with 99.25 percentile in 2014 and was awarded a GATE fellowship by the Government of India. His MTech thesis was based on the investigation of nano-wear in Mg-based composites in collaboration with the National University of Singapore, Singapore. He has a keen interest in working on the tribological characterization of interacting surfaces. He has experience of more than eight years as Assistant Professor and a Graduate Student Researcher to work with tribological systems, and their surface and wear characterization. He has gained hands-on experience working with sophisticated experimental equipment in the past eight and half years and has utilized his experimental skills to successfully complete more than ten research projects with reputed industrial partners such as Chevron Corporation, SpaceX, Chevron-Phillips, and Lawrence Livermore National Laboratory and academic collaborators such as the National University of Singapore, the Indian Institute of Technology Kanpur, California State University in Chico, Illinois State University in Normal, Illinois. He has utilized his exposure to demanding and challenging projects by publishing four peer-reviewed articles in peer-reviewed journals and presenting his work in five international conferences and publishing articles in two conference proceedings. He has worked as Book Editor and Book Proposal Reviewer for CRC Press, Taylor & Francis USA, and reviewed two book proposals, edited one book, and authored six book chapters. He has enthusiasm to learn about advanced research and he frequently volunteers as a peer reviewer for more than six high-impact journals (*Journal of Magnesium and Alloys* (IF = 17.6), *Lubricants* (IF = 3.584), *Journal of Composites Science* (IF = 3.3), *Tribology Transactions* (IF = 2.056), *Inorganic Chemistry Communications* (IF = 3.8), and *Materials Today: Proceedings* (CS = 3.2), and has been recognized as a Certified Peer Reviewer by Elsevier and MDPI. He has been selected as the Outstanding Reviewer twice by the *Journal of Magnesium and Alloys* (17.6 impact factor) due to his extraordinary contributions to the journal. He is also a member of several professional societies such as The American Society of Mechanical Engineers (ASME), The Society of Tribologists and Lubrication Engineers (STLE), The American Ceramic Society (ACerS), The Association for Iron and Steel Technology (AIST), The Minerals, Metals & Materials Society (TMS), and ASM International (ASM). His research contributions have been acknowledged by STLE and he has been selected as the Technical Editor for *Tribology & Lubrication Technology* (*TLT*) magazine by STLE. He is also the recipient of several awards and honors, including the 2024 Houston Section Scholarship by STLE. He is currently involved in research that involves tribology, mechanical behavior of materials, material characterization, mechanical testing, mechanical properties, microstructures, nanomaterials, material characteristics, and materials and materials processing.

Avinash Kumar is currently working as Assistant Professor in the Department of Mechanical Engineering at the Indian Institute of Information Technology, Design and Manufacturing (IIITD&M) Kancheepuram, Chennai (an institute of national importance under the Ministry of Education, Government of India). He is Visiting Assistant Professor at Stanford University, California, USA. Here, he is working on two projects, i.e., (1) "theoretical and experimental quantification of uncertainty in enzyme kinetics parameter of CRISPR – Cas12 and Cas13" and (2) "desalination of water using novel microfluidic-based mixers in FCDI (Flow-electrode Capacitive Deionization)". His research interests include micro/nanofabrication, laser machining, and surface engineering for microfluidics, bio-fluidics, MEMS/micro/bio-devices, and healthcare. He has published more than 14 international journals, 3 books, 12 chapters, and 11 international conference papers. His Google Scholar ID is eiJdMdoAAAAJ, ORCID ID is 0000-0002-8071-5748, Scopus ID is 56564412000, and Web of Science Researcher ID is P-6124-2018. He is a guest editor in *MDPI: Bioengineering* (international journal) and *JoVE* (*Journal of Visualization Experiments*). He is a review editor in *Frontiers in Lab on a Chip Technologies – Medical Diagnostics* (international journal). He is a reviewer in the *ASME Journal of Solar Energy Engineering*, the *Journal of Engineering Applications of Computational Fluid Mechanics*, the *ASME Journal of Heat Transfer*, the *Journal of Physics of Fluids*, and many more. He worked as a Post-Doctoral Fellow at the Indian Institute of

Technology Kanpur until 2019. His post-PhD research experience includes Early Post-Doctoral Fellow in the Department of Mechanical Engineering, Indian Institute of Technology Delhi. He also worked as Assistant Professor (TEQIP Faculty) in a World Bank and MHRD (Government of India) project (NPIU) for a semester (2018) at THDC Institute of Hydropower Engineering and Technology, Tehri (Uttarakhand). He obtained his PhD from the Department of Mechanical Engineering, Indian Institute of Technology Delhi in 2018. He received his Bachelor of Engineering in Mechanical Engineering from Rajiv Gandhi Proudyogiki Vishwavidyalaya, Bhopal, in 2010 and MTech degree in Mechanical Engineering from the Indian Institute of Technology Kanpur, India, in 2012. He was Research Associate at the Bio-MEMS and Micro-fluidics Laboratory, Indian Institute of Technology Kanpur, India, from August 2012 to November 2012. He worked as Research Associate with Prof. G.K. Ananthsuresh and Prof. Ashitava Ghoshal at Robert Bosch Centre for Cyber-Physical Systems, Mechanical Engineering Department, Indian Institute of Science Bangalore for a year (2013). He qualified for Indian Engineering Services in 2010. He qualified GATE-2010 with 98.20% in 2010. His PhD thesis was nominated for the best thesis award by the foreign and Indian examiners. He was selected for the India–Sri Lanka youth exchange program 2017, organized by the Ministry of Youth Affairs and Sports, Government of India. He was awarded the DST-International Travel Fellowship in 2018 to visit Hawaii, USA. He is currently working on the various research projects by CSIR and DST-GOI. He had coordinated and organized many conferences and workshops at IIITD&M Kanchipuram, Chennai; IIT Delhi; IIT Kanpur; and AIIMS Delhi. He has delivered many talks in Webinars, Faculty Development Programs, and workshops in institutes like (a) B.S. Abdur Rahman Crescent Institute of Science and Technology, Vandalur, Chennai; (b) Sathyabama Institute of Science & Technology, Chennai; (c) SRM University Chennai; (d) VIT Vellore; and (e) Indira Gandhi Delhi Technical University for Women, New Delhi. He has examined many MTech and PhD students at SRM University, Chennai. He has guided more than 20 BTech, MTech, and PhD students for their theses.

Ashwani Kumar holds a PhD in Mechanical Engineering, specializing in Mechanical Vibration and Design. Currently serving as Senior Lecturer in Mechanical Engineering (Gazetted Officer Group B) at the Technical Education Department of Uttar Pradesh, India, under the Government of Uttar Pradesh, he assumed this role in December 2013. Prior to this, Dr Kumar worked as Assistant Professor in the Department of Mechanical Engineering at Graphic Era University, Dehradun, India, with an NIRF Ranking of 55, from July 2010 to November 2013. With over 13 years of combined experience in research, academia, and administration, Dr Kumar has undertaken various roles, including Coordinator for AICTE-Extension of Approval, Nodal Officer for PMKVY-TI Scheme (Government of India), Internal Coordinator for the CDTP scheme (Government of Uttar Pradesh), Industry Academia Relation Officer, Assistant Centre Superintendent (ACS) in the Institute Examination Cell, and Zonal Officer for conducting the Joint Entrance Examination (JEE-Diploma), as well as Sector Magistrate for State Assembly Election held in Uttar Pradesh. As an academician and researcher, Dr Kumar serves as the Series Editor for five book series published by CRC Press (Taylor & Francis USA) and Wiley Scrivener Publishing, titled "Advances in Manufacturing, Design and Computational Intelligence Techniques", "Renewable and Sustainable Energy Developments", "Smart Innovations and Technological Advancements in Mechanical and Materials Engineering", "Solar Thermal Energy Systems", and "Computational Intelligence and Biomedical Engineering". Furthermore, he is the Guest Editor of a special issue titled "Sustainable Buildings, Resilient Cities, and Infrastructure Systems" in *Buildings* (ISSN: 2075-5309, I.F. 3.8). Dr Kumar holds the position of Editor-in-Chief for the *International Journal of Materials, Manufacturing and Sustainable Technologies* (*IJMMST*, ISSN: 2583-6625) and serves as the Editor of the *International Journal of Energy Resources Applications* (*IJERA*, ISSN: 2583-6617). Additionally, he acts as a Guest Editor and Editorial Board Member for eight international journals and is a Review Board Member for 20 prestigious international journals indexed in SCI/SCIE/Scopus, including *Applied Acoustics, Measurement*, *JESTEC*, *AJSE*, *SV-JME*, and *LAJSS*. Dr Kumar has a prolific publication record, with more than 100 articles in journals, book chapters, and conferences. He has authored, co-authored, and edited more than 30 books in mechanical, materials, and renewable energy engineering. Holding three patents, he is recognized for his contributions to academia and research, earning the title of Best Teacher for excellence in both fields. Actively engaged in academic pursuits, Dr Kumar has successfully supervised 15 BTech, MTech, and PhD theses and serves as an external doctoral committee member of SRM University, New Delhi. His current research interests encompass AI and ML in mechanical engineering, smart materials and manufacturing techniques, thermal energy storage, building efficiency, renewable energy harvesting, and sustainable mobility. Dr Kumar's Orcid is 0000-0003-4099-935X and the Google Scholar web link is https://scholar.google.com/citations?hl=en&user=KOILpEkAAAAJ.

Contributors

Amit J. Agrawal
Department of Petrochemical Technology, Laxminarayan Innovation Technological University, Nagpur-440033, Maharashtra, India

Pratibha Agrawal
Department of Applied Chemistry, Laxminarayan Innovation Technological University, Nagpur-440033, Maharashtra, India

Shivaji Suryabhan Aher
Department of Automobile Engineering, Shivajirao S. Jondhle Polytechnic Asangaon, Shahapur, Maharashtra, India

Arun Kumar Bambam
Department of Mechanical Engineering, Indian Institute of Information Technology Design and Manufacturing (IIITD&M) Kancheepuram, Chennai-600127, Tamil Nadu, India

Pramod Belkhode
Laxminarayan Innovation Technological University, Nagpur-10, Maharashtra, India

Amit Choudhari
Mechanical Engineering Department, Cleveland State University, Cleveland, OH 44115, USA

Nusrat Chowdhury
University of Illinois Urbana-Champaign, Urbana, IL, USA

Sakib Faisal
South Dakota State University, Brookings, SD, USA

Kishor Kumar Gajrani
Department of Mechanical Engineering, Indian Institute of Information Technology Design and Manufacturing (IIITD&M) Kancheepuram, Chennai-600127, Tamil Nadu, India

Yatika Gori
Department of Mechanical Engineering, Graphic Era Deemed to be University, Dehradun, Uttarakhand, India

Kavitha Sharanappa Gudadur
School of Interdisciplinary Design and Innovation (SIDI), Indian Institute of Information Technology Design and Manufacturing (IIITD&M) Kancheepuram, Chennai-600127, Tamil Nadu, India

Ashish Kumar Gupta
School of Mechanical and Aerospace Engineering, Oklahoma State University, Stillwater, OK 74078, USA

Arpita Johari
ABES Engineering College, Ghaziabad, Uttar Pradesh, India

Vijay Y. Karadbhajne
Department of Oil Technology, Laxminarayan Innovation Technological University, Nagpur-440033, Maharashtra, India

Chandra Kishore
Department of Mechanical Engineering, Graphic Era Deemed to be University, Dehradun, Uttarakhand, India

Abhishek Kumar
Mike Walker '66 Department of Mechanical Engineering, Texas A&M University, College Station, TX 77843-3123, USA

Ashwani Kumar
Technical Education Department, Kanpur, Uttar Pradesh, India

Avinash Kumar
Department of Mechanical Engineering, Indian Institute of Information Technology Design and Manufacturing (IIITD&M) Kancheepuram, Chennai-600127, Tamil Nadu, India

Sandeep Kumar
Department of Mechanical Engineering, Indian Institute of Information Technology, Kancheepuram, Chennai, Tamil Nadu, India

Anjan Mahapatra
Department of Mechanical Engineering, Indian Institute of Information Technology Design and Manufacturing (IIITD&M) Kancheepuram, Chennai-600127, Tamil Nadu, India

Upendra Maurya
Pimpari Chinchwad College of Engineering, Pune, Maharashtra, India

Pravin P. Patil
Department of Mechanical Engineering, NMIMS Deemed to be University Mumbai, Maharashtra, India

Bipasa Bimalendu Patra
Department of Electrical Engineering, Padmashri Dr V B Kolte College of Engineering Malkapur, Maharashtra, India

Neha Singh Raghuvanshi
ABES Engineering College, Ghaziabad, Uttar Pradesh, India

Vinay Saini
University of Calgary, Canada

Mudit Saxena
ABES Engineering College, Ghaziabad, Uttar Pradesh, India

Sandeep Soni
Department of Mechanical Engineering, S.V. National Institute of Technology, Surat-395007, Gujarat, India

Sivasakthivel Thangavel
Global College of Engineering and Technology (GCET), Muscat, Sultanate of Oman

Nikhil Thoppil
Muthoot Institute of Technology and Science, Kochi, Kerala, India

Richa Tiwari
Laxminarayan Innovation Technological University, Nagpur-440033, Maharashtra, India

M.R. Tyagi
Former Head, Tribology Division, Indian Institute of Petroleum, Dehradun, Uttarakhand, India

Prameet Vats
Department of Mechanical Engineering, Indian Institute of Information Technology Design and Manufacturing (IIITD&M) Kancheepuram, Chennai-600127, Tamil Nadu, India

Acknowledgments

We are grateful to CRC Press (Taylor & Francis) and their dedicated editorial team for their crucial role in publishing *Performance Characterization of Lubricants*. Their insightful suggestions, unwavering support, and commitment to excellence significantly enhanced the book's overall quality.

We also extend sincere appreciation to the numerous contributors and reviewers. Their illuminating perspectives enriched every chapter of this comprehensive resource. Their collective expertise undeniably contributed to the book's depth and breadth. *Performance Characterization of Lubricants* begins by providing a solid foundation in the fundamental concepts of lubricants, exploring their composition, properties, and synthesis methods. It then highlights the science of lubrication chemistry, linking it to various industrial applications such as electric vehicles, aerospace, and biomedical fields. Finally, the book explores how lubricant performance characterization can address critical issues related to lubricant development by applying interface science and fundamental chemistry for suitable device development in extreme environments.

Editors
Abhishek Kumar, Avinash Kumar, and Ashwani Kumar

Chapter 1

A comprehensive review on introduction to lubrication chemistry

Amit Choudhari, Abhishek Kumar, Ashish Kumar Gupta, Avinash Kumar, and Ashwani Kumar

1.1 INTRODUCTION

Lubrication is a fundamental process across diverse industries, aiming to reduce friction and wear between surfaces in relative motion by introducing a lubricant, which can be solid, liquid, or gaseous [1]. Beyond merely reducing friction, lubrication also aids in dissipating heat generated by frictional forces, thereby safeguarding the contacting surfaces from damage [2]. Its significance extends across sectors such as automotive, manufacturing, aerospace, and marine, where it plays a multifaceted role in enhancing equipment performance, ensuring energy efficiency, and protecting against corrosion [3]. Additionally, lubrication contributes to noise reduction, which is essential for regulatory compliance and ensuring safety and comfort in various applications [4]. Lubrication ensures product quality, cleanliness, and regulatory compliance with specialized formulations meeting stringent requirements in specialized industries such as food processing [5], pharmaceuticals [6], and electronics manufacturing [4, 7–9]. Overall, lubrication extends beyond friction reduction, impacting energy efficiency, equipment longevity, environmental protection, and product quality assurance, making lubrication chemistry pivotal in optimizing performance, driving technological advancements, and ensuring operational excellence as industries continue to evolve [10].

The study of lubrication chemistry is paramount across diverse industries due to its multifaceted contributions [11, 12]. Delving into the intricacies of lubrication chemistry allows engineers to optimize equipment performance through an informed selection of lubricants and additives tailored to specific applications [12]. This meticulous optimization minimizes friction, wear, and heat generation, enhancing machinery and engines' reliability, efficiency, and longevity. Understanding lubrication chemistry facilitates the reduction of energy consumption by mitigating frictional losses [13, 14]. By carefully selecting lubricants with appropriate viscosity and additives, engineers can enhance energy efficiency, leading to cost savings and a diminished environmental footprint [3, 15]. Practical lubrication chemistry prevents equipment failure by mitigating wear and corrosion on critical

DOI: 10.1201/9781003535447-1

components. Adopting proactive lubrication strategies enables industries to reduce downtime, maintenance costs, and the risk of catastrophic failures, ensuring operational reliability. Additionally, lubrication chemistry contributes to environmental sustainability by promoting biodegradable and environmentally friendly lubricants, which minimize the release of harmful substances into the environment, thereby mitigating pollution and ecological impact [16]. Knowledge of lubrication chemistry is instrumental in meeting regulatory requirements and ensuring compliance with industry standards and specifications [3]. By selecting lubricants that meet or exceed regulatory standards, industries can uphold product safety, quality, and environmental responsibility. Research and innovation in lubrication chemistry drive technological advancements in materials science, nanotechnology, and tribology. Exploring novel lubricant formulations, additives, and lubrication techniques contributes to developing more efficient, durable, and environmentally sustainable technologies [17].

Adetunla et al. [18] explained in detail that innovation in lubrication chemistry drives technological advancements across industries, from materials science to nanotechnology and tribology [19, 20]. Research into novel lubricant formulations such as by Singh et al., additives by Minami et al. [21], and lubrication techniques contribute to developing more efficient, durable, and environmentally sustainable technologies [22]. This interdisciplinary approach underpins various facets of industrial operations, encompassing equipment performance optimization, energy efficiency, and environmental protection [23]. Regulatory compliance and meeting industry standards further emphasize the importance of lubrication chemistry in ensuring operational excellence and sustainability [24].

To provide a comprehensive understanding of lubrication chemistry, this review paper is structured to explore various aspects, from fundamental principles to practical applications and recent advancements. Each section delves into different facets of lubrication chemistry, laying the groundwork for subsequent discussions on lubricant compositions, degradation mechanisms, and analysis of lubrication regimes. Additionally, the review paper examines lubrication chemistry in practice, addressing challenges and future perspectives. By elucidating the significance of lubrication chemistry, this paper aims to contribute to industry knowledge, inspire further research, and drive technological innovation across diverse sectors.

1.2 FUNDAMENTALS OF LUBRICATION SCIENCE

1.2.1 Tribology: The science of friction, wear, and lubrication

Tribology is indeed the science and engineering of interacting surfaces in relative motion, including the study of friction, wear, and lubrication

[25]. It encompasses various disciplines such as mechanical engineering, materials science, surface science, and physics. The field is concerned with understanding the mechanisms of friction and wear to improve mechanical systems' performance, reliability, and longevity [26]. Friction is the resistance to motion when two surfaces are in contact, and wear refers to the gradual removal of material from one or both surfaces due to this interaction. Lubrication involves the introduction of a lubricant between surfaces to reduce friction and wear, thus enhancing efficiency and preventing damage [27]. Tribology has applications in numerous industries, including automotive, aerospace, manufacturing [28], and biomedical engineering [29–34]. Engineers and researchers in tribology develop lubricants, coatings, and materials that can withstand high-stress conditions and minimize energy losses due to friction. Additionally, they study the effects of friction and wear on machine components to improve their design and maintenance strategies [35, 36] and to understand the wear and mechanism of the prosthetics installed in the body [37, 38].

Nanoparticles have garnered significant attention in the field of lubrication due to their unique ability to facilitate various lubrication mechanisms [39]. In prior experimental research conducted by relevant scholars, the lubrication mechanisms of nanoparticles have been categorized into five distinct types: the sliding effect, rolling effect, polishing effect, mending effect, and exfoliation [39, 40]. These mechanisms are illustrated in Figure 1.1 [41]. However, it is important to note that these classifications are primarily speculative and lack theoretical backing at the atomistic scale. Some controversy exists regarding the validity of these mechanisms. For instance, Chou et al. [41] raised doubts about whether diamond nanoparticles functioned as ball bearings (rolling effect) during lubrication. Furthermore, the lubrication mechanisms of nanoparticles can vary depending on the lubrication state, and due to the small scale of nanoparticles and lubricant films,

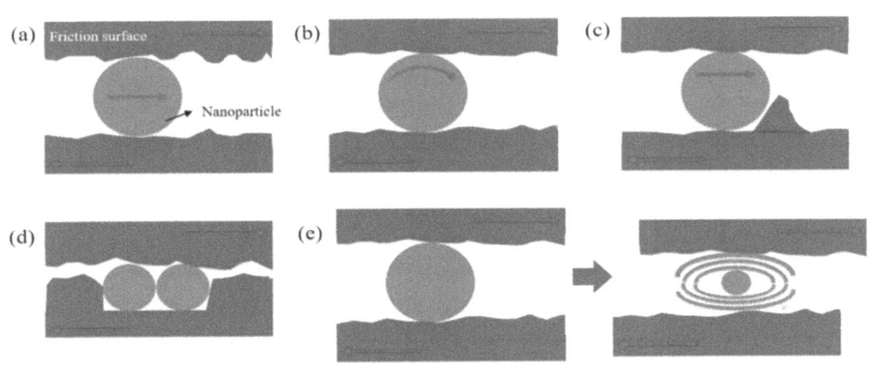

Figure 1.1 Schematic representations illustrating how nanoparticles facilitate various lubrication mechanisms, including (a) sliding, (b) rolling, (c) polishing, (d) mending, and (e) exfoliation [41].

the mechanisms derived from experimental results are often broad and indistinct [42]. The rolling effect involves nanoparticles acting as miniature rolling elements, decreasing friction and wear. Nanoparticles contribute to the polishing effect of smooth surface irregularities, enhancing the surface finish and reducing friction [43]. In the mending effect, nanoparticles repair surface damage and restore surface integrity, prolonging component life. Hence, nanoparticles facilitating exfoliation enhance the separation of contacting surfaces, preventing surface adhesion and reducing friction. These issues are challenging to address experimentally, necessitating the use of additional simulation and calculation methods. Fortunately, molecular dynamics (MD) simulation [44] methods serve as valuable complementary tools to experimental investigations of the tribological behavior of nanofluids. MD simulations offer unique insights into the friction mechanisms of nanolubricants under shear at the nanoscale, helping to fill the gaps left by experimental methods. In boundary and mixed lubrication conditions, where there is typically insufficient liquid lubricant at the contact interface to prevent direct surface contact of the friction pair, numerous MD simulations have been conducted to explore the behavior of nanoparticles in confined systems under dry friction conditions [45]. This exploration is crucial for understanding how the movement of nanoparticles influences their lubrication performance as additives. Understanding how nanoparticles influence lubrication mechanisms provides valuable insights for developing advanced lubrication solutions tailored to specific engineering applications. By harnessing the capabilities of nanoparticles, engineers and researchers can optimize friction reduction, wear protection, and overall system performance [46].

1.2.2 Surface interactions in lubrication

Table 1.1 outlines the various surface interactions in lubrication and their respective influences on friction and wear, providing insights into lubrication mechanisms. Understanding these surface interactions is essential for designing effective lubrication strategies to minimize friction and wear in mechanical systems. By considering factors such as surface roughness, material properties, lubricant characteristics, and environmental conditions, engineers can optimize lubrication effectiveness and prolong the lifespan of components [47, 48]. Solid–solid interactions involve direct contact between two solid surfaces, leading to increased friction and wear [49]. Surface roughness is critical, as rougher surfaces tend to interlock more, increasing adhesion and friction. Additionally, material properties are significant, with harder materials causing more wear on softer ones [50]. Furthermore, the normal load applied determines the contact area and, consequently, the friction experienced. Solid–liquid interactions occur between the lubricant and solid surfaces, resulting in reduced friction and wear [51]. Lubricant molecules adsorb onto solid surfaces, forming a protective film

Table 1.1 The surface interaction in lubricants and their influence on friction and wear

Interaction	*Description*	*Lubrication effect*	*Additional information*
Solid–solid interactions	Direct contact between two solid surfaces	Increases friction and wear	Roughness: Rougher surfaces have higher interlocking and adhesion, leading to more friction. Material properties: Harder materials tend to cause more wear on softer materials. Normal load: Higher load increases contact area and friction.
Solid–liquid interactions	Interactions between lubricant and solid surfaces	Reduces friction and wear	Adsorption: Lubricant molecules form a protective film on the solid surfaces, separating them. Shear strength: The lubricant's ability to resist shearing force during movement. Wetting: The lubricant's affinity for the solid surfaces, thus affecting film uniformity.
Solid–gas interactions	Interactions between gas and solid surfaces	Can have minimal or negative lubrication effects	Air entrainment: Trapped air between surfaces reduces lubrication effectiveness. Environmental factors: Humidity and temperature can affect both lubricant and surface properties (e.g., hotter environments might require higher viscosity lubricants).

that separates them, thus minimizing friction. The shear strength of the lubricant is crucial in resisting shearing forces during movement, ensuring effective lubrication [52]. Wetting, or the lubricant's affinity for solid surfaces, also influences film uniformity and, subsequently, friction reduction. Solid–gas interactions involve interactions between gas and solid surfaces, which can have minimal or negative lubrication effects [53]. Air entrainment, where trapped air between surfaces reduces lubricant effectiveness, is a common issue. Environmental factors such as humidity and temperature can also impact both the lubricant and surface properties [54]. For instance, hotter environments may necessitate higher viscosity lubricants to maintain optimal lubrication performance.

Figure 1.2 illustrates the lubrication mechanism of an oil-in-water (O/W)-based lubricant containing TiO_2 nanoparticles [55]. The depiction showcases how the TiO_2 nanoparticles interact with the surfaces of contacting asperities in a tribological system. The lubrication mechanism involves the reduction of friction and wear between two contacting surfaces by introducing a lubricant between them. This lubricant forms a thin film or layer that separates the surfaces, preventing direct metal-to-metal contact [55,

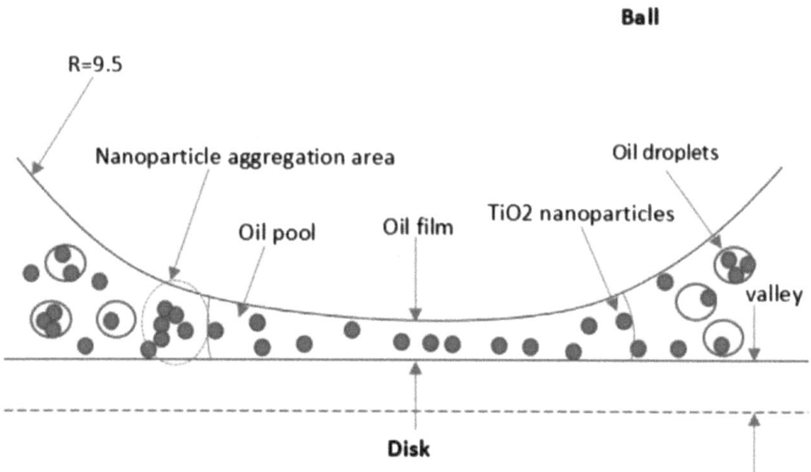

Figure 1.2 The mechanism by which the lubricant, composed of oil-in-water and containing TiO_2 nanoparticles, operates [55].

56]. The lubricant film reduces friction by allowing the surfaces to slide or roll more smoothly against each other. Additionally, it provides a protective barrier, minimizing wear by absorbing and dispersing forces and reducing surface damage. Hence, the lubrication mechanism aims to improve the efficiency, longevity, and reliability of mechanical systems by minimizing friction and wear between moving components [27, 47].

These nanoparticles, synthesized through various methods, such as vaporization and precipitation, are known to improve the load-carrying capacity of the lubricant. Upon incorporation into the lubricant, TiO_2 nanoparticles form a uniform tribochemical film on the surface of asperities [55]. This film reduces friction and wear by providing a protective layer between the contacting surfaces, thereby mitigating direct metal-to-metal contact and minimizing surface damage. Studies have indicated that even small volume fractions of TiO_2 additives in lubricant oil can significantly reduce the coefficient of friction (COF), highlighting the effectiveness of TiO_2 nanoparticles in enhancing lubrication performance [55, 57, 58]. The surface interaction of the lubricants in the vacuum condition varies depending on the vacuum inside the vessels or equipments. Hence, the design of the vacuum [59, 60] is very important from the point of view of selecting lubricants.

1.2.3 Types of lubricants

Table 1.2 shows a comprehensive breakdown of various types of lubricants, accompanied by illustrative examples, descriptive summaries, and their

Table 1.2 The classification of lubricants

Type of lubricant	*Examples*	*Description*	*Applications*	*Additional information*
A. Solid lubricants	Graphite, molybdenum disulfide (MoS_2), polytetrafluoroethylene (PTFE)	Solid particles with low shear strength that form a lubricating film	High-temperature applications (e.g., brakes), extreme pressure environments (e.g., gearboxes), and low-maintenance sliding surfaces	It can be applied as dry films, sticks, or powders. It may not be suitable for high-speed applications due to limited film formation.
B. Liquid lubricants	Mineral oils, synthetic oils, vegetable oils	Low-viscosity fluids that reduce friction between surfaces	General-purpose lubrication (e.g., engine oil, hydraulic fluids). Automotive engines and transmissions Industrial machinery. Food processing equipment (vegetable oils)	Mineral oils are the most common type but can break down at high temperatures. Synthetic oils offer better performance at high and low temperatures and resist oxidation. Vegetable oils are becoming more popular due to their biodegradability.
C. Semi-solid lubricants (greases)	Lithium grease, calcium grease, silicone grease	Thickened liquid lubricants that stay in place and provide long-lasting lubrication	Automotive chassis lubrication (e.g., wheel bearings) Industrial machinery lubrication (e.g., gears, bearings) Food processing equipment (using food-grade greases)	Greases consist of a base oil (liquid lubricant) thickened with soap or clay. The type of soap used determines the grease's properties, such as its water resistance and high-temperature performance. Silicone greases are suitable for high and low temperatures and offer good resistance to water washout.
D. Gaseous lubricants	Air, nitrogen	Gases are used to create a thin film that reduces friction	High-speed machinery (e.g., air bearings) Precision equipment (e.g., hard disk drives) Pneumatic systems (e.g., air tools)	Gaseous lubricants are clean and non-contaminating. They are unsuitable for high-load applications as they cannot provide a strong enough lubricating film. Often used in conjunction with other types of lubricants, such as solid lubricants for bearings in air turbines.

respective applications across diverse industries. Solid lubricants, such as graphite and MoS_2 [61, 62], represent a critical category known for directly applying onto surfaces. These substances find extensive use in high-temperature environments and under extreme pressure conditions, offering robust protection against friction and wear. However, despite their effectiveness, solid lubricants may not be suitable for all applications, necessitating a thorough understanding of their limitations and capabilities [63].

1.2.3.1 Solid lubricants

Solid lubricants encompass substances like graphite and MoS_2, which are applied directly onto surfaces to reduce friction and wear [64]. These materials possess unique properties that make them particularly effective in high-temperature environments and under extreme pressure conditions. Graphite, for instance, forms a slippery layer between surfaces due to its layered structure, allowing for smooth sliding motion and preventing direct contact between mating components [65]. Similarly, MoS_2, with its lamellar structure, provides excellent lubrication by shearing into thin layers under pressure, thereby minimizing friction and protecting surfaces from damage [66]. These solid lubricants find extensive applications in industries such as aerospace, automotive, and manufacturing, where traditional liquid lubricants may fail to perform adequately due to the harsh operating conditions. Additionally, solid lubricants offer advantages such as prolonged service life and reduced maintenance requirements, making them an attractive choice for critical components subjected to extreme environments [67]. Recent advancements in solid lubricant research have focused on enhancing the performance and versatility of these materials through techniques such as nanomaterial engineering and surface modification. Nanoparticles of solid lubricants, for example, have shown promise in improving the tribological properties of conventional solid lubricants by providing additional lubrication mechanisms, such as rolling and exfoliation, at the nanoscale. Moreover, surface treatments and coatings have been developed to enhance the adhesion and durability of solid lubricants, further expanding their applicability in demanding industrial settings. By exploring novel synthesis methods and tailoring the properties of solid lubricants at the molecular level, researchers aim to unlock new possibilities for enhancing the efficiency and reliability of mechanical systems across various sectors [68].

1.2.3.2 Liquid lubricants

Liquid lubricants, including mineral oils, synthetic oils, and vegetable oils, are pivotal in mitigating friction and wear in a wide range of mechanical systems [69, 70]. Mineral oils derived from petroleum serve as ubiquitous lubricants in automotive, manufacturing, and marine industries due to their excellent lubricating properties and affordability [71]. Conversely, synthetic oils are engineered fluids with precise properties tailored to meet specific performance requirements. They are ideal for high-performance engines, aerospace applications, and precision equipment where superior lubrication is essential. Moreover, vegetable oils, sourced from plants, offer environmentally friendly alternatives for lubrication, particularly in applications where biodegradability and sustainability are paramount, such as food processing and environmentally sensitive machinery [72]. Recent research efforts in liquid lubricant development have focused on enhancing lubricant performance and sustainability through innovations in additive technology and base oil formulation. Additives such as friction modifiers, anti-wear agents, and viscosity modifiers are incorporated into lubricant formulations to improve lubricity, reduce friction, and enhance wear protection [70]. Additionally, advancements in base oil technology, including developing synthetic base oils and bio-based alternatives, aim to address environmental concerns and regulatory requirements while maintaining or improving lubricant performance. By continuously refining lubricant formulations and exploring novel additives and base oils, researchers strive to meet the evolving demands of modern machinery and promote sustainable lubrication practices for a greener future [73].

1.2.3.3 Semi-solid lubricants (greases)

Semi-solid lubricants, commonly known as greases, represent a versatile class of lubricants that find widespread application in various industries and mechanical systems [74, 75]. Greases comprise a base oil thickened with additives to form a viscous or semi-solid consistency, providing superior lubrication and protection against friction and wear [76]. Examples of greases include lithium grease [77], silicone grease [78], and calcium-based greases [79], each tailored to specific operating conditions and application requirements. Lithium grease is one of the most widely used types of grease due to its excellent water resistance, high-temperature stability, and versatility across a wide range of applications [80]. It is commonly employed in automotive chassis lubrication, wheel bearings, and general machinery lubrication [81]. Silicone grease, on the other hand, offers exceptional resistance to extreme temperatures, making it suitable for high-temperature applications such as oven conveyor chains, electrical connectors, and rubber seals [82]. Calcium-based greases are preferred for applications requiring corrosion protection and water resistance, such as marine equipment, agricultural machinery, and outdoor equipment [83]. Recent advancements in grease technology have focused on improving the performance and sustainability of greases through innovations in thickener technology and additive formulations. Nanotechnology, for example, has enabled the development of nano-thickened greases with enhanced mechanical stability, shear stability, and load-carrying capacity. Incorporating advanced additives such as friction modifiers, extreme pressure agents, and antioxidants aims to further enhance the lubricating properties and durability of greases in demanding applications [84].

1.2.3.4 Gaseous lubricants

Gaseous lubricants represent a unique category that utilizes gases, such as air and nitrogen, to reduce friction and wear between moving surfaces [85]. Unlike solid or liquid lubricants, gaseous lubricants rely on the presence of a thin gas film to separate contacting surfaces and minimize direct metal-to-metal contact. This phenomenon, known as hydrodynamic lubrication, occurs when the fluid film thickness exceeds the surface roughness, resulting in a fluid barrier that prevents friction and wear [86, 87]. The lubricating properties of gaseous lubricants depend on factors such as pressure, temperature, and surface roughness. At high speeds and low loads, gaseous lubricants can effectively reduce friction and wear by maintaining a continuous gas film between moving surfaces. However, supplementary lubrication may be required under extreme conditions or in applications with high loads to prevent surface damage and ensure optimal performance [86]. Gaseous lubricants have focused on understanding the fundamental mechanisms of gas film formation and its influence on friction and wear at the nanoscale. Molecular dynamics simulations and computational fluid dynamics analyses have provided valuable insights into the dynamics of gas molecules at the interface of sliding surfaces, shedding light on the factors governing gas film stability and lubricating performance. Experimental studies have explored novel techniques for enhancing gas film formation, such as surface texturing and coating modifications, to improve the efficiency and reliability of gaseous lubrication in high-speed and precision applications.

1.3 LUBRICANT COMPOSITIONS

In the realm of lubricant compositions, base oils serve as the primary component, dictating the lubricating fluid's fundamental properties and performance characteristics. Table 1.3 represents a summary of the lubricantbase oil categories, which are hydrocarbon, synthetic, and bio-based oils. Base oils play a pivotal role in lubricant formulations, with hydrocarbon-based oils providing versatility and affordability, synthetic base oils offering advanced performance capabilities, and bio-based oils presenting sustainable alternatives with reduced environmental footprint. Understanding the properties and characteristics of each base oil type is essential for formulating lubricants tailored to specific application requirements and performance standards. Lubricants' compositions may vary depending on the temperature of the oil, hence an efficient cooling system [88, 89] is required.

Table 1.3 A brief summary of the comparison of lubricant-based oil categories

Category	*Description*	*Chemical formulas (examples)*	*Advantages*	*Disadvantages*
Hydrocarbon-based oils	The most common type is derived from petroleum refining	Mixtures of various alkanes (paraffin) and cycloalkanes (naphthene) with varying chain lengths	Good lubrication properties Affordable	Limited high-/low-temperature performance Not readily biodegradable
Synthetic base oils	Engineered lubricants offer superior performance	Polyalphaolefins (PAO) Esters: R-COO-R' (complex depending on the acid and alcohol) Polyglycols (PAG): HO(-CH_2-CH_2-O-)	Excellent high-/low-temperature performance High resistance to oxidation and shearing Long life	Higher cost compared to hydrocarbon oils It may not be compatible with all seal materials
Bio-based oils	Environmentally friendly alternatives derived from renewable sources	Triglycerides (fats and vegetable oils): vary depending on the source	Biodegradable Non-toxic	Lower performance compared to some synthetics in extreme conditions Higher volatility

Hydrocarbon-based oils, derived from crude oil through refining processes, constitute a substantial portion of commercially available lubricants [90]. These oils encompass various hydrocarbon molecules, ranging from simple linear chains to more complex branched and cyclic structures. Their composition influences key lubricant properties such as viscosity, oxidation stability, and volatility. Hydrocarbon-based oils offer excellent lubrication properties across a wide range of temperatures and operating conditions, making them suitable for diverse industrial applications [91]. Synthetic base oils, on the other hand, are engineered fluids synthesized through chemical processes such as polymerization, oligomerization, or hydrogenation. Synthetic base oils exhibit superior thermal stability, oxidative resistance, and viscosity index compared to their hydrocarbon counterparts. They offer precise control over molecular structure and properties, allowing for tailor-made lubricants optimized for specific applications. Synthetic base oils are commonly utilized in high-performance engines, precision machinery, and extreme operating environments where superior lubrication performance is paramount [92]. In recent years, bio-based oils have emerged as environmentally friendly alternatives to traditional petroleum-derived lubricants. These oils are derived from renewable biomass sources such as vegetable oils, animal fats, or microbial sources. Bio-based oils offer biodegradability, lower toxicity, and reduced environmental impact compared to conventional lubricants. However, challenges such as oxidative stability, cold-flow properties, and compatibility with existing infrastructure must be addressed to achieve widespread adoption in industrial applications [93].

1.4 LUBRICATION ADDITIVES

Adding additives to lubricants is crucial in enhancing their performance by imparting specific functionalities and improving overall efficacy in reducing friction and wear in mechanical systems. Additives are carefully selected and incorporated into lubricant formulations to address various tribological challenges and optimize lubrication performance [94–96]. The science behind lubricant additives involves a multifaceted approach aimed at addressing specific tribological challenges and optimizing lubrication performance. By understanding the underlying mechanisms of action and interactions between additives and lubricant-based oils, researchers and engineers can develop tailored lubricant formulations that meet the demands of modern machinery and ensure reliable operation in various operating conditions. Figure 1.3 illustrates how the addition of various kinds of additives improves the functionality of the lubricants. One key mechanism by which additives improve lubricant performance is by forming protective films or layers on the surfaces of contacting components. These films act as barriers, reducing direct metal-to-metal contact and minimizing friction and wear [97–106]. For example, anti-wear additives such as zinc dialkyldithiophosphate (ZDDP)

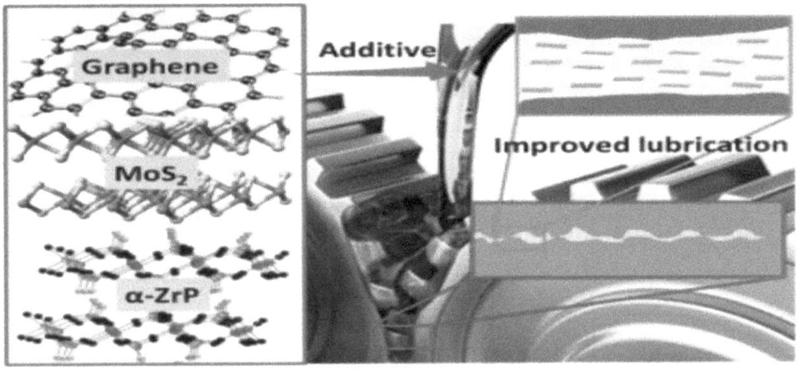

Figure 1.3 Illustrates the utilization of 2D nanosheets as additives in lubricants, contributing to enhanced lubrication performance [109].

[107] form sacrificial films on metal surfaces, providing a protective shield against surface asperities and preventing wear under high-load conditions. Similarly, friction modifiers like molybdenum disulfide (MoS_2) or graphite reduce friction by forming low shear boundary layers that facilitate smooth sliding between surfaces [108].

Another important function of additives is to enhance lubricant stability and longevity by inhibiting oxidation and degradation [110]. Oxidation inhibitors, such as hindered phenols or amine-based antioxidants, scavenge reactive oxygen species and free radicals, preventing the formation of harmful oxidation products that can degrade the lubricants and increase friction and wear. Additionally, detergents and dispersants are added to lubricants to keep contaminants and debris suspended in the fluid, preventing sludge formation and deposits that can interfere with lubricant flow and impair performance [111]. They also act as viscosity modifiers, which are commonly used additives that improve the viscosity–temperature relationship of lubricants, ensuring consistent lubricant film thickness over a wide range of operating temperatures. These additives alter the rheological properties of the lubricant, reducing viscosity at low temperatures to facilitate cold-start lubrication and increasing viscosity at high temperatures to maintain film integrity and prevent metal-to-metal contact [112].

Table 1.4 shows the various kinds of additives that are used to enhance the properties of the lubricants, along with the basic function of each additive. Adding additives to lubricants significantly enhances their performance by imparting specific functionalities and improving overall efficacy in reducing friction and wear in mechanical systems [113]. Each type of additive serves a distinct function, and understanding the underlying mechanisms behind their actions is crucial for optimizing lubrication performance. Anti-wear additives, such as zinc dialkyldithiophosphate (ZnDTP) [114] or molybdenum disulfide (MoS_2), function by forming sacrificial films on

Table 1.4 Functional additives in lubricants: mechanisms and benefits

Additive	*Function*	*Description*	*Benefits*
Anti-wear (AW) additives	Reduce wear on metal surfaces	Form a sacrificial film on metal surfaces, absorbing wear during contact. Often contain phosphorus (ZnDTP) or sulfur (MoS_2) compounds.	Minimize wear and tear, extending equipment life Improve fuel efficiency by reducing friction.
Extreme pressure (EP) additives	Enhance lubrication under high pressure and shock loads	Activate under extreme pressure, forming a protective film on metal surfaces. Often contain sulfur (sulfurized olefins) or phosphorus (chlorinated phosphates) compounds.	Protect components from damage during high-pressure operation. Reduce friction under heavy loads.
Antioxidants	Inhibit oil degradation	Neutralize free radicals that cause oil oxidation, thickening, and breakdown. Often contain phenolic antioxidants or hindered amine light stabilizers (HALS).	Extend oil life, reducing the need for frequent oil changes. Maintain optimal lubrication performance over time.
Corrosion inhibitors	Prevent rust and corrosion	Form a protective film on metal surfaces, hindering the interaction between metal and corrosive elements (e.g., water, acids). Often contain organic compounds with polar functional groups.	Protect metal components from rust and corrosion, extending equipment lifespan. Reduce maintenance costs associated with corrosion damage.
Viscosity modifiers (VM)	Maintain consistent viscosity across a wide temperature range	Thicken oil at high temperatures and reduce thinning at low temperatures, ensuring proper lubrication. Often consist of long-chain polymers that expand and contract with temperature changes.	Improve cold-start performance by ensuring adequate oil flow at low temperatures. Maintain optimal oil film thickness at high operating temperatures.
Pour point depressants (PPD)	Lower the oil's pour point	Reduce the temperature at which the oil becomes too thick to flow freely. Often contain copolymers that prevent wax crystals in the oil from forming large aggregates at low temperatures.	Improve cold-start performance by allowing easier oil circulation during engine startup. Reduce wear and tear during cold starts by ensuring proper lubrication.

metal surfaces. These films absorb wear during contact, minimizing metal-to-metal interactions and reducing friction. ZnDTP and MoS_2 compounds exhibit high affinity for metal surfaces, allowing them to adhere and provide effective lubrication under high-load conditions. By reducing wear and tear on metal surfaces, anti-wear additives extend equipment life and improve fuel efficiency by lowering frictional losses [115]. Extreme pressure (EP) additives activate under high pressure and shock load conditions, forming protective films on metal surfaces to prevent metal-to-metal contact and mitigate wear [116]. Compounds such as sulfurized olefins or chlorinated phosphates interact with metal surfaces to create durable boundary layers that withstand extreme pressure conditions. By protecting components from damage during heavy-duty operation, EP additives contribute to the overall reliability and longevity of mechanical systems. Antioxidants play a crucial role in inhibiting oil degradation by neutralizing free radicals that cause oxidation, thickening, and breakdown of the lubricant [117–125]. Phenolic antioxidants or hindered amine light stabilizers (HALS) effectively scavenge reactive oxygen species, preserving the integrity of the lubricant and maintaining optimal lubrication performance over time [126]. By extending oil life and reducing the need for frequent oil changes, antioxidants contribute to overall cost savings and equipment reliability. Corrosion inhibitors form protective films on metal surfaces, hindering the interaction between metal and corrosive elements such as water and acids. Organic compounds with polar functional groups effectively shield metal components from rust and corrosion, thereby extending equipment lifespan and reducing maintenance costs associated with corrosion damage. Viscosity modifiers maintain consistent viscosity across a wide temperature range by thickening oil at high temperatures and reducing thinning at low temperatures [127]. Long-chain polymers expand and contract with temperature changes, ensuring proper lubrication under varying operating conditions. Viscosity modifiers enhance overall lubrication effectiveness and equipment reliability by improving cold-start performance and maintaining optimal oil film thickness. Pour point depressants lower the oil's pour point by preventing wax crystals from forming large aggregates at low temperatures. Copolymers effectively inhibit wax crystallization, allowing easier oil circulation during engine startup and reducing wear and tear on components. By improving cold-start performance and ensuring proper lubrication, pour point depressants improve equipment reliability and longevity [128].

The depicted method and mechanism of surface protection are visually represented in Figure 1.4 [129]. This schematic illustrates the preparation process of nickel (Ni) nanoparticles modified graphene oxide (GO) and elucidates the mechanism underlying the formation of a protective film. Ni nanoparticles are modified with GO in the preparation process, likely through a chemical or physical interaction, resulting in a composite material. This modification process might involve techniques such as chemical reduction or physical adsorption, leading to the deposition of Ni

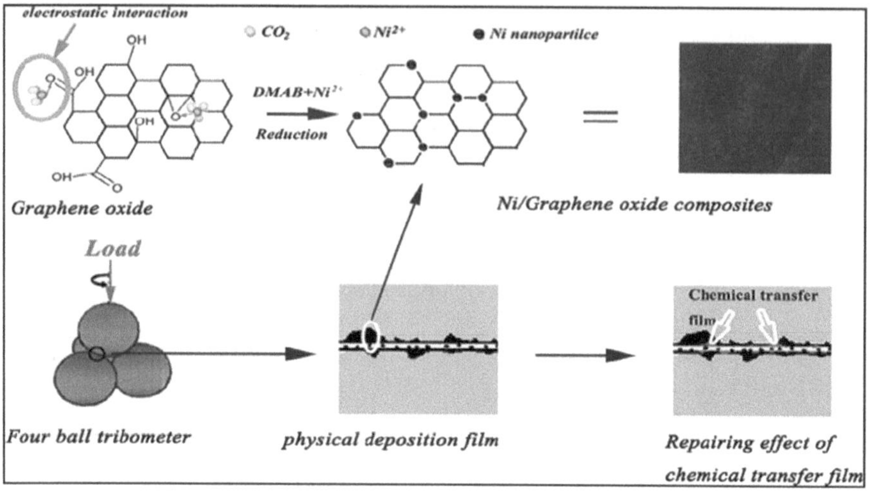

Figure 1.4 Illustrates the process of preparing nickel (Ni) nanoparticles modified graphene oxide (GO) and the mechanism involved in the formation of a protective film [129].

nanoparticles onto the GO sheets [129]. The protective film's formation mechanism involves the interaction between the modified Ni nanoparticles and the substrate surface. Upon application of the lubricant containing these additives, the modified Ni nanoparticles adhere to the surface of the contacting components. This adherence likely occurs through van der Waals forces, electrostatic interactions, or chemical bonding, depending on the nature of the substrate and the functional groups present on the modified nanoparticles [130]. As the surfaces move relative to each other during operation, the modified Ni nanoparticles act as active agents, forming a protective film on the surface. This film serves as a barrier between the contacting surfaces, preventing direct metal-to-metal contact and reducing friction and wear. The protective film is likely to be composed of a combination of Ni nanoparticles, graphene oxide sheets, and possibly other organic molecules present in the lubricant [20, 131, 132].

1.5 LUBRICATION REGIME ANALYSIS

1.5.1 Friction and wear characteristics in different lubrication regimes

Figure 1.5(a, b) illustrates the superior performance of liquid-like graphene as a lubricant additive compared to pure water and graphene oxide (GO). The graph showcases the remarkable reduction in friction and wear achieved by liquid-like graphene, with the most optimal tribological

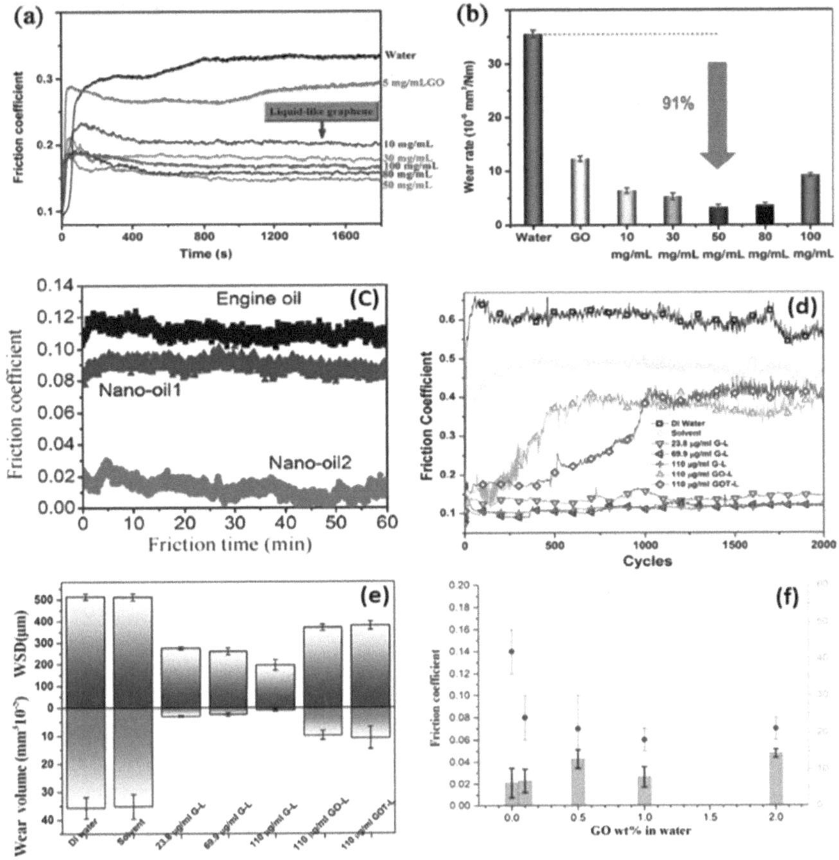

Figure 1.5 (a, b) Comparing the effectiveness of liquid-like graphene as an additive to pure water and GO additive in terms of performance [133]; (c) decrease in friction achieved by solar-exfoliated graphene when compared to engine oil, specifically nano-oil 1 at a concentration of 0.0125 mg/ml and nano-oil 2 at a concentration of 0.025 mg/ml [135]; (d, e) the frictional behavior of dispersed exfoliated graphene in deionized (DI) water with the aid of Triton X-100 [136]; and (f) demonstrates the decrease in friction and wear observed at the contact interface between DLC and stainless steel with the presence of graphene oxide (GO) in water [146].

performance observed at a medium concentration of 50 mg/ml in water [133]. This suggests that liquid-like graphene can be effectively dispersed in various solvents and exhibits exceptional friction and wear reduction properties. Compared to GO, liquid-like graphene demonstrates superior effectiveness in mitigating friction and wear, making it a promising candidate for lubricant formulations. Additionally, Ye et al. [134] conducted experiments modifying fluorinated graphene to enhance its dispersity in water, resulting in a significant reduction in wear rate compared to pure water. This highlights the potential of modified graphene compounds as effective lubricant additives. Figure 1.5c depicts the significant reduction in friction achieved by solar-exfoliated graphene compared to engine oil, as indicated by nano-oil 1 at a concentration of 0.0125 mg/ml and nano-oil 2 at 0.025 mg/ml [135]. The innovative method of exfoliating graphite oxide using focused solar electromagnetic radiation effectively removes oxygen from the original graphite oxide, imparting hydrophobic properties to the product and enabling easier dispersion in oil-based lubricants. Eswaraiah et al. [135] successfully dispersed and stabilized solar-exfoliated graphene in engine oil for one month, leading to considerable enhancement in engine oil tribological properties. This enhancement occurs by preventing direct surface contact, as the ultrathin graphene sheets penetrate into the contact, filling micro- and nanogaps in the rubbing surfaces to form a continuous lubricating film, thereby reducing friction and wear. Figure 1.5d illustrates the tribological performance of dispersed exfoliated graphene in deionized (DI) water with the assistance of Triton X-100 surfactant. Graphene, in its exfoliated form, tends to be hydrophobic, making it difficult to disperse in water [136]. However, the addition of surfactants, such as Triton X-100, provides a simple and effective solution to this issue. Triton X-100 contains both a hydrophobic tail and a hydrophilic chain: the hydrophobic tail attaches to the graphene, while the hydrophilic chain remains in the water. This configuration results in steric repulsion forces between the hydrophilic ends of the surfactant molecules, preventing the stacking and aggregation of graphene and ensuring its stable dispersion in water. Experimental findings demonstrated by Liang et al. [136] show that the inclusion of Triton X-100 in lubricants containing graphene additives led to significant improvements in tribological performance. Specifically, there was an 81.3% reduction in coefficient of friction (COF) and a 61.8% reduction in wear scar diameter, as depicted in Figure 1.5e. This indicates that Triton X-100 facilitates the effective dispersion of graphene in water-based lubricants, resulting in enhanced lubrication performance. Figure 1.5f illustrates the significant reduction in friction and wear achieved by the addition of graphene oxide (GO) in water-based lubricants. GO, a precursor necessary for graphene production, contains multiple carbon–oxygen functional groups and can be dissolved in water without requiring additional hydrophilic treatments [137–145]. Elomaa et al. [146] conducted experiments employing GO as an additive in water and studied its tribological performance between diamond-like carbon (DLC) and stainless steel surfaces. They observed a notable 57% decrease in the coefficient of friction (COF) compared to the control (pure water), indicating the effective lubricating properties of GO. Additionally, Kinoshita et al. [147] observed exceptionally low COF of approximately 0.05 at a tungsten carbide on stainless steel contact, with no evident wear even after 60,000 rubbing test cycles.

1.5.2 Role of surface roughness and asperity contact in lubrication regime determination

In exploring the dynamics of lubrication regimes, surface roughness and the intricacies of asperity contact play pivotal roles in determining the lubrication regime. Surface roughness, characterized by irregularities and deviations from an ideal flat surface, influences the interaction between lubricants and contacting surfaces. These irregularities create micro-asperities, which act as contact points between surfaces and significantly impact the lubrication behavior [148]. Understanding the role of surface roughness is essential for delineating the transition between different lubrication regimes, such as boundary, mixed, and hydrodynamic lubrication. Asperity contact, arising from the geometric irregularities present on contacting surfaces, exerts a profound influence on lubrication regime determination [149]. In boundary lubrication, where direct metal-to-metal contact occurs, the interaction between asperities governs the friction and wear characteristics. The effectiveness of lubricants in reducing friction and wear depends on their ability to form protective films or tribochemical reactions at these contact points. A mixed lubrication regime, characterized by a combination of fluid film lubrication and boundary interactions, further underscores the significance of asperity contact in determining overall friction and wear behavior [150]. Investigating the role of surface roughness and asperity contact in lubrication regime determination requires a comprehensive approach that integrates experimental studies, computational modeling, and theoretical analyses. Experimental techniques such as profilometry and atomic force microscopy (AFM) enable precise characterization of surface topography and measurement of surface roughness parameters [151]. Computational simulations, including finite element analysis (FEA) and molecular dynamics (MD), provide valuable insights into the complex interactions between lubricants, surfaces, and contacting asperities [152]. Theoretical models based on fundamental principles of fluid dynamics and solid mechanics further contribute to elucidating the underlying mechanisms governing lubrication regime transitions. A thorough understanding of the role of surface roughness and asperity contact is crucial for optimizing lubricant performance, mitigating friction and wear, and enhancing the efficiency and reliability of mechanical systems. By delving into the intricacies of these factors, researchers can advance the field of lubrication science and pave the way for the development of innovative lubrication solutions tailored to diverse engineering applications [153, 154].

1.6 ADVANCES IN LUBRICATION CHEMISTRY

1.6.1 Molecular dynamics (MD) simulation for lubrication

MD simulations offer a powerful tool for investigating the lubrication behavior and mechanisms of liquid lubricants at the atomic level, addressing challenges that are difficult to tackle experimentally. By subjecting lubricants to shear forces through bulk and confined MD simulations, researchers can gain insights into their flow and friction behavior between friction pairs. These simulations enable observing processes such as adsorption, film formation, and chemical reactions between lubricants and friction surfaces [155]. In the realm of fluid confined between solid surfaces and subjected to shear loads, the strength of wall–fluid adhesion emerges as a critical factor influencing lubricant tribological performance. MD simulations provide access to essential parameters like friction force, surface coverage of lubricants, and adsorption kinetics of molecules on friction surfaces [160]. These parameters help evaluate lubricant performance and mechanisms. For instance, Zhou et al. [161] investigated the lubrication performance of various molecules against ultra-high-molecular-weight polyethylene (UHMWPE) through MD simulations. By examining the interactions between lubricants and UHMWPE surfaces at the molecular level, they uncovered the adsorption behavior of lubricants in water and their correlation with surface coverage. Their findings suggested that the coefficient of friction decreased with increasing surface coverage, as Brownian motion facilitated lubricant molecules' approach to the surface, enabling adsorption via van der Waals and electrostatic forces. Similarly, Ta et al. [162] utilized MD simulations with the COMPASS force field to explore the lubrication performance of aqueous copolymer lubricants in the mixed lubrication regime. By analyzing the relationship between friction force, normal force, and system height during sliding, they inferred the formation of a lamellar structure by copolymer molecules when confined between iron surfaces. This physisorption on the surface contributed to lubrication. Moreover, they observed how the aqueous fluid underwent compression under pressure, leading to the gradual upward movement of the upper surface during friction. The plastic deformation of asperities was linked to concentration and loading pressure, with wear behavior showing linear dependence on the transferred atom number between contacting asperities [41, 163]. Hence, MD simulations provide invaluable insights into lubrication mechanisms and tribochemistry by elucidating molecular interactions and behaviors that underpin lubricant performance. These simulations complement experimental approaches and offer a deeper understanding of lubrication phenomena at the atomic scale, paving the way for advancements in lubrication science and technology. Table 1.5 shows the summarized version of various MD analyses conducted by the researchers.

1.6.2 Green lubricants

Green lubricants, often referred to as environmentally friendly lubricants or bio-based lubricants, are lubricants that are formulated using renewable, sustainable, and biodegradable materials. These lubricants aim to minimize environmental impact while providing adequate lubrication and

Table 1.5 Summarizes representative research endeavors employing molecular dynamics (MD) simulations to investigate various types of lubricants

Lubricant	*MD simulation details*	*Key findings*	*Additional information*	*Reference*
Aqueous triblock copolymer on TiO_2 (rutile) surface	• Simulation domain: 120 × 120 × 311 Å³. • Fixed rigid layers (NVE ensemble). • Mobile parts controlled at 300 K (Nosé–Hoover NVT).	• Pluronics (copolymer) physically bonds to TiO_2 surface, with stronger adhesion for polypropylene oxide (PPO) segments. • Higher concentration lubricant increases effective shear rate, leading to a non-slip behavior.	• Pluronics are versatile lubricants used in various applications like drug delivery and coatings. • Non-slip behavior indicates excellent lubrication properties.	[156]
Polytetrafluoroethylene (PTFE) sliding against Cu layer under water lubrication	• Optimization and annealing (Nosé thermostat NVT ensemble, 300–500 K). • Radial distribution function (RDF) analysis.	• Water molecules accumulate at the interface, forming a lubricating layer that improves wear resistance. • Wear mechanism transitions from mixed to elastohydrodynamic lubrication.	• Water lubrication plays a crucial role in enhancing PTFE performance. • Elastohydrodynamic lubrication signifies a pressure-dependent lubricating film.	[157]
Hydrocarbon lubricant (C_5H_{10}) on amorphous carbon films (a-C)	• ReaxFF and AIREBO potential functions. • Initial relaxation at 300 K (NVE ensemble).	• Unsaturated C_5H_{10} molecules weakly interact with a-C surface through intermolecular forces, not chemical bonds. • Low friction results from the self-passivation of a-C and the formation of a hydrodynamic lubricating film by C_5H_{10}.	• ReaxFF and AIREBO are powerful force fields for simulating complex materials like a-C. • Self-passivation suggests the a-C surface modifies itself to reduce friction.	[158]
Acetaldehyde molecules (CH_3CHO) as lubricant between α-Al_2O_3 (0001) surfaces	• First-principles MD (FPMD) simulations. • Systems with varying acetaldehyde molecule density.	• At low pressure, acetaldehyde molecules remain structurally intact. • High pressure triggers tribochemical reactions, forming oligomers (chains of molecules) that reduce friction through a slip mechanism. • Friction coefficient as low as 0.034 achieved.	• FPMD offers high accuracy for simulating interactions at the atomic level. • Tribochemical reactions can be beneficial for lubrication under extreme pressure conditions.	[159]

protection to machinery and equipment [164]. From a scientific perspective, green lubricants are designed to meet specific criteria that make them environmentally preferable compared to traditional petroleum-based lubricants [165–173]. This includes using base oils derived from renewable sources such as vegetable oils, animal fats, or synthetic esters, which have lower toxicity and biodegradability than conventional mineral oils [174]. Additionally, green lubricants may incorporate additives derived from natural sources or synthetic compounds that are less harmful to the environment. These additives help improve the lubricant's performance characteristics, such as viscosity, thermal stability, and anti-wear properties, while reducing friction and minimizing energy consumption. In terms of their environmental impact, green lubricants offer several advantages. They have lower carbon footprints than petroleum-based lubricants due to using renewable feedstocks and reduced dependence on fossil fuels. Moreover, their biodegradability ensures that they break down more readily in the environment, reducing the risk of pollution and ecosystem harm [22]. Green lubricants represent a promising avenue for reducing the environmental impact of lubrication processes across various industries while maintaining high performance and reliability in machinery and equipment. Continued research and development in this field are crucial for advancing the adoption of sustainable lubrication solutions and mitigating environmental concerns associated with traditional lubricants [181]. Under the category of green lubricants, the following are the two main aspects to consider.

1.6.2.1 Biodegradable lubricants

Biodegradable lubricants are designed to break down naturally in the environment over time, minimizing their impact on ecosystems. They are typically derived from renewable resources and formulated to degrade into harmless byproducts. Biodegradable lubricants offer an environmentally friendly alternative to traditional petroleum-based lubricants, reducing pollution and ecological damage [16, 69, 175].

1.6.2.2 Sustainable lubricant production methods

These refer to the techniques and processes used to manufacture lubricants in an environmentally sustainable manner. They involve minimizing resource consumption, reducing waste generation, and adopting renewable energy sources in lubricant production facilities. Sustainable production methods aim to decrease the environmental footprint of lubricant manufacturing while ensuring efficient and reliable production processes. By implementing sustainable practices, lubricant manufacturers can contribute to environmental conservation and mitigate the ecological impact of lubricant production [177–178].

1.6.3 Self-healing lubricants

These are lubricants engineered to repair minor damage or wear autonomously within the lubrication system. Through innovative formulations, self-healing lubricants can detect and mend small cracks or defects that may develop during operation, thereby extending the lifespan of mechanical components and improving overall system reliability. The self-healing properties of these lubricants are often achieved through the incorporation of microcapsules containing reactive agents or polymers capable of filling in gaps upon rupture, restoring the integrity of the lubricant film [180].

1.6.4 Responsive lubricants

These lubricants are designed to adapt their properties in response to changes in operating conditions or environmental factors. By leveraging advanced materials or additives, responsive lubricants can optimize their viscosity, film thickness, or friction coefficient based on real-time data such as temperature, pressure, or shear rate. This responsiveness allows the lubricant to provide tailored lubrication performance, enhancing efficiency and minimizing wear under varying operational scenarios. Responsive lubricants represent a promising avenue for optimizing tribological performance in dynamic systems, offering potential benefits in terms of energy efficiency, component longevity, and overall system performance [180–182].

1.7 CHALLENGES AND FUTURE PERSPECTIVES

In the landscape of lubrication chemistry, emerging challenges warrant attention and innovative solutions. These challenges encompass various facets, including stringent environmental regulations and escalating performance demands in extreme operating conditions. Environmental regulations pose a significant challenge to lubricant manufacturers and users alike, necessitating the development of lubricants that comply with increasingly stringent environmental standards [183]. Additionally, the escalating performance demands in extreme conditions, such as high temperatures, pressures, and speeds, require lubricants that can withstand and function optimally under such rigorous circumstances. Addressing these emerging challenges in lubrication chemistry requires a multifaceted approach involving the development of novel lubricant formulations, advancements in manufacturing processes, and adherence to regulatory requirements [184].

Looking ahead, the future directions of lubrication research offer promising avenues for innovation and advancement. Two key focal points for future research endeavors include computational modeling of lubricant behavior and the integration of lubrication with other cutting-edge technologies, such as tribology and materials science. Computational modeling holds immense potential for gaining deeper insights into the complex interactions between lubricants, surfaces, and operating conditions. Researchers can elucidate fundamental mechanisms governing lubricant behavior, predict performance outcomes, and optimize lubricant formulations for specific applications by employing sophisticated computational tools and simulation techniques. Furthermore, integrating lubrication research with other interdisciplinary fields, such as tribology and materials science, opens up new possibilities for synergistic advancements. Collaborative efforts across these domains can foster innovation in lubricant design, surface engineering, and materials development, leading to the creation of next-generation lubrication solutions with enhanced performance, durability, and sustainability [185].

1.8 CONCLUSION

In conclusion, this comprehensive review has provided an in-depth exploration of lubrication science, covering fundamental principles, types of lubricants, lubricant compositions, additives, lubrication regime analysis, advances in lubrication chemistry, and future perspectives. Tribology, the science of friction, wear, and lubrication, forms the cornerstone of lubrication science. Understanding surface interactions and the various types of lubricants, including solids, liquids, greases, and gases, is crucial for effective lubrication system design and operation. Lubricant compositions are vital in determining lubricant performance, with base oils and additives working synergistically to achieve optimal lubrication. Additionally, lubrication additives enhance lubricant properties such as viscosity, thermal stability, and wear resistance. The importance of lubrication regime analysis cannot be overstated, as it aids in selecting the most suitable lubricant for specific operating conditions. Friction and wear characteristics across different lubrication regimes, along with the role of surface roughness and asperity contact, are critical factors to consider. Advancements in lubrication chemistry, such as molecular dynamics (MD) simulations, offer valuable insights into lubricant behavior at the molecular level. Moreover, the emergence of green lubricants, including biodegradable options and sustainable production methods, reflects a growing commitment to environmental sustainability in the lubrication industry. Exciting innovations like self-healing lubricants and responsive lubricants present promising avenues for improving lubrication system performance and longevity. However, challenges remain, including optimizing lubricant performance under extreme conditions and ensuring compatibility with emerging technologies. Looking ahead, continued research, innovation, and collaboration are essential to address these challenges and unlock the full potential of lubrication science in advancing technological progress and sustainability. By fostering interdisciplinary partnerships and leveraging cutting-edge technologies, one can drive innovation and propel lubrication science into a new era of efficiency, reliability, and environmental responsibility.

REFERENCES

1. Mang, T. and W. Dresel, *Lubricants and Lubrication.* John Wiley & Sons, 2007.
2. Mang, T., et al., *Lubrication.* John Wiley & Sons, 2011.
3. Boyde, S., *Green Chemistry*, 2002. **4**(4): p. 293–307.
4. Kajdas, C., et al., *Industrial Lubricants.* Chemistry and Technology of Lubricants, 2010, p. 239–292.
5. Lawate, S., *Lubricant Additives.* CRC Press, 2017, p. 485–502.
6. Sander, J., *Environmentally Considerate Lubricants.* ASTM International, 2014.
7. Chudnovsky, B.H., *Lubrication of Electrical Contacts.* IEEE, 2005.
8. Gupta, A.K., et al., *Crystals*, 2022. **12**: p. 624.
9. Gupta, A.K., et al., *ACS Applied Materials and Interfaces*, 2020. **12**(40): p. 45558–45563.
10. Jones, M.H. and D. Scott, *Lubrication and Wear.* Vol. 8. Elsevier, 1983.
11. Korcek, S., et al., *Industrial Lubrication and Tribology*, 2000. **52**(5): p. 209–220.
12. Hossain, K.R., et al., *Macromolecular Materials and Engineering*, 2023. **308**(11): p. 2300147.
13. Wang, Z., et al., *Energies*, 2021. **14**(20): p. 6656.
14. Cosimbescu, L., et al., *ACS Applied Materials & Interfaces.* 2018. **10**(1): p. 1317–1323.
15. Pirro, D.M., et al., *Lubrication Fundamentals, Revised and Expanded.* CRC Press, 2016.
16. Zainal, N.A., et al., *Renewable and Sustainable Energy Reviews.* 2018, **82**: p. 80–102.
17. Taylor, R.I., et al., *Tribology and Interface Engineering Series.* Elsevier, 2005, p. 565–572.
18. Adetunla, A., et al., *E3S Web Conference.* Vol. 391. EDP Sciences, 2023. https://doi.org/10.1051/e3sconf/202339101014
19. Roy, D., et al., *Materials Letters*, 2020. **271**: p. 127780.
20. Chakraborty, S., et al., *Materials Letters*, 2019. **257**: p. 126739.
21. Minami, I., *Applied Sciences*, 2017. **7**(5): p. 445.
22. Mannekote, J.K., et al., *Renewable and Sustainable Energy Reviews*, 2018. **81**: p. 1787–1801.
23. Ghosh, S. and P.V. Rao, *Journal of Cleaner Production*, 2015. **100**: p. 17–34.
24. Nosonovsky, M. and B. Bhushan, *The Royal Society Publishing*, 2010. **368**: p. 4677–4694.
25. Czichos, H., *Tribology: A Systems Approach to the Science and Technology of Friction, Lubrication, and Wear.* Vol. 1. Elsevier, 2009.
26. Bhushan, B., *Tribology: Friction, Wear, and Lubrication.* The Engineering Handbook, 2000, p. 80–120.
27. Wakelin, R.J., *Annual Review of Materials Science*, 1974. **4**(1): p. 221–253.

28. Choudhari, A., et al., *Lecture Notes in Mechanical Engineering*. Springer. doi: 10.1007/978-981-15-4485-9_21
29. Sahoo, P. and S.K. Das, Tribology – a tool for mechanical and industrial engineering, in: *Mechanical and Industrial Engineering: Historical Aspects and Future Directions*. Springer, 2022, p. 1–37.
30. Menezes, P.L., et al., *Tribology for Scientists and Engineers*. Springer, 2013.
31. Kumar, A., et al., Biotribology, in: *Application of Biotribology in Biomedical System*. Springer, 2024.
32. Gupta, A.K., et al., Composites for drug eluting devices: Emerging biomedical applications, in: *Application of Biotribology in Biomedical System*, Springer, 2024.
33. Kumar, A., et al., Introduction to biotribology: A science of surface interaction, in: *Application of Biotribology in Biomedical System*. Springer, 2024.
34. Avinash et al., Tribological hurdles in biomedical manufacturing: A comprehensive examination, in: *Application of Biotribology in Biomedical System*. Springer, 2024.
35. Neale, M. and M. Gee, *A guide to wear problems and testing for industry*. William Andrew, 2001.
36. Budinski, K.G., *Guide to Friction, Wear, and Erosion Testing*. ASTM International, 2007.
37. Choudhari, A., et al., Wear and friction mechanism study in knee and hip rehabilitation: A comprehensive review, in: *Application of Biotribology in Biomedical System*. Springer, 2024.
38. Chowdhury, N., et al., Characteristics of hydrogel properties in the advancement of biotribology, in: *Application of Biotribology in Biomedical System*. Springer, 2024.
39. Shafi, W.K. and M.S. Charoo, *Tribology-Materials, Surfaces & Interfaces*, 2021. **15**(1): p. 201–254.
40. Gulzar, M., et al., *Journal of Nanoparticle Research*, 2016. **18**: p. 11–25.
41. Chou, C.-C. and S.-H. Lee, *Wear*, 2010. **269**(111–112): p. 7571–762.
42. Zhai, W. and K. Zhou, *Advanced Functional Materials*, 2019. **29**(28): p. 1806395.
43. Gupta, H., *Study of Rheological and Tribological Properties of Metal Oxide Nanoparticles Blended Lubricating Oil*. 2021.PhD Thesis University of Petroleum and Energy Studies Dehradun, India, Department of Mechanical Engineering . https://shodhganga.inflibnet.ac.in/handle/10603/371018
44. Essajai, R., et al., *Results in Physics*, 2019. **15**: p. 102576.
45. Miura, K. and S. Kamiya, *Europhysics Letters*, 2002. **58**(4): p. 610.
46. Opia, A.C., et al., Nano-particles Additives as a Promising Trend in Tribology: A Review on Their Fundamentals and Mechanisms on Friction and *Wear Reduction*. 2021.*EVERGREEN Joint Journal of Novel Carbon Resource Sciences & Green Asia Strategy*. (8) 4: pp 777–798. https://doi.org/10.5109/4742121
47. Gong, H., et al., *Composites Part B: Engineering*, 2020. **202**: p. 108450.
48. Amiril, S.A.S., et al., *Journal of Cleaner Production*, 2017. **168**: p. 1571–1589.
49. Mohan, S. and A. Mohan, *Anti-Abrasive Nanocoatings*. Elsevier, 2015, p. 3–22.
50. Singer, I.L., *Langmuir*, 1996. **12**(19): p. 4486–4491.

51. Chen, K., et al., *The International Journal of Advanced Manufacturing Technology*, 2022. **122**(3): p. 1115–1141.
52. Höglund, E., *Wear*, 1999. **232**(2): p. 176–184.
53. Nguyen, A., et al., *Encyclopedia of Surface and Colloid Science*. Citeseer, 2007.
54. Chen, Z., et al., *Lubricants*, 2018. **6**(3): p. 74.
55. Xia, W., et al., *Wear*, 2018. **396**: p. 162–171.
56. Jacobson, B., *Tribology International*, 2000. **33**(3–4): p. 205–210.
57. Wu, H., et al., *Tribology International*, 2017. **109**: p. 398–408.
58. Hettiarachchi, S.J., *Enhancing Engine Oil Performance using Nanoparticles and Bio-Lubricants as Additives*, 2022.PhD Thesis London South Bank University School of Engineering. https://doi.org/10.18744/lsbu.92806
59. Gupta, A., et al., *Lecture Notes in Mechanical Engineering*. Springer. doi: 10.1007/978-981-13-2490-1_13.
60. Choudhari, A., et al., *Proceedings of International Conference on Intelligent Manufacturing and Automation: ICIMA 2022*. Springer. doi: 10.1007/978-981-19-7971-2_50.
61. Sunil, T., et al., *International Journal of Mechanical Engineering and Technology*, 2016. **7**: p. 193–199.
62. Asmoro, G., et al., *Role of Solid Lubricant (MoS2 and graphite) Variations on Characteristics of Brake Lining Composite*. IOP Publishing, 2018. doi: 10.1088/1757-899X/420/1/012022
63. Sahoo, A.K., et al., *International Journal of Machining and Machinability of Materials*, 2012. **12**(4): p. 297–307.
64. Scharf, T.W. and S.V. Prasad, *Journal of Materials Science*, 2013. **48**: p. 511–531.
65. Cho, M.H., et al., *Wear*, 2006. **260**(7–8): p. 855–860.
66. Vazirisereshk, M.R., et al., *Lubricants*, 2019. **7**(7): p. 57.
67. Kumar, R., et al., *Materials Today: Proceedings*, 2020. **28**: p. 1962–1967.
68. Reeves, C.J., et al., Tribology of solid lubricants, in: *Tribology for Scientists and Engineers: from Basics to Advanced Concepts*, 2013, p. 447–494.
69. Syahir, A.Z., et al., *Journal of Cleaner Production*, 2017. **168**: p. 997–1016.
70. Rahman, M.H., et al., *Lubricants*, 2021. **9**(8): p. 73.
71. Shafi, W.K., et al., *Tribology-Materials, Surfaces & Interfaces*, 2018. **12**(1): p. 27–43.
72. Sharma, B.K., et al., *Synthetics, Mineral Oils, and Bio-Based Lubricants*. CRC Press, 2020, p. 399–430.
73. Srivyas, P.D. and M.S. Charoo, *Materials Today: Proceedings*, 2019. **18**: p. 4773–4781.
74. Caines, A.J., et al., *Automotive Lubricants Reference Book*, Vol. 354. John Wiley & Sons, 2004.
75. Murugesan, R. and M. Raafat Youssef Afifi Elsisi, *Supporting Sustainable Selection for Lubricants in Automotive Industry: Constant Velocity Joint Grease as a Case Study*. 2023.Linköping University | Department of Management and Engineering Master's thesis, 30 credits| Sustainability Engineering and Management Spring 2023| LIU-IEI-TEK-A--23/04706--SE.
76. Rawat, S.S. and A.P. Harsha, *Automotive Tribology*, 2019: p. 147–182.
77. Delgado, M.A., et al., *Chemical Engineering Research and Design*, 2005. **83**(9): p. 1085–1092.

78. Wang, Q., et al., *Journal of Applied Polymer Science*, 2003. **89**(9): p. 2397–2399.
79. Kizhakkeappillil, J.B., *Proceedings of the Institution of Mechanical Engineers, Part C: Journal of Mechanical Engineering Science*, 2023. **237**(3): p. 708–716.
80. Ren, G., et al., *Friction*, 2021. **9**: p. 75–91.
81. Zhang, X., et al., *Industrial & Engineering Chemistry Research*, 2014. **53**(1): p. 182–188.
82. Gow, G., Lubricating grease, in: *Chemistry and Technology of Lubricants*. Springer, 2010, p. 411–432.
83. Lugt, P.M., *Tribology International*, 2016. **97**: p. 467–477.
84. Dixena, R.K., *Study on the Application of Additive Systems in Polymeric Lubricating Greases*, 2015.University of Petroleum and Energy Studies Dehradun, India.
85. Brunetière, N., *Journal of Tribology*, 2018. **140**(2): p. 021702.
86. Zhang, X., et al., *Journal of Tribology*, 2017. **139**(2): p. 021703.
87. Persson, B.N.J. and M. Scaraggi, *Journal of Physics: Condensed Matter*, 2009. **21**(18): p. 185002.
88. Talkar, S., A. Choudhari, and P. Rayar, *Proceedings of International Conference on Intelligent Manufacturing and Automation. Lecture Notes in Mechanical Engineering*. Springer, 2020. doi: 10.1007/978-981-15-4485-9_31.
89. Choudhari, A., et al., Designing and prototyping for conservation and effective utilization of waste heat from air conditioner, in: *IOP Conference Series: Materials Science and Engineering*, Vol. 1104, no. 1, p. 012007. IOP Publishing. doi: 10.1088/1757-899X/1104/1/012007.
90. Rizvi, S.Q.A., *Petroleum-Derived Hydrocarbon Base Oils*. ASTM International, 2019.
91. Ho, C.K., et al., *Renewable and Sustainable Energy Reviews*, 2019. **113**: p. 109261.
92. Katiyar, J.K., et al., *Industrial Tribology: Sustainable Machinery and Industry 4.0*. CRC Press, 2022.
93. Anwar, A. and A. Garforth, *Fuel*, 2016. **173**: p. 189–208.
94. Shahnazar, S., et al., *International Journal of Hydrogen Energy*, 2016. **41**(4): p. 3153–3170.
95. Leventini, S.D., et al., *Tribology Letters*, 2022. **70**(4): p. 127.
96. Kumar, A., et al., *Tribology Transactions*, 2018. **61**(5): p. 951–959.
97. Gupta, A.K., et al., Composites for drug eluting devices: Emerging biomedical applications, in: *Applications of Biotribology in Biomedical Systems*, Springer Press, Springer Nature Group, 2024.
98. Choudhari, A., et al., Wear and friction mechanism study in knee and hip rehabilitation: A comprehensive review, in: *Applications of Biotribology in Biomedical Systems*, Springer Press, Springer Nature Group, 2024.
99. Datta, A., et al., Advanced materials in biological implants and surgical tools, in: *Advanced Materials for Biomedical Applications*. CRC Press, 2022, p. 21–43.
100. Kumar, A., et al., Recent advancements and future trends in next-generation materials for biomedical applications, in: *Advanced Materials for Biomedical Applications*. CRC Press, 2022, p. 1–19.

101. Kumar, A., et al., Liquid slippage in confined flows: Effect of periodic micropatterns of arbitrary pitch and amplitude. ASME 2016 5th International Conference on Micro/Nanoscale Heat and Mass Transfer, MNHMT, 2016a.
102. Kumar, A., et al., *Physics of Fluids*, 2016. **28**: p. 082002.
103. Kumar, A., et al., Reduction of hydraulic friction in confined flows by laser texturing: Experiments and theoretical validation. ASME 2018 16th International Conference on Nanochannels, Microchannels, and Minichannels, ICNMM, 2018.
104. Kumar, A., et al., *Journal of Heat and Mass Transfer*, 2018, p. 140.
105. Kumar, A., et al., Liquid slippage in con fined flows: Effect of periodic micropatterns of arbitrary pitch and amplitude. ASME 2016 5th Micro/Nanoscale Heat and Mass Transfer International Conference (MNHMT2016). Biopolis, Singapore, 2016.140(1): 012403. https://doi.org/10.1115/1.4037363
106. Kumar, A., et al., Biomedical study of femur bone fracture and healing, in: *Advanced Materials for Biomedical Applications*. CRC Press, 2022, p. 235–250
107. Njiwa, P., et al., *Tribology Letters*, 2011. **44**: p. 19–30.
108. Winer, W.O., *Wear*, 1967. **10**(6): p. 422–452.
109. Xiao, H. and S. Liu, *Materials & Design*, 2017. **135**: p. 319–332.
110. Aguilar, G., et al., *Chemistry and Technology of Lubricants*. Springer, 2010, p. 107–152.
111. Ahmed, N.S., et al., *Tribology-Lubricants and Lubrication*. InTech, 2011, p. 249–268.
112. Braun, J., *Additives. Lubricants and Lubrication*, 2017, p. 117–152. doi:10.1002/9783527645565.ch6.
113. Baderna, D., et al., *Lubricants and Additives: A Point of View. Global Risk-Based Management of Chemical Additives I: Production, Usage and Environmental Occurrence*. Springer, 2012, p. 109–132.
114. Equey, S., et al., *Tribology International*, 2008. **41**(11): p. 1090–1096.
115. Spikes, H., *Lubrication Science*, 2008. **20**(2): p. 103–136.
116. Farng, L.O. and L. Rudnick, *Lubricant Additives: Chemistry and Applications*. Routledge, 2003, p. 223–257.
117. Heidari, M., S. et al., Navigating the landscape: Cutting-edge biomedical manufacturing techniques, in: *Applications of Biotribology in Biomedical Systems*, Springer Press, Springer Nature Group, 2024.
118. Gour, R., et al., Animal tribology, in: *Applications of Biotribology in Biomedical Systems*, Springer Press, Springer Nature Group, 2024.
119. Kumar, A., et al., *Tribology Transactions*, 2018. **61**: p. 951–959.
120. Kumar, A., et al., *Biosensor Based Advanced Cancer Diagnostics*. Elsevier, 2022, p. 197–224.
121. Kumar, A., et al., *Trends in Fabrication of Polymers and Polymer Composition*. AIP Publishing, 2022, p. 1–3.
122. Kumar, A., et al., Clinically available/under trial drugs and vaccines for treatment of SARS-COV-2, in: *Computational Approaches for Novel Therapeutic and Diagnostic Designing to Mitigate SARS-CoV2 Infection*. Elsevier, 2022, pp. 451–488.
123. Kumar, A., et al., *ACS Applied Bio Materials*, 2022. **5**: p. 2046–2068.
124. Kumar, A., et al., Physics of laser—matter interaction in laser-based manufacturing, in: *Laser-based Technologies for Sustainable Manufacturing*. CRC Press, 2023, p. 45–54.

125. Kumar, A., et al., *Lubricants*, 2023. **11**, p. 64.
126. Amit Ranjan, N., *Syntheses of New Butylated Hydroxytoluene Analogues and Their Potential Application as Antioxidant for Synthetic Lubricant Oil*, 2019. https://myto.upm.edu.my/find/Record/u1130871/Details.
127. Martini, A., et al., *Tribology Letters*, 2018. **66**: p. 1–14.
128. Stambaugh, R.L. and B.G. Kinker, *Chemistry and Technology of Lubricants*. Springer, 2010: p. 153–187.
129. Meng, Y., et al., *ACS Applied Materials & Interfaces*, 2015. 7(21): p. 11604–11612.
130. Van Oss, C.J., et al., *Journal of Colloid and Interface Science*, 1986. **111**(2): p. 378–390.
131. Sun, J. and S. Du, *RSC Advances*, 2019. **9**(69): p. 40642–40661.
132. Gupta, A.K., et al., *International Journal of Engineering Research and Applications*, 2016. **6**(4): p. 68–73.
133. Yang, J., et al., *Tribology International*, 2017. **105**: p. 118–124.
134. Ye, X., et al., *ACS Applied Materials & Interfaces*, 2016. 8(11): p. 7483–7488.
135. Eswaraiah, V., et al., *ACS Applied Materials & Interfaces*, 2011. **3**(11): p. 4221–4227.
136. Liang, S., et al., *Carbon*, 2016. **96**: p. 1181–1190. https://www.sciencedirect.com/science/article/pii/S0008622315303870?via%3Dihub.
137. Kumar, A., et al., *Proceedings of Institution of Mechanical Engineers. Part E: Journal of Process Mechanical Engineering*, 2022: p. 09544089221110980.10.1063/9780735423916
138. Kumar, S., et al., *Trends in Fabrication of Polymers and Polymer Composition*. AIP Publishing Books, 2022, p. 10–11. https://pubs.aip.org/books/monograph/53/chapter-abstract/20635222/Importance-of-Chemically-Treated-Natural-Fibers-in?redirectedFrom=fulltext.
139. Leventini, S.D., et al., *Tribology Letters*, 2022. **70**: p. 127, 2022.
140. Mehra, A.K., et al., *Composites Communications*, 2021. **25**: p. 100732.
141. Parihar, A., et al., *Advanced Biology*, 2023. **7**: p. 2200285. https://onlinelibrary.wiley.com/doi/abs/10.1002/adbi.202200285
142. Parihar, A., et al., *Regenerative Engineering and Translational Medicine*, 2021, p. 1–27. https://link.springer.com/article/10.1007/s40883-021-00219-w.
143. Parmar, V., et al., *Journal of Laser Applications*, 2018: p. 30. https://pubs.aip.org/lia/jla/article-abstract/30/3/032001/700091/Oxidation-facilitated-antimicrobial-ability-of?redirectedFrom=fulltext.
144. Pal, A., et al., Tribological measurements of human skin, in: *Applications of Biotribology in Biomedical Systems*. Springer Press, Springer Nature Group, 2024. https://link.springer.com/9783031583261.
145. Kumar, A., et al., Tribological hurdles in biomedical manufacturing: A comprehensive examination, in: *Applications of Biotribology in Biomedical Systems*. Springer Press, Springer Nature Group, 2024.
146. Elomaa, O., et al., *Diamond and Related Materials*, 2015. **52**: p. 43–48.
147. Kinoshita, H., et al., *Carbon*, 2014. **66**: p. 720–723.
148. He, Y., et al., *Applied Surface Science*, 2022. **585**: p. 152501.
149. Holmes, M.J.A., et al., *Journal of Tribology*, 2005. **127**(1): p. 61–69.
150. Brandão, J.A., et al., *Tribology International*, 2012. **47**: p. 50–61.
151. Esayanur, M.S., et al., *Journal of Adhesion Science and Technology*, 2005. **19**(8): p. 611–626.

152. Vakis, A.I., et al., *Tribology International*, 2018. **125**: p. 169–199.
153. Baumli, P., et al., *Advances in Colloid and Interface Science*, 2021. **287**: p. 102329.
154. Kumar, A., *Materials Today: Proceedings*, 2021. **42**: p. 884–892.
155. Donato, M.T., et al., *Journal of Physical Chemistry B*, 2024. **128**(10): p. 2559–2568. https://doi.org/10.1021/acs.jpcb.3c08397
156. Ta, T.D., et al., *Tribology International*, 2019. **130**: p. 170–183.
157. Song, J. and G. Zhao, *Tribology International*, 2019. **136**: p. 234–239.
158. Li, X., et al., *Computational Materials Science*, 2019. **161**: p. 1–9.
159. Haw, S.M. and N.J. Mosey, *The Journal of Physical Chemistry C*, 2012. **116**(3): p. 2132–2145.
160. Ewen, J.P., et al., *Tribology Letters*, 2021. **69**(1): p. 24.
161. Zhou, Y., et al., *Carbon*, 2015. **86**: p. 132–138.
162. Ta, T.D., et al., *ACS Applied Materials & Interfaces*, 2016. **8**(8): p. 5641–5652.
163. Liu, D., et al., *Chemical Engineering Science*, 2023. **270**: p. 118541.
164. Salih, N. and J. Salimon, *Biointerface Research in Applied Chemistry*, 2021. **11**(5): p. 13303–13327.
165. Kumar, A. et al., Introduction to biotribology: A science of surface interaction, in: *Applications of Biotribology in Biomedical Systems.* Springer Press, Springer Nature Group, 2024.
166. Chowdhury, N., et al., Characterization of hydrogel properties in the advancement of biotribology, in: *Applications of Biotribology in Biomedical Systems.* Springer Press, Springer Nature Group, 2024.
167. Pathak, A., et al., Application of laser technology in the mechanical and machine manufacturing industry, in: *Laser-based Technologies for Sustainable Manufacturing.* CRC Press, 2023, p. 107–155.
168. Saurabh Gupta Ruchika Saini AKPS, *International Journal of Recent Technology and Engineering*, 2020. **8**: p. 4569–4573.
169. Shrivastava Prateek, S.R., et al., Investigation of torsional rod to minimize vibration in automobile using ANSYS. 2020 1st International Conference on Innovative Research in Applied Science, Engineering and Technology, 2020, p. 1–6.
170. Singh, M., et al., Capillary as a liquid diode. *Physical Review Fluids*, 2020. **5**, p. 102101.
171. Singh, RK, et al., *Microsyst Technology*, 2014b. **20**: p. 101–111.
172. Vats, P., K.K. Gajrani and A. Kumar, Laser-based additive manufacturing, in: *Laser-based Technologies for Sustainable Manufacturing.* CRC Press, 2023, p. 67–83.
173. Dias, D., et al., *Lubricants,* 2024. **12**(4): p. 114.
174. Sharma, B.K. and G. Biresaw, *Tribology & Lubrication Technology*, 2010. **66**(1): p. 32.
175. Mehra, A.K., et al., *Composites Communications*, 2021. **25**: p. 100732.
176. Shah, R., et al., *Lubricants*, 2021. **9**(2): p. 21.
177. Pichler, J., et al., *Green Chemistry Letters and Reviews*, 2023. **16**(1): p. 2185547.
178. Kumar, A., et al., Recent advancements and future trends in next-generation materials for biomedical applications, in: *Advanced Materials for Biomedical Applications*, CRC Press, 2022, p. 1–19.

179. Nosonovsky, M. and P.K. Rohatgi, *Biomimetics in Materials Science: Self-Healing, Self-Lubricating, and Self-Cleaning Materials*. Vol. 152. Springer Science & Business Media, 2011.
180. Qian, H., et al., *Surface and Coatings Technology*, 2021. **406**: p. 126742.
181. Gupta, A.K., et al., *Materials Performance Characterization*, 2020. **9**(1): p. 25–35.
182. Roy, D., et al., *Philosophical Magazine*, 2022. **102**(3): p. 189–209.
183. Cecilia, J.A., et al., *Processes*, 2020. **8**(3): p. 257.
184. Tung, S.C. and M.L. McMillan, *Tribology International*, 2004. **37**(7): p. 517–536.
185. Narayana Sarma, R. and R. Vinu, *Lubricants*, 2022. **10**(4): p. 70.

Chapter 2

Sustainable green lubricants

Anjan Mahapatra, Prameet Vats,
Arun Kumar Bambam, Avinash Kumar,
and Kishor Kumar Gajrani

2.1 INTRODUCTION

Lubricants have played a vital role in human history, with water and air recognized as the earliest known forms and the other types of well-known lubricants are classified as shown in Figure 2.1. In natural joints, synovial fluid serves not only as a natural lubricant but also as a biochemical medium, thus facilitating the transport of nutrients and regulatory cytokines [1]. While conventional lubricants have been the cornerstone of machinery maintenance for years, offering crucial protection and ensuring smooth operation in diverse industries, they predominantly consist of mineral-based oils derived from petroleum. Despite their cost-effectiveness and widespread availability, these traditional lubricants pose significant environmental challenges, contributing to various ecological issues [2].

The escalating global demand for energy, primarily derived from fossil fuels, has heightened the need for petroleum-based products, including lubricants. As a response, bio-based lubricants, referred to as bio-lubricants, have gained prominence. These lubricants are readily biodegradable and safe for the environment, addressing concerns related to ecotoxicity and reducing their ecological footprint. The incorporation of innovative additives, such as antioxidants and friction modifiers, enhances their performance without compromising environmental considerations [3].

Industries, including automotive and manufacturing, where lubricants are indispensable, are increasingly adopting bio-lubricants as part of their commitment to sustainability. These lubricants not only align with stringent regulations but also reflect a corporate emphasis on environmental responsibility, considering factors like waste reduction and circular economy principles. Certification programs like the European Ecolabel and the USDA BioPreferred program play a crucial role in guiding consumers and businesses toward environmentally conscious lubricant choices [4].

As the demand for green and sustainable alternatives continues to grow across diverse industries, bio-lubricants are poised to play an increasingly vital role in shaping a more environmentally responsible and sustainable industrial landscape. The shift toward these eco-friendly lubricants signifies

DOI: 10.1201/9781003535447-2

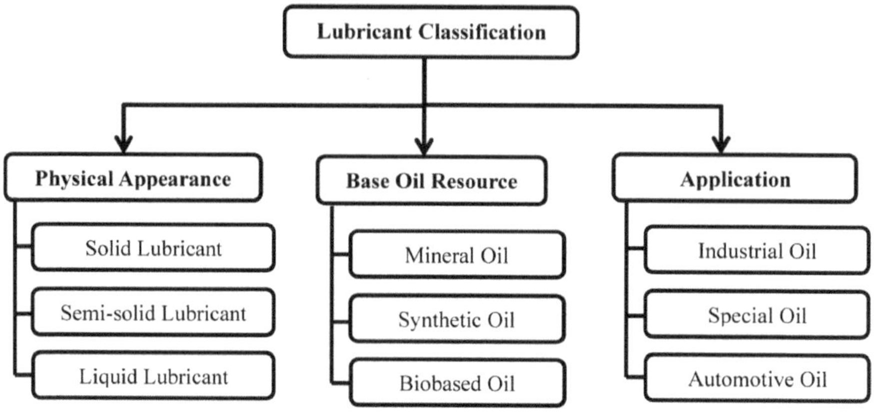

Figure 2.1 Basic classification and application of lubricants [1].

a broader commitment to environmental conservation, climate change mitigation, and resource efficiency, ultimately contributing to a greener and more sustainable future [5].

The chapter concludes with a forward-looking exploration of the dynamic landscape of on-shape research and development efforts. It underscores the imperative to fine-tune the efficiency and cost-effectiveness of sustainable lubricants as they evolve into essential catalysts for cultivating a more environmentally responsible and resilient industrial future.

2.1.1 Objectives

The core objectives of sustainable lubricants are deeply rooted in the broader aim of developing lubrication solutions that are both environmentally conscious and socially beneficial. These objectives center around minimizing environmental impact by prioritizing biodegradability and utilizing renewable, eco-friendly feedstocks [6]. Sustainable lubricants seek to address issues such as pollution, soil contamination, and resource depletion associated with traditional lubricants. Energy efficiency is a pivotal goal, as these lubricants are specifically formulated to reduce friction and wear, ultimately improving the overall efficiency of mechanical systems. Furthermore, sustainable lubricants aim to achieve regulatory compliance, aligning with stringent environmental standards to promote responsible industrial practices [7]. They also strive to foster corporate responsibility, positively influencing public perceptions and meeting the increasing demand for environmentally conscious products. Moreover, the scope of sustainable lubricants extends to promoting innovation and research. Continuous efforts in research and development are essential to refining formulations, exploring cutting-edge technologies, and enhancing the overall performance of these lubricants [8]. This dynamic approach ensures that sustainable lubricants

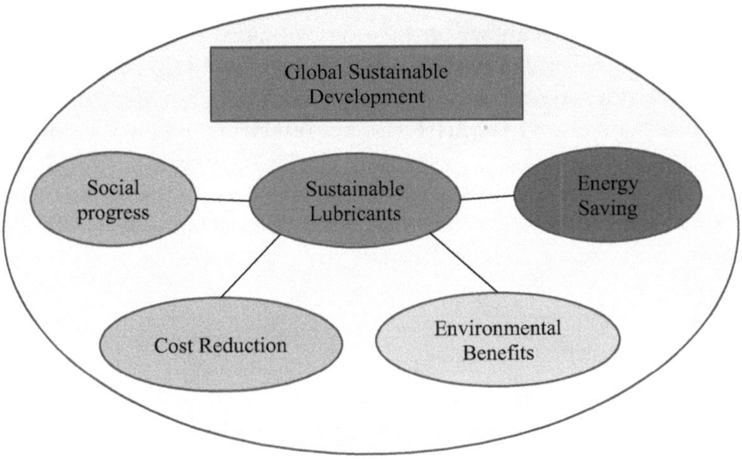

Figure 2.2 Role of sustainable lubricants in sustainable development.

remain at the forefront of technological advancement, adapting to emerging challenges and evolving sustainability standards.

Reduced toxicity and fewer health hazards are integral to creating a safer working environment, aligning with broader goals of occupational health and safety. Recognizing the critical importance of the well-being of individuals interacting with lubricants, from production to application, is a key aspect of sustainability [9]. The adoption of sustainable lubricants also encompasses a corporate responsibility dimension. The positive public perception associated with environmentally friendly products becomes a goal beyond technical specifications, as using sustainable lubricants enhances a company's image by demonstrating a commitment to environmental responsibility [6, 9]. Figure 2.2 shows how the proper selection of lubricants can help in meeting sustainable goals.

2.1.1.1 *Resource conservation*

In the development of sustainable lubricants, resource conservation aims to address concerns related to the depletion of non-renewable resources. Resource conservation aligns with broader sustainability goals, emphasizing the need to transition from finite fossil fuels to renewable and bio-based alternatives. Promotion of circular economy, enhancement of resource efficiency, and diversification of feedstock resources lead to addressing the geopolitical risks [10].

Moreover, their formulation emphasized on extended equipment lifespan through reduced friction and wear, diminishing the need for frequent replacements, thus conserving materials and minimizing waste generation. Regulatory compliance is integral to resource conservation, adding recyclability as an emerging dimension in the resource conservation narrative of sustainable lubricants. Some formulation is designed to be more readily recyclable, allowing for the recovery and reuse of lubricant components. This approach not only minimizes waste generation but also supports a circular economy model, where materials are reintroduced into the production cycle, further reducing the reliance on virgin resources. Additionally, the energy efficiency inherent in sustainable lubricants aligns with the broader efforts to conserve energy resources, while their stringent environmental regulations underscore a commitment to responsible industrial practices that contribute to resource conservation [11]. In essence, sustainable lubricants form a cornerstone in the pursuit of a more environmentally sustainable and resource-efficient future.

2.1.1.2 Biodegradability and eco-friendly disposal

Biodegradability and eco-friendly disposal practices are fundamental components of sustainable lubricants, reflecting a dedicated commitment to environmental responsibility throughout their lifecycle. Biodegradable lubricants possess the capability to naturally break down, preventing persistent harm when released into the environment. Additionally, a focus on eco-friendly disposal, including recycling and aligning with ecological waste management principles, forms a comprehensive strategy to minimize the ecological footprint associated with lubricant use. This emphasis on responsible end-of-life scenarios aims to reduce environmental burdens linked to lubricant disposal, contributing to a more sustainable waste management system [12]. By combining biodegradability with eco-friendly disposal practices, sustainable lubricants not only demonstrate effective performance during use but also actively support closed-loop systems, fostering environmentally sound practices and aligning with broader sustainability objectives for a resilient and responsible industrial future [13].

2.1.1.3 Reducing emissions

Integral to the environmental sustainability goals, sustainable lubricants aim to minimize harmful substance release during application, addressing air quality concerns and supporting population reduction initiatives. This commitment aligns with broader objectives such as health and safety, regulatory compliance, climate change mitigation, and overall environmental performance improvement. A key strategy involves choosing bio-based or synthetic base oils from renewable sources, reducing the carbon footprint compared to mineral oil-based lubricants. Advanced additive technologies, including friction modifiers and antioxidants, enhance lubricant performance, reduce wear, and minimize harmful by-products, contributing to a cleaner and more sustainable operating environment [14]. Sustainable lubricants play a crucial role in minimizing the environmental impact of industrial operations, offering a pathway to cleaner, more efficient, and environmentally conscious practices as industries globally transition toward a more sustainable future.

2.2 ENVIRONMENTAL IMPACT OF CONVENTIONAL LUBRICANTS

Conventional lubricants, long-standing workhorses in machinery operations, undeniably contribute to the efficient functioning of various industrial sectors. However, the environmental toll extracted by these lubricants raises significant concerns, prompting a critical evaluation of their life cycle and consequences [15]. Primarily derived from fossil fuels, conventional lubricants undergo processes from crude oil extraction to refining and blending, intensifying their ecological impact as shown in Figure 2.3 [16]. The ramifications extend across different facets, emphasizing the urgent need for a transition to sustainable alternatives.

2.2.1 Life cycle analysis

The life cycle of conventional lubricants commences with the extraction of base oils from crude petroleum. This initial phase involves processes that release greenhouse gas emissions, disrupt habitats, and pose the risk of oil spills. The extraction and transportation of crude oil contribute to climate change, habitat degradation, and potential ecological disasters. Furthermore, the refining process, aimed at transforming base oils into usable lubricants, intensifies the environmental impact. Air pollution, resulting from oil refining, further exacerbates the ecological consequences associated with conventional lubricants [17].

Figure 2.3 Environmental impact of conventional lubricants (adapted from Ref. [16]).

2.2.2 Resource depletion

Resource depletion is a significant concern associated with the life cycle of conventional lubricants. The manufacturing process, spanning crude oil extraction to refining and blending, intensifies resource depletion due to its high energy demand, particularly from non-renewable sources. The reliance on finite fossil fuel resources exacerbates environmental challenges linked to energy-intensive manufacturing, raw material extraction, limited lifecycle considerations, and a global supply chain [14, 18].

2.2.3 Greenhouse gas emissions

Conventional lubricants contribute significantly to greenhouse gas (GHG) emissions throughout their life cycle, from raw material extraction to disposal. Energy-intensive activities like drilling, refining, and transportation during extraction release carbon dioxide (CO_2) and pollutants, thus contributing to GHG emissions. The rising demand for lubricants intensifies the environmental impact of additive production. Transportation across the supply chain further amplifies emissions, highlighting the interconnectedness of the lubricant supply chain. During the operational phase, lubricants accumulate contaminants during machinery use, leading to additional emissions during combustion. Incineration of conventional lubricants releases CO_2, and their decomposition during disposal may yield methane (CH_4) if not managed responsibly [19]. This intricate web of GHG emissions emphasizes the urgency of transitioning to sustainable alternatives. Derived from renewable feedstocks, sustainable lubricants offer a promising solution with a lower carbon footprint and improved biodegradability, mitigating long-term environmental impact.

2.2.4 Disposal challenges

The disposal phase of conventional lubricants presents further environmental challenges. Improper disposal methods, including incorrect recycling or disposal in landfills, can cause soil and water contamination. Conventional lubricants often persist in the environment, contributing to pollution and posing risks to ecosystems and human health. Sustainable lubricants, on the other hand, offer an eco-friendlier alternative. Being biodegradable and potentially recyclable, they minimize pollution risks associated with conventional lubricants, contributing to a circular economy model [20].

2.3 GREEN LUBRICANTS

Green lubricants, also known as environmentally friendly or sustainable lubricants, signify a transformative shift in the lubrication industry, placing a strong emphasis on ecological responsibility and resource conservation. Unlike traditional lubricants derived from non-renewable fossil fuels, green lubricants utilize base oils sourced from renewable and sustainable raw materials such as plant oils, animal fats, and synthetic esters [14]. One key feature of green lubricants is their biodegradability, engineered to naturally break down into environmentally benign components. This characteristic not only reduces the risk of long-term environmental harm but also aligns with circular economy principles, promoting a closed-loop system where components can be recovered, reused, or safely returned to the environment without causing harm (see Fig. 2.4) [21].

Energy efficiency is another crucial aspect of green lubricants. With superior lubricating properties, they minimize friction and wear in machinery, leading to enhanced efficiency and energy savings. This contributes to improved overall system performance, addressing the global imperative to optimize energy consumption and reduce environmental impact. Contrary to misconceptions, green lubricants often exhibit superior lubricating properties compared to conventional counterparts, ensuring effective operation across various applications. While there may be a higher upfront cost, the total cost of ownership, considering factors like extended equipment life, reduced maintenance, and improved energy efficiency, makes green lubricants a financially prudent choice over the long term. Consumer awareness and evolving market trends are pivotal in driving the adoption of green

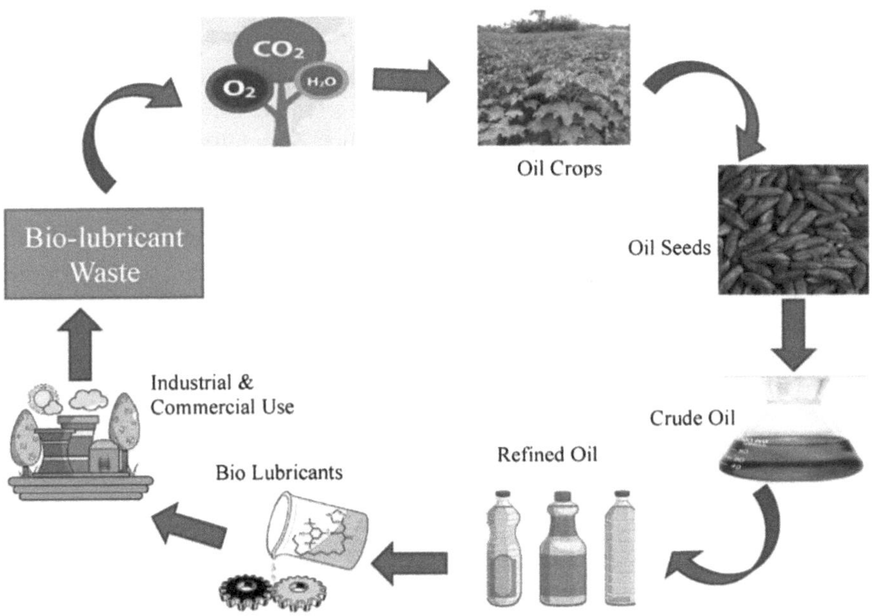

Figure 2.4 Lifecycle of green lubricants [21].

lubricants. As consumers become more environmentally conscious, there is a growing demand for products and services that align with sustainable values. Lubricant manufacturers are increasingly incorporating sustainability into their marketing strategies, emphasizing the environmental benefits of their products. Certifications and standards, such as those from the European Ecolabel and the USDA BioPreferred program, guide the industry toward sustainable practices [22]. Adherence to these standards fosters a competitive landscape where lubricant manufacturers strive to meet and exceed them, driving continuous improvement in sustainability practices.

In essence, green lubricants represent a multifaceted approach to addressing environmental challenges in the lubrication industry, embodying circular economy principles, renewable feedstocks, performance advancements, economic viability, and global adoption trends. The evolution of green lubricants reflects a crucial stride toward a more environmentally conscious and responsible future, requiring collaborative efforts, regulatory support, and a collective commitment to balancing ecological integrity with operational efficiency [8].

2.4 RECENT DEVELOPMENTS IN SUSTAINABLE GREEN LUBRICANTS

Recent developments in sustainable green lubricants have witnessed notable strides in the pursuit of environmentally friendly and high-performance lubrication solutions. Technological advancements have played a pivotal role in enhancing the performance of green lubricants. Manufacturers are investing in innovative formulations and additives such as nanoparticles and ionic liquids that elevate lubricant efficiency, thermal stability, and overall effectiveness (see Figure 2.5) [23]. The integration of synthetic

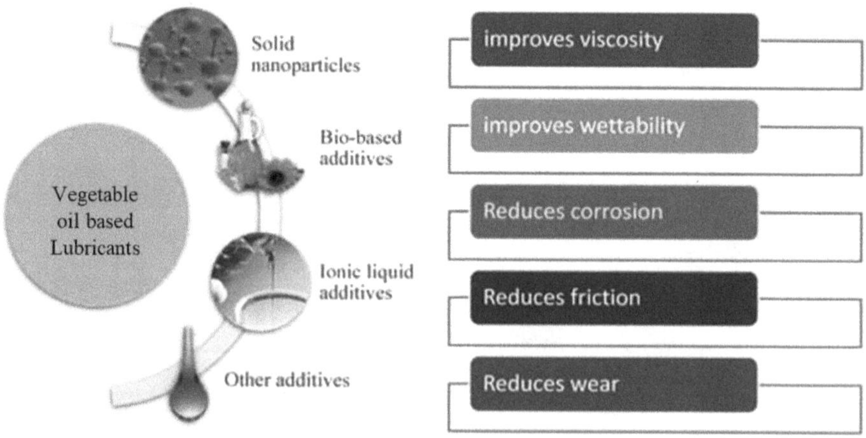

Figure 2.5 Effect of lubricant additive [23].

esters, polyalkylene glycols (PAGs), and other advanced compounds has contributed to the development of lubricants that not only meet but often exceed the performance benchmarks set by traditional petroleum-based counterparts.

2.4.1 Bio-based lubricants

Biofluids, as an innovative subset of lubricants, represent a forward-thinking and environmentally friendly solution in lubrication technology. Sourced from bio-based materials such as rapeseed, sunflower, soybean, and palm oil, these lubricants underscore a commitment to sustainability, addressing environmental concerns associated with traditional fossil fuel-derived lubricants. The utilization of renewable feedstocks supports sustainable practices, contributing to the reduction of carbon footprints in lubricant production. One of the key advantages of biofluids lies in their natural ability to break down into environmentally benign components, aligning seamlessly with sustainability objectives and minimizing the ecological impact associated with their use. This characteristic is particularly crucial in applications where lubricants may encounter soil or water, reducing the risk of long-term environmental harm. Beyond their eco-friendly attributes, biofluids exhibit superior lubricating properties, including exceptional film strength, wear protection, and oxidative stability. This versatility makes them well-suited for a diverse range of applications, ensuring optimal performance while meeting stringent environmental standards [24]. In essence, biofluids play a pivotal role in steering industries toward a more environmentally conscious future. Their superior performance, coupled with biodegradability and alignment with sustainability goals, positions them as key contributors to the transition toward eco-friendly and responsible industrial practices.

2.4.2 Synthetic ester lubricants: exceptional performance for demanding applications

Synthetic ester lubricants, crafted through esterification processes, stand out as high-performance solutions for demanding applications across industries. Known for exceptional thermal stability, these lubricants excel in extreme conditions, benefiting sectors like aviation and automotive racing where precision is vital. Despite higher initial costs, the long-term advantages—extended equipment life, enhanced efficiency, and reduced maintenance make synthetic ester lubricants a cost-effective investment. This class of lubricants represents a pinnacle in technology, seamlessly combining superior performance with environmental responsibility [25]. Their resistance to thermal breakdown ensures optimal reliability, positioning them as the preferred choice for critical applications.

2.4.3 Ionic fluids: unprecedented versatility and environmental benignity

Ionic fluids, also known as ionic liquids, represent a distinctive class of substances garnering attention for their unique properties and versatile applications. Composed entirely of ions, typically organic cations and organic or inorganic anions, ionic fluids offer low volatility and remain in liquid form over a wide range of temperatures. Their non-volatile nature reduces the risk of emissions, contributing to a lower environmental impact compared to traditional solvents [26]. The exceptional solvation properties of ionic fluids, allowing them to dissolve a wide array of substances, make them effective solvents for various chemical processes. This versatility has led to applications in catalysis, chemical synthesis, and extraction processes, thus showcasing their utility in green and sustainable chemistry practices [27]. Despite challenges such as potential toxicity and higher production costs, the tenable nature of ionic fluids positions them as promising candidates for applications ranging from chemistry and engineering to energy and environmental science.

2.4.4 Nanofluids

Nanofluids, a forefront innovation in lubrication technology, extend the trajectory toward environmentally conscious solutions, akin to biofluids. Integrating nanoparticles into conventional lubricants, typically at the nanometer scale, nanofluids enhance thermal conductivity, wear resistance, and stability, contributing to more efficient and durable lubrication. This not only reduces the need for frequent lubricant replacements but also minimizes the environmental impact associated with disposal. Notably, the improved heat transfer capabilities of nanofluids align with the global goal of optimizing energy consumption and decreasing the environmental footprint related to excessive energy use in various applications. Beyond performance enhancements, nanofluids play a role in resource conservation by extending the lifespan of equipment and machinery through reduced friction and wear, thereby lessening the demand for raw materials and resources [28, 29]. While considering potential challenges, such as the environmental impact of nanoparticle production and disposal, ongoing research aims to ensure that nanofluids remain aligned with sustainability goals, offering a futuristic and eco-friendly dimension to lubrication practices.

2.4.5 Applications and performance of green lubricants

Green lubricants, renowned for their eco-friendly formulations, have become integral across industries, driven by a global shift toward sustainability. In automotive applications, these lubricants enhance fuel efficiency,

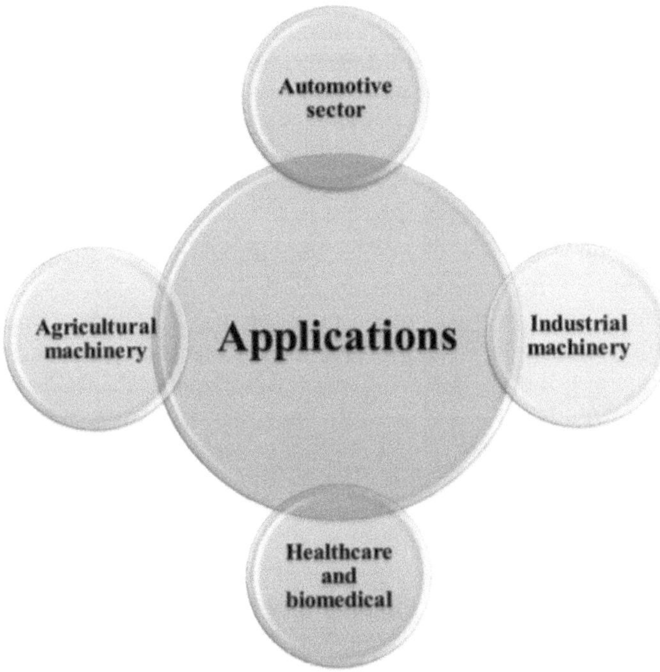

Figure 2.6 Application of green lubricants.

reduce emissions, and contribute to the overall performance of vehicles. In industrial manufacturing, they play a vital role in machinery lubrication, ensuring smooth operations while aligning with sustainability goals. Their reach extends to sensitive domains like healthcare, where biocompatible biofluids contribute to the safety of medical equipment. The performance benefits of green lubricants include biodegradability, renewable feedstocks, reduced emissions, enhanced equipment life, and safety advantages [30]. As industries prioritize environmentally conscious practices, green lubricants emerge as key contributors to both operational excellence and ecological responsibility. Figure 2.6 shows the potential application of green lubricants. The potential application of green lubricants includes sectors such as the automotive sector, industrial machinery, agricultural machinery, and healthcare and biomedical sectors.

2.5 CHALLENGES AND FUTURE PERSPECTIVES

Although advancements in sustainable green lubricants have brought about notable improvements in lubrication technology, presenting an eco-friendly alternative to traditional lubricants, these innovations face several challenges that need to be addressed for widespread adoption. One significant hurdle lies in meeting or surpassing the performance standards set by conventional lubricants. Green lubricants must demonstrate efficacy, stability, and durability under diverse operating conditions to gain industry-wide acceptance. Additionally, the economic viability of these sustainable alternatives poses a challenge [31, 32]. Production costs, often higher than those of conventional lubricants, necessitate ongoing efforts to optimize production processes, explore cost-effective feedstocks, and enhance competitiveness. Compatibility issues with existing materials, technologies, and lubricants present another challenge.

Furthermore, the limited availability of bio-based feedstocks, such as plant oils, underscores the importance of developing a consistent and sustainable supply chain. Future perspectives in sustainable green lubricants pivot on addressing these challenges through strategic approaches. Research and development efforts will be crucial for fine-tuning formulations and overcoming performance limitations. Innovations in production processes, efficient sourcing of feedstocks, and scaling up production are key to achieving economic viability. Regulatory support plays a pivotal role, with governments incentivizing the use of sustainable lubricants through policies and standards. Increasing public awareness about the environmental benefits and long-term advantages of green lubricants is essential for driving demand and acceptance. Moreover, integrating sustainable lubricants into the broader concept of a circular economy, where recycling and responsible disposal practices are emphasized, holds promise for enhancing overall sustainability. A commitment to innovation, economic viability, regulatory support, and environmental responsibility will collectively shape the trajectory of sustainable green lubricants toward mainstream adoption and contribute to a more sustainable and resilient future in lubrication technology.

2.6 CONCLUSIONS

In conclusion, the global shift toward sustainability has instigated significant strides in lubrication technology, with a particular emphasis on the evolution of green lubricants. Recent advancements in green lubricants encompass the utilization of bio-feedstocks, such as plant-based oils and waste cooking oils, serving as eco-friendly substitutes for traditional lubricants. The integration of bio-additive technology, coupled with the burgeoning field of nanotechnology featuring nanoparticles like graphene and nanoclay additives, not only enhances sustainability but also improves lubricant properties, boosting efficiency, and extending equipment lifespan. The focus extends to developing lubricants with heightened biodegradability and establishing closed-loop systems for recovery and recycling, aligning with the principles of a circular economy. This chapter provides a comprehensive overview of recent developments in sustainable lubricant technologies, spotlighting environmentally friendly alternatives that address performance, biodegradability, and resource efficiency. As industries continue to prioritize eco-conscious practices, these advancements herald a more sustainable future for lubrication technologies, where environmental responsibility converges with enhanced performance and efficiency.

REFERENCES

1. Soodoo, N., Bouzidi, L., Narine, S. S. (2023). Fundamental structure–function relationships in vegetable oil-based lubricants: A critical review. *Lubricants*, *11*(7), 284.
2. Bambam, A. K., Dhanola, A., Gajrani, K. K. (2022). Machining of titanium alloys using phosphonium-based halogen-free ionic liquid as lubricant additives. *Industrial Lubrication and Tribology*, *74*(6), 722–728.
3. Salem, A., Hopkins, C., Imad, M., Hegab, H., Darras, B., Kishawy, H. A. (2020). Environmental analysis of sustainable and traditional cooling and lubrication strategies during machining processes. *Sustainability*, *12*(20), 8462.
4. Nagendramma, P., Pandey, A., Singh, A. P., Ray, A. Tribological aspects of alternative materials in the manufacture of green lubricants for industrial applications. In: *Tribology in Sustainable Manufacturing* (pp. 125–144). CRC Press.
5. Gul, M., Masjuki, H. H., Kalam, M. A., Zulkifli, N. W. M., Mujtaba, M. A. (2020). A review: role of fatty acids composition in characterizing potential feedstock for sustainable green lubricants by advance transesterification process and its global as well as Pakistani prospective. *Bio Energy Research*, *13*, 1–22.
6. Mannekote, J. K., Kailas, S. V., Venkatesh, K., Kathyayini, N. (2018). Environmentally friendly functional fluids from renewable and sustainable sources—a review. *Renewable and Sustainable Energy Reviews*, *81*, 1787–1801.
7. Bambam, A. K., Alok, A., Rajak, A., Gajrani, K. K. (2023). Tribological performance of phosphonium-based halogen-free ionic liquids as lubricant additives. *Proceedings of the Institution of Mechanical Engineers, Part J: Journal of Engineering Tribology*, *237*(4), 881–893.
8. Shah, R., Woydt, M., Zhang, S. (2021). The economic and environmental significance of sustainable lubricants. *Lubricants*, *9*(2), 21.
9. Li, J., Wu, Y. (2014). Lubricants in pharmaceutical solid dosage forms. *Lubricants*, *2*(1), 21–43.
10. Pinheiro, C. T., Quina, M. J., Gando-Ferreira, L. M. (2021). Management of waste lubricant oil in Europe: A circular economy approach. *Critical Reviews in Environmental Science and Technology*, *51*(18), 2015–2050.
11. Miller, S. A., Landis, A. E., Theis, T. L., Reich, R. A. (2007). A comparative life cycle assessment of petroleum and soybean-based lubricants. *Environmental Science and Technology, 41*(11), 4143–4149.
12. Balakrishnan, M., Sacia, E. R., Sreekumar, S., Gunbas, G., Gokhale, A. A., Scown, C. D., Toste, F. D., Bell, A. T. (2015). Novel pathways for fuels and

lubricants from biomass optimized using life-cycle greenhouse gas assessment. *Proceedings of the National Academy of Sciences*, *112*(25), 7645–7649.
13. Cecilia, J. A., Ballesteros Plata, D., Alves Saboya, R. M., Tavares de Luna, F. M., Cavalcante, C. L., Rodríguez-Castellón, E. (2020). An overview of the biolubricant production process: Challenges and future perspectives. *Processes*, *8*(3), 257.
14. Boyde, S. (2002). Green lubricants. Environmental benefits and impacts of lubrication. *Green Chemistry*, *4*(4), 293–307.
15. Pal, M., Singhal, S. (2000). Environmentally adapted lubricants, Part I. An overview. *Journal of Synthetic Lubrication*, *17*(2), 135–143.
16. Sayed, K., Baloo, L., Sharma, N. K. (2021). Bioremediation of total petroleum hydrocarbons (TPH) by bioaugmentation and biostimulation in water with floating oil spill containment booms as bioreactor basin. *International Journal of Environmental Research and Public Health*, *18*(5), 2226.
17. Raimondi, A., Girotti, G., Blengini, G. A., Fino, D. (2012). LCA of petroleum-based lubricants: state of art and inclusion of additives. *International Journal of Life Cycle Assessment*, *17*, 987–996.
18. Singh, G., Aggarwal, V., Singh, S. (2020). Critical review on ecological, economical and technological aspects of minimum quantity lubrication towards sustainable machining. *Journal of Cleaner Production*, *271*, 122185.
19. Wong, V. W., Tung, S. C. (2016). Overview of automotive engine friction and reduction trends—effects of surface, material, and lubricant-additive technologies. *Friction*, *4*, 1–28.
20. Bartz, W. J. (1998). Lubricants and the environment. *Tribology International*, *31*(1–3), 35–47.
21. Joshi, J. R., Bhanderi, K. K., Patel, J. V. (2023). A review on bio-lubricants from non-edible oils—recent advances, chemical modifications and applications. *Journal of the Indian Chemical Society*, *100*(1), 100849.
22. Khadem, M., Kang, W. B., Kim, D. E. (2024). Green tribology: A review of biodegradable lubricants—properties, current status, and future improvement trends. *International Journal of Precision Engineering and Manufacturing Green Technology*, 11. 565–583.
23. Rahman, M. H., Warneke, H., Webbert, H., Rodriguez, J., Austin, E., Tokunaga, K., Menezes, P. L. (2021). Water-based lubricants: Development, properties, and performances. *Lubricants*, *9*(8), 73.
24. Cermak, S. C., Evangelista, R. L., Jackson, M. A., Compton, D. L., Knothe, G., Laszlo, J. A., Evans, K. O. (2014). Biobased lubricants and 17 functional products from cuphea oil. *Surfactants in Tribology, 4*(4), 443.
25. Eisentraeger, A., Schmidt, M., Murrenhoff, H., Dott, W., Hahn, S. (2002). Biodegradability testing of synthetic ester lubricants––effects of additives and usage. *Chemosphere*, *48*(1), 89–96.
26. Bambam, A. K., Vats, P., Gajrani, K. K. (2023). Wetting behaviour of phosphonium-based halogen-free ionic liquids as lubricant additives on various metal surfaces. *Materials Today: Proceedings*.https://doi.org/10.1016/j.matpr.2023.06.066
27. Bambam, A. K., Dhanola, A., Gajrani, K. K. (2023). A critical review on halogen-free ionic liquids as potential metalworking fluid additives. *Journal of Molecular Liquids*, 380, 121727.

28. Dubey, V., Sharma, A. K., Vats, P., Pimenov, D. Y., Giasin, K., Chuchala, D. (2021). Study of a multicriterion decision-making approach to the MQL turning of AISI 304 steel using hybrid nanocutting fluid. *Materials*, *14*(23), 7207.
29. Gajrani, K. K., Ram, D., Sankar, M. R., Dixit, U. S., Suvin, P. S., Kailas, S. V. (2017). Machining of hardened AISI H-13 steel using minimum quantity eco-friendly cutting fluid. *International Journal of Additive and Subtractive Materials Manufacturing*, *1*(3–4), 240–256.
30. Chidambaram, C. T., Kumar, S., Martin, V., Joseph, P. V., Vignesh, S., Kumar, S. V. (2023). Green lubrication in manufacturing. *Tribology in Sustainable Manufacturing*, 83–109.
31. Bambam, A. K., Gajrani, K. K. (2023). Challenges in achieving sustainability during manufacturing. In *Sustainable Materials and Manufacturing Technologies* (pp. 108–124). CRC Press.
32. Srinivas, M. S., Gajrani, K. K., Udayakumar, A., Sankar, M. R. (2021). Sustainable machining of Cf/SiC ceramic matrix composite using green cutting fluids. *Procedia CIRP*, 98, 151–156.

Chapter 3

Tribology of green lubricants

Environmental benefits and impacts

Richa Tiwari, Pratibha Agrawal,
Amit J. Agrawal, and Vijay Karadbhajane

3.1 INTRODUCTION

Lubricants are substances that exist in quasi-solid, solid, and gaseous or in liquid state and are capable of improving the efficiency. In order to protect and separate the surfaces, lubricants set up a layer between interconnecting surfaces and also provide a reduction in the risk of failures, transport or dissolve the foreign particles, precluding rust or corrosion, dissipating the heat, sealing clearances, carrying away debris and contaminants, and maintaining the machine function reliable [1]. In tribology, lubricants play a significant role and are formed by a mixture of base oil and chemical additives. Generally, lubricants contain ~90% mineral base oil and ~10% chemical additive for the lubrication of mechanical parts. These lubricants are of low cost as compared to other lubricants and have good oxidation stability (quite stable) [2]. The overall utilization of mineral lubricants is close to 41 million tons, which is increasing accordingly and is expected to rise ~1.5% yearly in the next few years, which also increases their disposal issues since these lubricants are poorly biodegradable and raise environmental concerns. The automotive lubricant as well as the automotive fuel industry because of the use of conventional fuel faces environmental directives and energy security [3]. Several studies have shown the hazardous effect of mineral lubricants on living beings, and Chand et al. [4] suggested that prolonged contact with skin causes various skin infections from mild to acute type such as skin discoloration, eczema, itching, skin thickening, and dermatitis. These lubricants contain PAHs (polycyclic aromatic hydrocarbons) are the organic compounds consisting of two/more fused aromatic rings and on burning produce gases containing hydrocarbons, CO_2, SO_2, and CO, which on inhalation results in acute illness of living beings. Shankar et al. [5] studied the effect of lubricating oil on marine systems and reported that lube oil is applied to marine engines for the reduction of wear and friction and elimination of high temperature to induce high power output and efficiency. After the spilling of lube oil into the sea, it embarks to expand onto the surface of the sea and increases the atmospheric processes namely dispersion, dissolution, sedimentation, emulsification, and

 DOI: 10.1201/9781003535447-3

evaporation, thereby affecting the marine environment. It has been found that morphological defects mainly tail fin fold defects were elevated with higher frequency. The factors related to atmospheric spills/disposal, global warming, and pollution can be attenuated by the utilization of bio-based compounds, namely vegetable oil (edible/non-edible).

Therefore, this issue makes the researchers find an ecofriendly substitute for mineral lubricants, and hence, in this context, green lubricants that lessen wear/abrasion, are energy-efficient, and have good biodegradability and bioavailability are efficient substitutes. Vegetable oils that are easily available have low ecological toxicity and can be utilized in making chemicals that can be produced from mineral oil-based substances. Thus, the substances obtained from edible/non-edible vegetable oils have applications in different areas such as in the fuel industry (biofuels), in paints and coatings, in the food industry, in adhesives, in plasticizers, in construction materials, and in pharmaceutical industries. Further, the lubricity properties of vegetable oils enable them to make lubricant substances in the tribology industry [6]. These oils impart various advantages such as viscousness and increased lubricity, better shear stability, reduced volatility, and high biodegradability as compared to synthetic and mineral oils, and are tremendously used for the production of green lubricants on a large scale. Therefore, this chapter discussed various types of lubricants such as mineral lubricants, synthetic lubricants, bio-lubricants, and feedstocks for the production of green lubricants. Non-edible oils as a raw material are used for manufacturing green lubricants, and their properties are also discussed. Further, an analysis of the catalyst involved in the conversion of vegetable oil into green lubricants and the feasibility of mahua oil as a base oil for the production of bio-lubricants and their properties are also discussed. The preparation of green lubricants from vegetable oil reduces the price as compared to mineral oil lubricant since fossil fuel reserves are decreasing. Hence, this chapter provides a detailed analysis and production of green lubricants and their environmental effects.

3.2 TYPES OF LUBRICANTS

The parts of machinery are lubricated by materials usually liquids (~ solids or semi-solids) known as lubricants, which extended their lifespan while reducing friction. Various chemical additives and base fluids are utilized for the formulation of lubricants, and based on the class of base fluids, lubricants are classified into different categories namely mineral or petroleum-based synthetics such as poly-alkylene glycols, silicones, poly-alpha-olefins, synthetic esters, and green or bio-lubricants [7].

The complex nature of mineral oils contains hydrocarbons (C_{20}–C_{50}), which consist of waxes (linear alkanes), paraffins (branched alkanes), naphthenes (alicyclic), aromatic, olefinic species, and heteroatoms (mostly sulfur). Mineral oils are easily obtainable in different spans of viscosities as compared to natural oils, low-priced, and are stable. However, the drawback associated with mineral oils is that they have different properties due to their origin as they are procured from various fields, are less biodegradable, and pollute the environment by liberating toxic components. Sulfur compounds and heavy metals are released into the atmosphere by additives such as zinc dialkyl acrylates and acrylates, which are generated from crude oil resources and the low molecular weight compounds present in it causing a reduction in its flash point. However, globally mineral-based lubricants are extensively used since they have more oxidation stability [8].

Another class of lubricants are synthesized chemically known as synthetic lubricants containing enhanced properties and have many applications namely marine, aviation, automobile, metal-working, and industrial. Synthetic lubricants namely poly-alpha olefins (PAOs) are widely used in industries. The ester group enhances the lubricant property by providing fluidity at low temperatures, decreases the volatility at higher temperature, and confers the metal surface affinity. Monoesters are usually utilized as an additive or base stock in formulation of the lubricants for hydraulic oils, greases, engines, and metal-working fluids, and similarly, polyol esters [trimethylolpropane (TMP) trioleate, pentaerythritol (PE), and NPG (neopentylglycol) esters] are utilized in hydraulic oils, automotive oils, textile lubricants, metal-working fluids, and greases as an additive or base stocks since these are thermally stable [9]. However, the drawbacks related to synthetic lubricants are these are costlier than lubricants derived from mineral oils, their biodegradability is lower than the lubricants based on vegetable oils, more toxic, have lower friction resistance, and so on. Despite the price of bio-lubricants obtained from vegetable oils, these are favorable alternatives to mineral and synthetic oil-based lubricants and have characteristic properties such as good lubricity, high flash point, high viscosity index, high biodegradability, low aquatic toxicity, and superior anticorrosion properties. The vegetable oil structure indicates that it is amphiphilic with polar end groups and a long fatty chain. Vegetable oils derived from plant seeds contain phospholipids and galactolipids (polar lipids), 91–95% triacylglycerols, diacylglycerols, monoacylglycerols, and free fatty acids. The additives obtained from vegetable oil work optimally at higher pressures and show superlative tribological characteristics [10]. The viscosity index of non-edible oil is approximately 221 while that of mineral oil is 89–101, and it was estimated that the lubricants based on vegetable oils due to high viscosity index functioned well at higher temperatures by keeping the lubricating coating unscathed. Further, the fire resistance properties of lubricants were resolute by fire and flash points and these lubricants tend to decrease the operating temperature and increase the energy savings (by 16%). Therefore, these lubricants are biodegradable having lower sulphur and higher heat content. Lubricants based on vegetable oil can be applied as hydraulic oils,

Table 3.1 Applications of various vegetable oils [11]

Sr. No.	*Vegetable oil*	*Applications*
1	Palm oil	Rolling lubricant and grease
2	Linseed oil	Lacquers, stains, and varnishes
3	Jojoba oil	Lubricant applications and cosmetic industry
4	Soybean oil	Hydraulic oil, pesticides, plasticizers, disinfectants, and printing inks
5	Cuphea oil	Cosmetics and motor oil
6	Canola oil	Tractor transmission fluids, hydraulic fluids, and metal-working oils

grease, multi-purpose oils, and engine oils. Some of the vegetable oils along with their applications are shown in Table 3.1.

3.3 OVERVIEW OF FEEDSTOCKS FOR GREEN LUBRICANTS

Green lubricants (bio-lubricants) are a class of lubricants that are produced from natural feedstocks, namely non-edible and edible vegetable oils and fats from animals that are ecofriendly, innocuous, and renewable and are generally prepared from vegetable oils, which have unique characteristics because of their chemical composition as compared to mineral oils. Lubricants obtained from vegetable oils usually have a viscosity index and a higher flash point and show good lubrication properties; however, fatty acid contents present in them reduce their thermal/oxidation properties. Different methods (transesterification, epoxidation, and hydrogenation) can reduce the degree of unsaturation in the oil, thereby enhancing the tribological characteristics making it suitable for lubrication [12]. Vegetable oils are obtained from different types of crops depending on geographical and climate factors. Soybean oil, rapeseed oil, sunflower oil, coconut oil, cottonseed oil, palm oil, karanja oil, jatropha oil, etc., are vegetable oils that can be utilized for the production of green lubricants. Generally, vegetable oils are triglycerides containing a long chain of fatty acids that has unsaturated and unconjugated carbon atoms, which gives the oil low hydrolytic stability, low fluidity, and low bio-resistance. Table 3.2 shows the composition of fatty acids present in different types of edible oils.

Thus, from Table 3.2, it was estimated that with the knowledge of different fatty acids in edible oils (vegetable oils), the characteristics of the green lubricant can be improved by applying different methods. However, non-edible vegetable oils, namely castor, rice bran, jatropha, neem, linseed, and mahua oil, are precedence in the generation of bio-lubricants or biofuels as compared to edible oils since these are economical. In view of cultivation, non-edible crops are of interest in the production of green lubricant since there is little or no effect on the world food prices as well as no effect

Table 3.2 Fatty acids composition present in vegetable oils [12, 13]

Edible oils	*Stearic*	*Linoleic*	*Palmitic*	*Linolenic*	*Oleic*
Soybean oil	5.3	63.6	6.1	4.8	20.4
Sunflower oil	5.4	66.6	6.2	—	21.2
Cottonseed oil	2.1	37.8	17.8	1.1	41.2
Canola oil	0.9	22.4	2.3	8.4	64.5
Corn oil	2.1	59.6	10.4	0.8	26.9
Peanut oil	2.9	30.1	9.8	—	64.6
Olive oil	2.2	16.3	14.2	1.9	63.8
Palm oil	4.8	10.2	42.2	—	40.9
Rapeseed oil	0.8	22.5	3.6	8.3	19.6
Coconut oil	2.4	1.1	9.2	—	7.3

on the cultivation of agricultural (edible) sources, and these crops can be easily cultured in the lesser fertile field as well as can be sowed nearby irrigation canals, railways, degraded forests, roads, and poverty-stricken regions. Therefore, non-edible oils emerging as a potential substitute for the production of bio-lubricants because these are less harmful, possess good biodegradability, and can reduce the problems associated with energy, debate related to food versus fuel and environment. Some of the non-edible oils suitable for bio-lubricant production are as follows.

3.3.1 Neem oil

The non-edible neem oil is produced from an evergreen tree named *Azadirachta indica* and the tree grows in different geological regions of India. In India, the environment is favorable for the growth of neem trees and the tree yields 38–40 kg of fruit each year and the seeds produce approximately 37–39% of oil, which is bitter in taste, dark brown color, and has strong odor and the average production of neem oil annually is about 30,000 tons. Neem oil consists of triglycerides, steroids (beta-sitosterol, stigmasterol, etc.), and a substantial number of triterpenoids of which a bioactive compound termed Azadirachtin whose content ranges from 300–2500 ppm based on the quality of the neem seeds (crushed) and extraction technology. Furthermore, neem oil consists of saturated fatty acids such as stearic acid and palmitic acid and also consists of polyunsaturated fatty acids namely linoleic and oleic acids, which makes it a suitable raw material for the production of lubricants [14].

3.3.2 Karanja oil

The evergreen karanja tree (*Pongamia pinnata*, poonga oil tree) is medium in size, hardy, semi-deciduous, drought-resistant, and perennial tree, which is native to tropical, other parts of India, and can also be found in parts of Japan, Australia, China, and Malaysia. The leguminous nitrogen-fixing tree generates seeds (oil-rich) containing 40–56% oil twice yearly, and the kernels of the tree are white in color and covered with reddish thin skin, per tree the kernel yield is ~8–24 kg and the composition of kernel (air-dried) contains oil (33.9%), moisture (19.1%), and protein (17.5%). The oil obtained from karanja seeds is non-edible, yellow-orange or brown in color, has bitter taste, is viscous in nature, and contains unsaturated fatty acids namely stearic, linoleic, palmitic, lignoceric, behenic, arachidic, and oleic acid in which oleic acid is 45–70% as major constituent, making it a feedstock for green lubricants. In India, approximately more than 151,000 tons of karanja seeds are produced from which approximately 51,000 tons of oil are extracted, which have applications in making soap, pesticides, and medicines and the oil cake can be utilized as a combustible and a fertilizer [15].

3.3.3 Castor oil

Ricinus communis commonly known as castor bean belongs to the Euphorbiaceae family and contribute about 58–60% of non-edible oil. The plant grows in various geographical areas and the oil is produced by extracting and expressing the castor plant seed and is inexpensive and ecofriendly. The average percentage of castor seed oil is about 45–59%, and globally, the annual production of oil is about 1.8–2.0 million tons. Castor oil is pale yellow in color, viscous, non-drying, and non-volatile with slightly acrid taste having imperceptible odor and better shelf life and does not become rancid until subjected to high temperature. Castor oil contains fatty acids such as oleic, palmitic, linolenic, ricinoleic, etc., and in all of the fatty acids, the oil contains approximately 70–90% of ricinoleic acid (monounsaturated fatty acid) whose structure constitutes the chemistry of castor oil and imparts the properties namely prominent miscibility and viscosity, low freezing point, and iodine content. However, other parameters such as hydroxyl groups, single location of unsaturation, and carboxylic groups also contribute toward the chemistry of castor oil [16]. The hydroxyl group present in it provides the characteristics of hydrogen bonding to the castor oil, which imparts high viscosity and makes it suitable for utilization in blending lubricants. The various physicochemical characteristics of non-edible oils are shown in Table 3.3.

3.4 CATALYSTS EMPLOYED IN GREEN LUBRICANTS PRODUCTION

Different catalysts have been employed for the production of green lubricants from natural sources. Some of them are as follows.

Table 3.3 Different physicochemical characteristics possessed by non-edible oils [17]

	Non-edible oils			
Properties	*Jatropha*	*Castor*	*Karanja*	*Neem*
Kinematic viscosity [(40°C), mm^2/s]	4.81	15.26	4.79	5.22
Density (kg/m^3)	880–939	900–954	921–936	885–918
Oxidation stability (110°C, h)	2.4	1.09	6.0	7.2
Cloud point (°C)	2.8	−13.3	9.0	14.3
Flash point (°C)	136–224	260	150	42

3.4.1 *p*-Toluene sulfonic acid

p-toluene sulphonic acid (*p*-TSA) is a solid acid catalyst and have been employed in organic synthesis as a catalyst in reactions such as esterification, etherification, alkylation, and acetylation since it is inexpensive, easily available and provide high catalytic activity. Therefore, Moreira et al. [18] synthesized green lubricant from moringa oil using *p*-toluene sulfonic acid (*p*-TSA) as a catalyst. The first step involves the preparation of moringa fatty acids in which 39 g of moringa oil was mixed with NaOH solution and heated with stirring at 363 K for 4 h. After that, extraction of ethyl acetate was carried out followed by cooling the solution at 323 K and neutralization with HCl. The extract was then dried and used for the preparation of moringa trimethylolpropane esters (MTMPE) and pentaerythritol esters (MPEE). For MTMPE, moringa fatty acid (19.7 g) and 24.6 ml of toluene were mixed with stirring at 408 K using 0.21 g of *p*-TSA catalyst for 6 h. The obtained mixture was then cooled and washed with H_2O at room temperature and then the organic layer was separated and dried and purified with the help of flash chromatography. Similarly, for MPEE, 14.9 g of moringa fatty acid was mixed with 1.36 g of pentaerythritol and then 0.146 g of the *p*-TSA catalyst was added to it, and the obtained ester was dried and collected. It was found from the experiment that in the presence of a *p*-TSA catalyst, approximately 98.5% and 96% conversion to MTMPE and MPEE was observed and thus the ester formed showed high a viscosity index of approximately 170.1, non-toxic and had good thermal stability and can be used as a lubricant.

3.4.2 Titanium iso-propoxide

Encinar et al. [19] utilized titanium the iso-propoxide catalyst for the conversion of rapeseed and castor oil into lubricant with various alcohols. The first stage of the process consists of the conversion of rapeseed and castor oil into FAME (fatty acid methyl esters), which involves the reaction of oils with methanol in the required ratio (6:1). Heating at a temperature of 338 K takes place with the formation of 3 moles of FAME and by-product (glycerol). In the second stage of the process, the production of green lubricant takes place in which the obtained FAME reacted with various alcohols such as 1-heptanol, 2-ethyl-1-hexanol, etc., in different ratios in the presence of titanium iso-propoxide catalyst by the transesterification method at different temperatures of 423 K, 433 K, and 444 K. It was observed that titanium iso-propoxide is effective in the production of green lubricant (yield 92–93%), and the lubricant shows higher flash points of 223–270°C as compared to mineral oil lubricant and low viscosity.

3.4.3 Amberlite IR-120

Rajwani et al. [20] utilized a blend of vegetable oil, i.e., DCO (dehydrated castor oil) and WRBO (waste rice bran oil) to synthesize green lubricant in the presence of a heterogeneous catalyst Amberlite IR-120. The process involves the formation of BOME (blended oil methyl esters) in which blended oil was mixed with methanol in the ratio of 1:12 and sodium methoxide and the mixture was heated in a flask at 338 K for 180 min. After the separation of methyl esters, it was then converted into blended oil fatty acid methyl ester epoxides by epoxidation method in which acetic acid, hydrogen peroxide, and BOME in the ratio of 0.5:1.5:1 and 15 wt% strong cationic ion-exchange resin, Amberlite IR-120, were employed. The product thus obtained was washed with water until the pH became neutral. The yield of the product obtained was 92–94%, and it was estimated by the characterization of the lubricant that it shows high thermal stability and improved pour point (−4.6°C) and can be utilized for hydraulic functions.

Further, the summary of the catalyst employed for the preparation of green lubricant is shown in Table 3.4.

3.5 MAHUA OIL AS FEEDSTOCKS

Mahua tree (*Madhuca indica*) grows in the vicinity of tropical regions in India and belongs to the Sapotaceae family having different species such as *Madhuca longifolia*, *Madhuca butyracea*, and *Madhuca latifolia*, and because of its extensive various applications, it is regarded as a principal tree for economic purpose. Mahua oil is produced from *Madhuca indica* seeds (kernel of seeds), which consists of elements namely calcium, magnesium, sodium, nitrogen, carbon, and phosphorous by solvent extraction process, is pale yellow in color, and at room temperature it exists in semi-solid form. During cold atmosphere, the oil is thrown out as it solidifies (buttery consistency) and the oil contains higher acidity. The triacylglycerol content in the oil is amphiphilic in nature, which makes it suitable for the production of functional fluid and lubricant (green lubricant) and substitute for mineral base oil. As reported by Kumar et al. [27], mahua oil is yellowish green and pale yellow in color having a typical taste, characteristic odor,

Table 3.4 List of catalyst applied for the conversion of vegetable oils into bio-lubricants

Vegetable oils	*Catalysts employed*	*Catalyst loading*	*Reaction condition*	*Conversion*	*Inference*	*Ref.*
Jatropha oil	NaOMe	0.81% (w/w)	T = 473 K t = 150 min	45–47%	Pour point = −6.3°C Viscosity index = ~182	[21]
Palm oil	$CaO_{15}SrO$	1.2% (w/w)	T = 453 K	80–83%	—	[22]
Castor oil	Dowex 50W-X8	9.8 wt%	T = 373 K	13–23%	Pour point = −38.6°C Viscosity index = 7.8	[23]
Castor oil	Perchloric acid	Not mentioned	t = 7 h	Not mentioned	Pour point = −18.7°C Flash point = 268°C	[24]
Palm kernel oil	Conventional KOH	5 wt%	T = 373 K, T = 4 h	Not mentioned, yield= 51%	Pour point = −9°C Viscosity index = 110	[25]
Palm and jatropha oil	NaOMe	0.9–1.0 wt%	T= 121–130°C, t = 4 h	Not mentioned, yield = 95–98%	Pour point = −3°C Viscosity index = 140	[26]

a specific gravity of 0.905, peroxide value 10.1, saponification value ~195, iodine value 62–70.5, and refractive index ~1.453–1.461, and the composition of fatty acids present in mahua oil is shown in Figure 3.1.

The oil is utilized for many purposes in industries such as for the production of creams (skin care), soaps and detergents, and vegetable butter, and in addition, it can further be utilized as a fuel in machines and engines. Karupannasamy et al. [28] utilized mahua oil for the preparation of lubricant. The process involves an emulsifier polysorbate-80 for the preparation of an oil emulsion mixture. This mixture is then blended with H_2O (1:20). The analysis of the tribological characteristics of the obtained emulsion indicates that there is a reduction in friction forces, coefficient of friction, and wear rate. Singh et al. [29] chemically modified the nature of mahua oil in three steps so as to convert it into a suitable bio-lubricant by first converting it into FAME and then with the use of trimethylolpropane (3.5:1 ratio), catalyst NaOMe at 393 K, and concluded that at different loads the tribological analysis showed an enhancement in anti-wear and frictional characteristics. Chaurasia et al. [30] utilized mahua oil for lubricant production by epoxidation method in the presence of copper oxide nanoparticles. In this method, a mixture of mahua oil (10.73 ml) and peracetic acid (37.13 ml) was heated at 293 K for 270 min, then decant into the water, kept for 60 min, and the obtained epoxide oil was separated. The obtained oil was then mixed with copper oxide nanoparticles in different proportions (0.2, 0.4, 0.8%), and subsequently Triton X-100 (surface modifier, 50 wt% vs nanoparticles) was utilized for the stabilization of nanoparticles in the resultant oil. The agitation of the mixture was carried out for 41 min with the help of an ultrasonic probe sonicator at 333 K. The obtained lubricant's

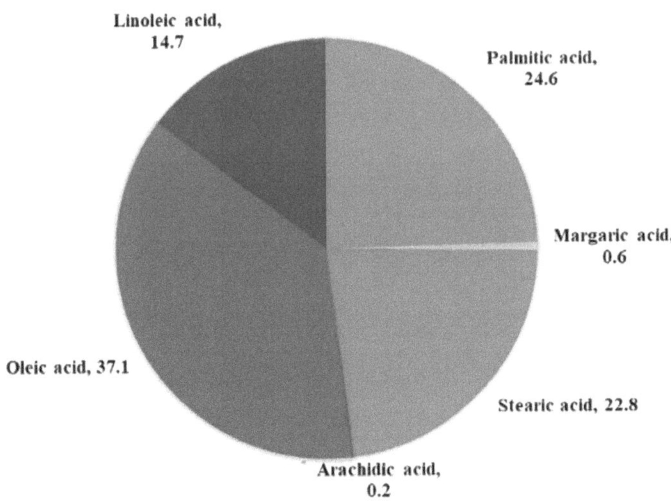

Figure 3.1 Mahua oil fatty acid composition [27].

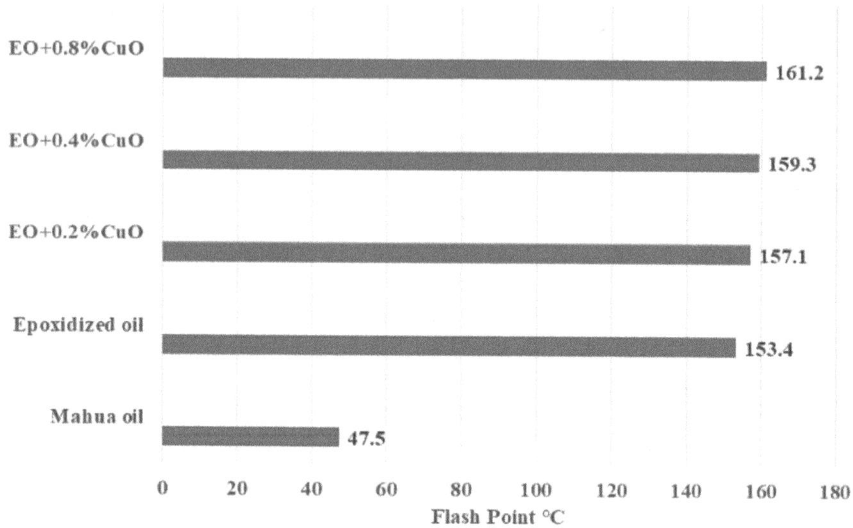

Figure 3.2 Flash point of various lubricants [30].

physicochemical characteristics showed an improvement in anti-wear characteristics and a decrease in friction coefficient and with the addition of nanoparticles (0.2, 0.4%) it showed enhanced lubrication properties.

As shown in Figure 3.2, it was observed that the flash point of the lubricants increases with an increase in copper oxide nanoparticle concentration, which increases the lubricant's upper operating temperature and therefore increases the lubricity property of the resultant epoxide oil.

3.6 CONVERSION OF MAHUA OIL TO DIETHYLENE GLYCOL (DEG)-BASED BIO-LUBRICANT

3.6.1 Materials

Feedstock mahua oil was acquired from domestic commerce and chemicals namely H_2SO_4 (98%), NaOMe, NaOH, CH_3OH, TMP (trimethylolpropane), DEG (diethylene glycol), PE (pentaerythritol), potassium permanganate, butyric anhydride, acetic anhydride, and C_2H_5OH were acquired from SD Fine Chemicals Pvt Ltd.

3.6.2 Synthesis of mahua oil FAME

Mahua oil contains a higher acid value of ~29.8% (>1%, required for transesterification), indicating a higher FFA (free fatty acid) content, and hence,

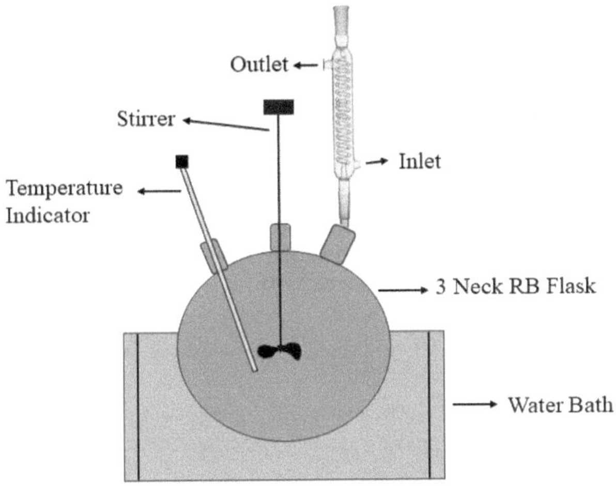

Figure 3.3 FAME synthesis setup.

the two-stage pretreatment process has been conducted to convert a higher level of FFA into FAME (fatty acid methyl esters) and thus reduced the concentration of FFA by <1%. The experimental setup consists of an RB flask (1L), which is assembled with a stirrer, a reflux condenser, and a temperature monitoring system. The flask was kept in a water bath at a temperature of 333–338 K as it is shown in Figure 3.3.

3.6.2.1 *Pretreatment process*

In this process, the addition of acid catalyst, i.e., H_2SO_4 (1.0%) to the reaction mixture which was comprised of methanol–oil mixture (0.34% v/v ratio) takes place with continuous stirring at a required temperature for 60 min. The mixture was then cooled and the layers of transesterified oil and water–alcohol were isolated with the use of a separating funnel. The acid value of esterified oil was determined (12.6 mg KOH/g), which is higher than the preferred value, and therefore a second-stage esterification process was carried out. The product thus obtained contains the acid value <2.0 mg KOH/g and is utilized for further processing.

3.6.2.2 *Transesterification process*

In this process, the treated oil having a low acid value was mixed with CH_3OH, alcohol (0.25% v/v), and base catalyst KOH (0.7% v/v) at a temperature of 333 K for 60 min, and the resultant was then kept in a separating funnel and after 12–15 h the upper layer of FAME was separated from the by-product glycerol and was collected in a flask.

3.6.3 Synthesis of DEG-based green lubricant

The refinement of obtained crude FAME from mahua oil was then treated with diethylene glycol for further transesterification process. The reaction occurs in the presence of catalyst NaOMe (second-stage transesterification process) in batches (200 ml). The process was carried out with diethylene glycol to FAME wt ratio of 1:3.5 and the catalyst amount 0.78% w/w (of total reactants) at 121°C temperatures for 2.5 h. Once the reaction was completed, the obtained product was kept for cooling and then tested for different physicochemical properties. Similar reactions were carried out by changing the wt ratio of glycerol to FAME in various batches, which is shown in Table 3.5.

3.7 CHARACTERIZATION OF DEG-BASED BIO-LUBRICANT

Bio-lubricant possesses properties corresponding to lubricants based on mineral oil and can be obtained by ASTM methods. The various physicochemical characteristics of bio-lubricants are summarized below.

i. **Viscosity:** The property of viscosity allows the operability and condition of the lubricant since if the viscosity is low, the oil film between two sliding or moving surfaces cannot persist, whereas if the viscosity is too high irrelevant erosion will occur [31].
ii. **Viscosity index:** The good lubricant must possess a consistency that will not change with temperature and can be applied under various conditions. If the lubricity of the liquid drops with rising temperature, then the liquid possesses a low viscosity index; however, if the lubricity of the liquid is slightly influenced by temperature, then it will have a high viscosity index [31].
iii. **Acid value:** Acid value measures the acidity of the lubricant that is responsible for causing the corrosion during the operation by the

Table 3.5 Effect of FAME:DEG ratio on synthesis of bio-lubricant

Batch No.	*DEG:FAME*	*NaOMe (catalyst) wt%*	*Stirring rate (rpm)*	*Reaction temperature (°C)*	*Inference*
1	1:2	0.78	184	121	Uniform quasi-solid
2	1:2.5	0.78	184	121	Presence of irregular lumps in liquid
3	1:3	0.78	184	121	Presence of infinitesimal lumps in liquid
4	1:3.5	0.78	184	121	Uniform viscous liquid

generation of acidic compounds. The good bio-lubricant is expected to have a minimum acid value in order to reduce the corrosion of machinery and equipment during use [32].

iv. **Flash and fire point:** The lubricating characteristics of the oil are not affected by the flash or fire point of the oil, but at a higher temperature, these properties are of significant value because during usage it will protect contrarily to hazards of fire if the oil has higher flash and fire point [33].

v. **Pour point:** Generally, lubricating oils applied in a machinery have low pour point for operating at low temperatures since stiffening of the oil will create blockage of the machinery. It was estimated that waxes present in the oil increase the pour point of lubricating oil [34].

The distinctive properties of the sample (mahua oil) and mahua oil methyl ester (MOME) are provided in Table 3.6, whereas those of DEG-based bio-lubricant are provided in Table 3.7.

vi. **Oxidation stability**

The oxidative stability property of bio-lubricant (green lubricant) is defined as its capacity to resist auto-oxidation, and it is an important characteristic

Table 3.6 Characteristic properties of mahua oil and mahua oil methyl ester

Sr. No.	*Properties*	*Mahua oil*	*MOME*	*Analytical method*
1	Density (g/ml)	0.91	0.853	ASTM-D1298
2	Viscosity at 40 (cSt)	37.19	4.46	ASTM-D445
3	Viscosity at 100 (cSt)	11.68	2.03	ASTM-D445
4	Viscosity index	325.5	446.2	ASTM-D2270
5	Pour point	17	0	ASTM-D97
6	Flash point	212	148	ASTM-D92
7	Fire point	222	166	ASTM-D92
8	Acid value	26.25	2.4	ASTM-D664

Table 3.7 Characteristics of DEG-based bio-lubricant

Sr. No.	*Characteristic properties*	*Commercial 4T Engine oil*	*Bio-lubricant (mahua DEG ester)*	*Analytical methods*
1	Viscosity at 100°C (cSt)	6.45	9.42	ASTM-D445
2	Viscosity at 40°C (cSt)	41.9	61.33	ASTM-D445
3	Density (g/ml)	0.96	0.94	ASTM-D1298
4	Viscosity index	100.59	134.14	ASTM-D2270
5	Acid value (mg KOH/g)	0.50	1.13	ASTM-D664
6	Pour point (°C)	−14.6	4.8	ASTM-D97

Table 3.8 Oxidation stability of mahua oil bio-lubricant

Sr. No.	*Properties*	*Mahua oil*	*DEG bio-lubricant*	*Commercial 4T oil*
1	Density (g/ml) before oxidation	0.91	0.94	0.96
2	Density (g/ml) after oxidation	0.921	0.974	0.987
3	Viscosity at 40°C, cSt (before oxidation)	37.19	61.33	45.3
4	Viscosity at 40°C, cSt (after oxidation)	66.39	78.82	52.3
5	Viscosity ratio	1.79	1.30	1.15

to increase the working life of bio-lubricant. The oxidation and degradation at higher temperatures result in the formation of rusts, froths, acids, and suspension of the oil film, which decrease the life span of the lubricant. Thus, a lubricant is expected to possess soundness toward oxidative degradation. The oxidative stability property of mahua oil bio-lubricant is provided in Table 3.8.

3.8 RESULTS AND DISCUSSION

It is known that while utilizing petroleum-based lubricants, the performance of the engine is influenced by oil thickening because of soot stacking; oil decomposition by oxidation and high-temperature activity; oil dilution because of fuel mixing (crankcase dilution); and deposit formation and wear because of the exhaustion of anti-wear additives. Consequently, the assessment of decomposition of bio-lubricants in the long run in terms of their oxidation stability and, furthermore, the impact of oil on motor execution and discharges are still to be perceived. The major part of the discussion is based on the experimental study and detailed analysis, which has been carried out as follows.

3.8.1 Effect on viscosity

Viscosity is the quantity that measures the fluid resistance to flow. Figure 3.4 shows that the viscosity of mahua oil is 37.19 and 11.68 cSt at 40 and 100°C which is very high. However, on conversion to methyl ester by transesterification process, a decrease in viscosity (4.46 cSt, 2.03 cSt) at two different temperatures has been observed which is again increased by the reaction with diethylene glycol during the preparation of bio-lubricants. From Figure 3.4, the viscosity of bio-lubricant (mahua DEG ester) is 61.33 cSt, 9.42 cSt at 40°C, 100°C, which is higher than commercial 4T Engine oil, which shows that as the chain length increases the viscosity also increases. Thus, bio-lubricant made from mahua oil shows better viscosity as compared to commercial 4T Engine oil.

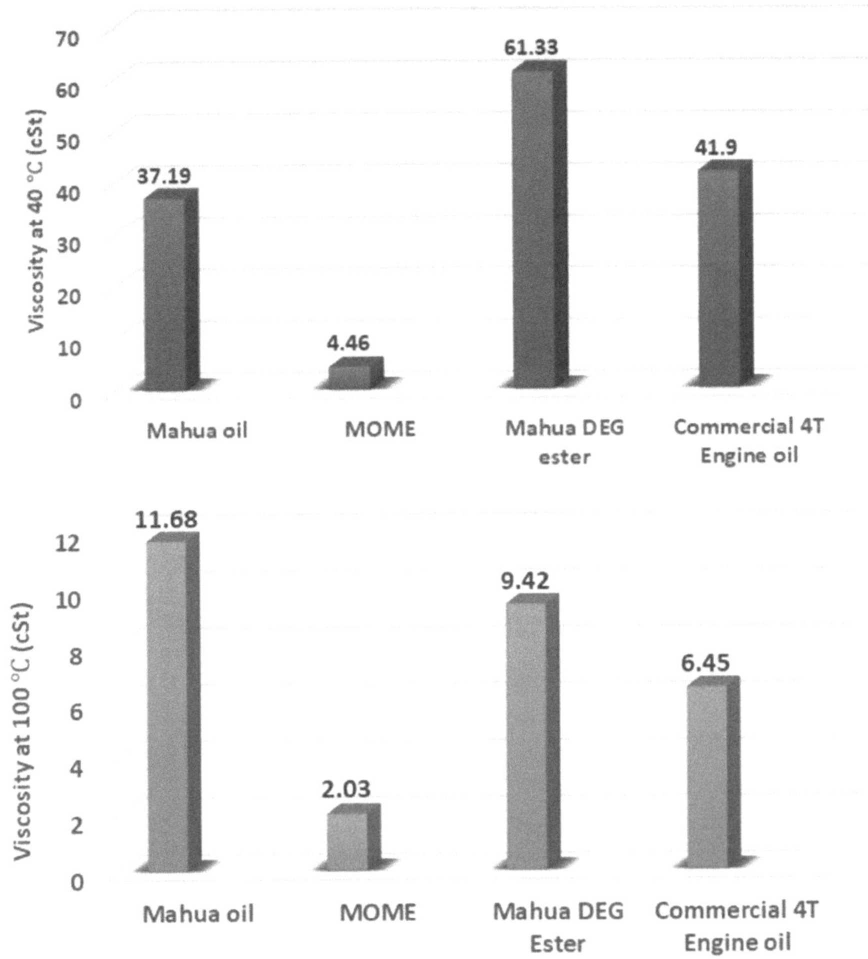

Figure 3.4 Viscosity of products of mahua oil and commercial 4T Engine oil at two different temperatures.

3.8.2 Effect on viscosity index

Viscosity index is an important property of lubricating oils since high viscosity index results in reduced lubricant consumption and wear, and according to Epelle et al. [35], lubricants containing high viscosity index (>129 cSt) are widely used for different applications. From Figure 3.5, it has been observed that the viscosity index (134.14 cSt) of bio-lubricant obtained from mahua oil was found to be higher than commercial 4T Engine oil (100.59 cSt). Furthermore, the viscosity index of methyl ester of mahua oil is 446.2 cSt and is higher than the corresponding oil (325.5 cSt) and the final product.

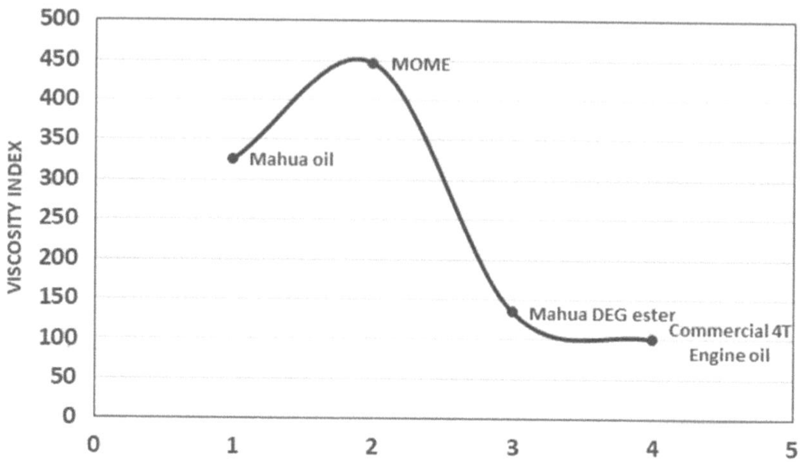

Figure 3.5 Viscosity index of different products of mahua oil and commercial 4T Engine oil.

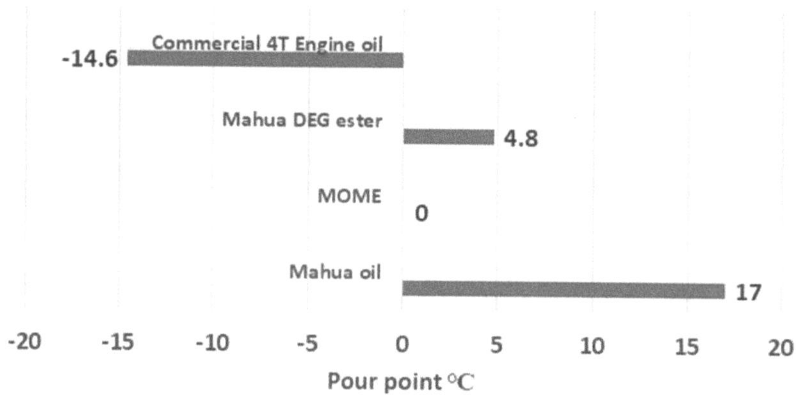

Figure 3.6 Pour point of products of mahua oil and 4T Engine oil.

3.8.3 Effect on pour point

Pour point is an essential property of vegetable oils for the production of bio-lubricant because at low temperature, the oil poorly flows which limits its utilization in the fluid industry. The chemical treatment of mahua oil improved its pour point from 17°C to 0°C as shown in the Figure 3.6, which is due to the chain length increase/removal of polyunsaturation. Generally, bio-lubricants are utilized in machines operating at low temperature, and hence, these lubricants must possess low pour point in order to abstain solidification. From Figure 3.6, it was estimated that the bio-lubricants obtained from mahua oil have a higher pour point as compared to commercial 4T Engine oil, but it can be lowered with the help of some additives.

3.8.4 Effect on other parameters

From Table 3.6, it has been observed that the acid value of mahua oil is very high and it cannot be directly employed for bio-lubricant production since increased acid value indicates the presence of free fatty acids (FFA) in higher amounts. However, acids in any form are not acceptable for lubricating oils, and thus the conversion of mahua oil into methyl ester by transesterification method lowers the acid content present in it since the decrease in acid value confirms the transformation of FFA into methyl esters. It is also estimated from Table 3.7 that the bio-lubricant produced from mahua oil contains acid value which is inacceptable according to commercial lubricants. Moreover, from Figure 3.7, the density of the obtained bio-lubricant (Mahua DEG Ester) is slightly lower than the commercial lubricant, but it can be maintained in the range by the addition of some additives.

3.8.5 Oxidation stability

The oxidation of bio-lubricant synthesized from mahua oil was accomplished at a higher temperature (200°C) for 360 min, and the change in density and viscosity was determined at 40°C before and after the process of oxidation. The variation in the values of viscosity and density is shown in Table 3.8. It was estimated from Table 3.8 that bio-lubricant (Mahua DEG Ester) has a somewhat higher viscosity ratio (1.30) as compared to commercial 4T Engine oil (1.15), which indicates the low oxidation stability of the final product.

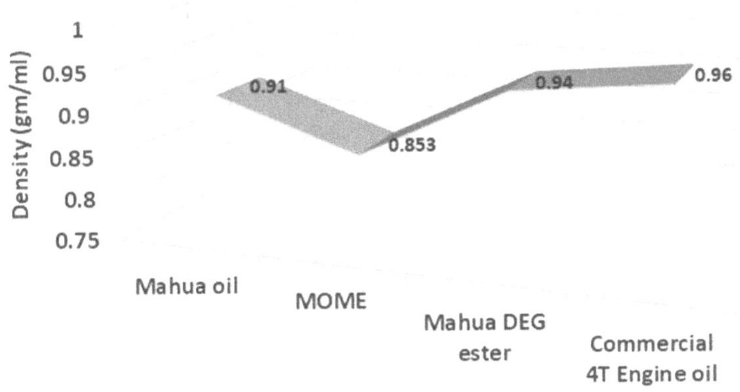

Figure 3.7 Density of products of mahua oil and commercial 4T Engine oil.

3.8.6 Effect of mahua oil bio-lubricant on marine life

Clarias gariepinus fishes were subjected to bio-lubricant (Mahua DEG Ester) and commercial oil (4 ml/20 L water) for four days. The morphology of gills indicates the mild clogging of gill tissues when exposed to mahua bio-lubricant, whereas severe clogging of gill tissues has been observed on exposure to commercial lubricant. Furthermore, calcification and acute inflammation of gill tissues also have been seen with a lower concentration of commercial lubricants. The folding of lamellae and irregular surface of lamellae also occurred while on exposure to mahua oil bio-lubricant no major changes are observed in the gills of the fishes, and hence mahua oil bio-lubricant appeared to be environmentally favorable.

3.9 CONCLUSION

Green lubricants are the class of lubricants that are generally made from vegetable/natural sources, and these are usually denoted frequently as environmentally benign, environmentally friendly, and environmentally acceptable. Nowadays, due to depleting petroleum reserves, day-to-day price hike, and environmental concern, green lubricants are an emerging technology in order to replace mineral lubricants. This chapter discussed about the types of lubricants based on the property of base oil used for their production, their environmental effects, and inedible vegetable oils as a feedstock for the production of green lubricants. In the automobile industry, the property of biodegradability of lubricants is an important factor and therefore the non-edible oils are more favorable as compared to mineral base oil since these oils are renewable, easily available, biodegradable in nature, and have low marine toxicity while reducing the cost. Various types of catalysts are used for the conversion of vegetable oils (non-edible) into bio-lubricants such as *p*-toluene sulfonic acid, titanium iso-propoxide, Amberlite IR-120, etc., which impart good yield and improvement in thermal stability, flash point, and pour point to the obtained bio-lubricant. This chapter represents the synthesis of bio-lubricants from mahua oil through different routes, namely esterification and transesterification utilizing diethylene glycol (DEG). A comparative analysis of the properties of obtained bio-lubricant with commercial 4T Engine oil has been carried out, which showed that the untreated mahua oil has increased viscosity as well as increased viscosity index, but it also contains high acid value, low thermo-oxidative stability, and high pour point, which restricts their direct use as bio-lubricants. Therefore, mahua oil is treated to reduce its acid value and improve pour point and then utilized for the synthesis of mahua oil bio-lubricants with the use of DEG at a weight ratio of 1:3.5 for glycol to oil. The viscosity index (134.14) of DEG-based mahua oil bio-lubricant is more than that of commercial 4T Engine oil (100.59), and it also has a higher pour point that can be improved by the addition of pour point depressants. Furthermore, the study on fish showed that the bio-lubricant (final product) obtained from mahua oil is ecofriendly and safe for marine life. Thus, this chapter showed the feasibility of mahua oil as a feedstock for the production of green lubricants that are less toxic, biodegradable, and ecofriendly. The large-scale cultivation of non-edible seeds enables the country to energy security, maintains the environmental balance, gives rise to green cover, decreases the dependency of the country on imported fuels and chemicals, and will provide the opportunity for employment in remote areas.

REFERENCES

1. E. J. Parente, J. P. C. Marques, I. C. Rios, J. A. Cecilia, E. Rodriguez-Castellon, F. M. T. Luna, C. L. Cavalcante *et al.*, "Production of biolubricants from soybean oil: Studies for an integrated process with the current biodiesel industry," *Chem. Eng. Res. Des.*, 2021, doi: 10.1016/j.cherd.2020.11.012.
2. S. Bekal and N. R. Bhat, "Bio-lubricant as an alternative to mineral oil for a CI engine—an experimental investigation with pongamia oil as a lubricant," *Energy Sources, Part A Recover. Util. Environ. Eff.*, 2012, doi: 10.1080/15567031003735303.
3. N. Salih and J. Salimon, "A review on new trends, challenges and prospects of ecofriendly friendly green food-grade biolubricants," *Biointerface Res. Appl. Chem.*, 2021, doi: 10.33263/briac121.11851207.
4. R. Chand and B. Kumar, "Oil and lubricant hazard effects on human health," *Int. J. Innov. Sci. Eng. Technol.*, 2017, https://www.researchgate.net/publication/317821507%0Awww.ijiset.com.
5. R. Shankar, J. H. Jung, A. Loh, J. G. An, S. Y. Ha, and U. H. Yim, "Environmental significance of lubricant oil: A systematic study of photooxidation and its consequences," *Water Res.*, 2020, doi: 10.1016/j.watres.2019.115183.
6. J. McNutt and Q. S. He, "Development of biolubricants from vegetable oils via chemical modification," *J. Ind. Eng. Che.*, 2016, doi: 10.1016/j.jiec.2016.02.008.
7. J. K. Katiyar, R. K. Sahu, and T. C. S. M. Gupta, "Types of lubricants," in: *Sustainable Lubrication*, 2022. doi: 10.1201/9781003201199-2.
8. P. G. Nikolakopoulos and D. A. Bompos, "Experimental measurements of journal bearing friction using mineral, synthetic, and bio-based lubricants," *Lubricants*, 2015, doi: 10.3390/lubricants3020155.
9. R. S. Kamal, N. S. Ahmed, and A. M. Nassar, "Synthesis and characterization of mixed esters as synthetic lubricants," *Pet. Coal*, 2017, 59, 736–746.
10. F. M. T. Luna, J. B. Cavalcante, F. O. N. Silva, and C. L. Cavalcante, "Studies on biodegradability of bio-based lubricants," *Tribol. Int.*, 2015, doi: 10.1016/j.triboint.2015.07.007.
11. F. J. Owuna, "Stability of vegetable based oils used in the formulation of ecofriendly lubricants—a review," *Egypt. J. Pet.*, 2020. doi: 10.1016/j.ejpe.2020.09.003.

12. N. K. Attia, S. A. El-Mekkawi, O. A. Elardy, and E. A. Abdelkader, "Chemical and rheological assessment of produced biolubricants from different vegetable oils," *Fuel*, 2020, doi: 10.1016/j.fuel.2020.117578.
13. A. N. Annisa and W. Widayat, "A review of bio-lubricant production from vegetable oils using esterification transesterification process," in *MATEC Web of Conferences*, 2018. doi: 10.1051/matecconf/201815606007.
14. M. Kareemullah, A. Afzal, K. F. Rehman, K. C. Vishwanath, H. Khan, M. Elahi, M. Saudager, A. R. Kaladgi *et al.*, "Preparation and physicochemical properties evaluation of epoxidized neem oil-based bio-lubricant," *Aust. J. Mech. Eng.*, 2023, doi: 10.1080/14484846.2021.1920141.
15. U. C. Sharma and S. Sachan, "Friction and wear behavior of karanja oil derived biolubricant base oil," *SN Appl. Sci.*, 2019, doi: 10.1007/s42452-019-0706-y.
16. P. R. Campos Flexa Ribeiro Filho, M. Rocha do Nascimento, S. S. Otaviano da Silva, F. M. Tavares de Luna, E. Rodríguez-Castellón, and C. Loureiro Cavalcante, "Synthesis and frictional characteristics of bio-based lubricants obtained from fatty acids of castor oil," *Lubricants*, 2023, doi: 10.3390/lubricants11020057.
17. N. P. Chauke, H. E. Mukaya, and D. B. Nkazi, "Chemical modifications of castor oil: A review," *Sci. Prog.*, 2019, doi: 10.1177/0036850419859118.
18. D. R. Moreira *et al.*, "Moringa polyesters as eco-friendly lubricants and its blends with naphthalenic lubricant," *Ind. Crops Prod.*, 2020, doi: 10.1016/j.indcrop.2020.112937.
19. J. M. Encinar, S. Nogales-Delgado, N. Sánchez, and J. F. González, "Biolubricants from rapeseed and castor oil transesterification by using titanium isopropoxide as a catalyst: Production and characterization," *Catalysts*, 2020, doi: 10.3390/catal10040366.
20. S. Rajwani and P. K. S. Yadav, "Synthesis of biolubricant via chemical modification from vegetable oil blend using heterogeneous catalyst," *Asian J. Chem.*, 2021, doi: 10.14233/ajchem.2021.23263.
21. M. F. M. Gunam Resul, T. I. Tinia, and A. Idris, "Kinetic study of jatropha biolubricant from transesterification of jatropha curcas oil with trimethylolpropane: Effects of temperature," *Ind. Crops Prod.*, 2012, doi: 10.1016/j.indcrop.2012.01.012.
22. C. T. Ivan-Tan, A. Islam, R. Yunus, and Y. H. Taufiq-Yap, "Screening of solid base catalysts on palm oil based biolubricant synthesis," *J. Clean. Prod.*, 2017, doi: 10.1016/j.jclepro.2017.02.027.
23. R. M. A. Saboya *et al.*, "Assessment of commercial resins in the biolubricants production from free fatty acids of castor oil," *Catal. Today*, 2017, doi: 10.1016/j.cattod.2016.02.020.
24. A. Rochmat, A. Dudayev Alfaruqi, Saefuri, Suaedah, and M. Nurtanto, "Synthesis biolubricant by esterification of castor oil (*Jatropha Curcas L*-oil) with chlorate acid catalyst," *J. Phys. Conf. Ser.*, 2020. doi: 10.1088/1742-6596/1573/1/012004.
25. M. B. Alang, M. K. Ndikontar, Y. M. Sani, and P. T. Ndifon, "Synthesis and characterisation of a biolubricant from cameroon palm kernel seed oil using a locally produced base catalyst from plantain peelings," *Green Sustain. Chem.*, 2018, doi: 10.4236/gsc.2018.83018.

26. E. K. Heikal, M. S. Elmelawy, S. A. Khalil, and N. M. Elbasuny, "Manufacturing of environment friendly biolubricants from vegetable oils," *Egypt. J. Pet.*, 2017, doi: 10.1016/j.ejpe.2016.03.003.
27. D. Kumar, V. K. Chibber, and A. Singh, "Physical and chemical properties of Mahua and Sal seed oils," *Adv. Intell. Syst. Comput.*, 2018, doi: 10.1007/978-981-10-5903-2_146.
28. D. K. Karupannasamy and R. Ruthuraraj, "Synthesis of eco-friendly machining lubricants and its tribological characteristics," *Int. J. Recent Technol. Eng.*, 2019, doi: 10.35940/ijrte.C4466.098319.
29. Y. Singh, N. K. Singh, A. Sharma, C. Kishore, and A. Raturi, "Madhuca Indica (Mahua): A novel feedstock for bio based lubricant application treated with trimethylolpropane and tribological analysis," *Aust. J. Mech. Eng.*, 2023, doi: 10.1080/14484846.2020.1842300.
30. S. K. Chaurasia, A. K. Sehgal, and N. K. Singh, "Improved lubrication mechanism of chemically modified Mahua (*Madhuca indica*) oil with addition of copper oxide nanoparticles," *J. Bio- Tribo-Corrosion*, 2020, doi: 10.1007/s40735-020-00387-2.
31. ASTM International, *ASTM D445 : Standard Test Method for Kinematic Viscosity of Transparent and Opaque Liquids (and Calculation of Dynamic Viscosity)*. ASTM International, 2019.
32. J. C. J. Bart, E. Gucciardi, and S. Cavallaro, "Lubricants: Properties and characteristics," *Biolubricants*, 2013. doi: 10.1533/9780857096326.24.
33. American Society for Testing and Materials, "Standard test method for flash and fire points by Cleveland Open Cup Tester ASTM D92," *Annu. B. ASTM Stand.*, 2018.
34. ASTM International: West Conshohocken, "ASTM D97-17b—standard test method for pour point of petroleum products," *Annu. B. ASTM Stand.*, 2017.
35. E. I. Epelle, A. J. Otaru, Y. O. Zubair, and O. J. Awele, "Improving the viscosity index of used lubricating oil by solvent extraction," *Int. Res. J. Eng. Technol.*, 2017, www.irjet.net.

Chapter 4

Eco-friendly lubricants

Neha Singh Raghuvanshi, Arpita Johari, Mudit Saxena, Ashish Kumar Gupta, Abhishek Kumar, Ashwani Kumar, Avinash Kumar, and Ashutosh Gupta

4.1 INTRODUCTION

Eco-friendly lubricants, also called environmentally friendly lubricants, encompass a range of formulations designed to mitigate environmental impact while ensuring efficient lubrication [1]. The call for environmental stewardship and sustainability has become urgent in today's global context. As industries confront the complexities of climate change, resource scarcity, and ecological degradation, the lubricant sector finds itself at the forefront of a profound evolution. The emergence of environmentally friendly and sustainable lubricants marks a significant departure toward more ecologically responsible practices within industrial lubrication [2]. These lubricants signify a dedicated effort to curb environmental impact while maintaining peak performance and efficiency across various applications. Characterized by their eco-conscious formulations and strict adherence to sustainability standards, they epitomize a comprehensive approach to lubrication that prioritizes environmental integrity throughout their lifecycle [3].

At the core of environmentally friendly lubricants lies a commitment to mitigating the adverse environmental consequences of traditional lubrication methods. Conventional lubricants, often derived from fossil fuels and laden with harmful additives, pose significant threats to ecosystems, human health, and the planet. Their production and usage contribute to air and water pollution, habitat destruction, and greenhouse gas emissions, exacerbating today's environmental challenges. In stark contrast, environmentally friendly lubricants prioritize renewable, biodegradable base oils sourced from sustainable feedstocks like vegetable oils, animal fats, or synthetic esters. By harnessing nature's resources, these lubricants minimize reliance on finite fossil fuel reserves and diminish the carbon footprint associated with their manufacturing and use [4].

Beyond their composition, environmentally friendly lubricants stand out for their eco-friendly performance characteristics and adherence to stringent sustainability benchmarks. Engineered to deliver outstanding lubrication and protection across diverse industrial applications, they undergo rigorous testing and certification to ensure compliance with environmental regulations

 DOI: 10.1201/9781003535447-4

and industry standards. These lubricants boast superior thermal stability, oxidation resistance, and wear protection properties, allowing them to endure the rigors of high-temperature, high-pressure conditions without compromising environmental standards. Furthermore, they were formulated to reduce the release of harmful emissions such as volatile organic compounds (VOCs), toxic chemicals, and greenhouse gases, thereby safeguarding both air quality and human health in industrial and residential settings [5].

In addition to their environmental advantages, environmentally friendly lubricants offer tangible sustainability and resource management benefits. Extending equipment lifespan, minimizing friction and wear, and improving energy efficiency optimize operational performance while reducing resource consumption throughout the lubrication cycle. Innovative technologies and advanced formulations empower industries to achieve higher productivity, reliability, and cost-effectiveness while simultaneously curbing waste generation and environmental impact [6]. Moreover, their biodegradable nature facilitates safe and responsible disposal, minimizing the buildup of hazardous pollutants in landfills and ecosystems. The embrace of environmentally friendly lubricants signifies a transformative shift toward a more sustainable and resilient future for industries worldwide. By fostering a sustainability and corporate responsibility culture, lubricant manufacturers, suppliers, and end-users can spearhead positive environmental change and promote the transition to a circular economy model. Through collaborative partnerships, research, and innovation, stakeholders across the lubricant value chain can harness the transformative potential of environmentally friendly lubricants to address pressing environmental challenges and cultivate a culture of environmental stewardship and sustainability [7–10]. As we stand on the threshold of a new era in industrial lubrication, adopting environmentally friendly lubricants heralds a paradigm shift toward greener, cleaner, and more sustainable practices that will define the industry's future for generations.

In summary, the need for eco-friendly lubricants has never been more urgent. As industries seek to reduce their environmental impact, eco-friendly lubricants offer a viable solution, providing excellent performance while contributing to sustainability goals. This chapter will explore the different types of eco-friendly lubricants, their characteristics, and their applications across various industries. Understanding their benefits and how they can be incorporated into industrial processes can pave the way for broader adoption of environmentally responsible practices. This shift ensures a sustainable future where operational efficiency and environmental stewardship coexist.

4.2 CLASSIFICATION OF ECO-FRIENDLY LUBRICANTS

Eco-friendly lubricants are a growing category designed to minimize environmental impact throughout their lifecycle. Unlike traditional lubricants derived from petroleum, eco-friendly options prioritize biodegradability, reduced toxicity, and the use of renewable resources. Table 4.1 shows the classification of eco-friendly lubricants.

1. **Lubricants:**
   This classification encompasses a variety of lubricants, each with its own strengths and limitations. Biodegradable lubricants, which are made from readily decomposable materials like vegetable oils or synthetic esters, offer excellent environmental disposal options and reduce pollution risks in case of spills. However, their performance might be lower in extreme temperatures or high pressures [3]. Bio-based lubricants formulated with significant renewable resources offer a more sustainable alternative to traditional options. Biodegradability can vary depending on the base stock, and performance characteristics may differ.
2. **Synthetic eco-friendly lubricants:**
   These man-made lubricants combine high performance with improved biodegradability compared to traditional synthetics. They may also use non-toxic additives. While offering excellent lubrication and reduced environmental impact, they can be more expensive than other options. Beyond these broad categories, there are more specific options like vegetable-based lubricants ideal for certain industries, bio-synthetic lubricants that combine biodegradability with exceptional performance, and water-based lubricants perfect for metal-working applications due to their cooling and lubricating properties [11]. The choice of eco-friendly lubricant ultimately depends on the specific application, considering factors like performance requirements, environmental impact, and cost.

4.2.1 Vegetable-based lubricants

These are derived from renewable sources like plants, fruits, and vegetables and offer eco-conscious alternatives to conventional petroleum-derived oils. These lubricants promote biodegradability, non-toxicity, and low volatility, thereby minimizing the emission of harmful substances into the environment. These are widely utilized in agriculture, forestry, marine, and food processing; vegetable-based lubricants align with industries where environmental considerations are paramount. Soybean oil lubricants, canola oil lubricants, sunflower oil lubricants, corn oil lubricants, and rapeseed oil lubricants are popular vegetable-based lubricants. These are suitable for applications such as hydraulic systems, metal-working fluids, and gear lubrication [12].

4.2.2 Bio-synthetic lubricants

Bio-synthetic lubricants blend renewable bio-based oils with synthetic additives to enhance performance and durability. Their exceptional

Table 4.1 A detailed classification of eco-friendly lubricants

Category	*Description*	*Key characteristics*	*Benefits*	*Limitations*
Biodegradable Lubricants	Made from readily decomposable natural oils (e.g., vegetable oils) or synthetic esters	High biodegradability (> 60% in 28 days by standard tests) Lower aquatic toxicity compared to mineral oils	Environmentally friendly disposal Reduced risk of pollution in case of spills	Lower performance in extreme temperatures or high pressures Shorter lifespan requiring more frequent changes
Bio-based lubricants	Formulated with a significant portion (often >25%) of renewable resources	Made from plant or animal-derived materials It may not be fully biodegradable	Reduced reliance on fossil fuels Potential for improved sustainability	Performance characteristics can vary depending on base stock Biodegradability may not be guaranteed
Synthetic eco-friendly lubricants	Man-made lubricants are designed for minimal environmental impact	High performance Improved biodegradability compared to traditional synthetics May use non-toxic additives	Can offer excellent lubrication in demanding applications Reduced environmental impact compared to mineral oils	Can be more expensive than traditional mineral or bio-based oils Newer technology's long-term environmental effects may require further study
Vegetable-based lubricants	Derived from renewable sources like soybeans, canola, sunflowers, corn, and rapeseed	Biodegradable Non-toxic Low volatility	Reduced environmental impact Suitable for agriculture, forestry, marine, and food processing industries	Lower performance in extreme temperatures or high pressures It may require more frequent changes
Bio-synthetic lubricants (combines bio-based and some synthetic eco-friendly)	A blend of bio-based oils and synthetic additives for improved performance	Biodegradable Excellent thermal stability and oxidation resistance Suitable for demanding applications	Minimizes environmental impact of spills Offers exceptional performance	Can be more expensive than some lubricants
Water-based lubricants	Primarily water with additives for lubrication	Outstanding biodegradability Free of harmful solvents and VOCs	Environmentally safe; Cooling and lubrication for metal-working	Limited lubrication performance in extreme applications
Polyalkylene glycol (PAG) lubricants	Synthetic lubricants offer superior thermal and lubricating properties	Biodegradable Non-toxic High thermal stability	Reduced environmental impact Suitable for extreme conditions	May not be compatible with all materials
Silicone-based lubricants	Silicone oil-based lubricants are known for inertness and wide temperature range	Biodegradable Non-toxic Excellent thermal stability	Long-lasting lubrication Reduced emissions	May not be suitable for all applications

thermal stability, oxidation resistance, and shear stability render them suitable for demanding applications in high-temperature and high-pressure environments. Engineered to degrade rapidly in soil and water, these lubricants mitigate environmental impact in scenarios involving spills or leaks. They find extensive use in hydraulic systems, gearboxes, chains, and environmentally sensitive areas like nature reserves and marine ecosystems. Bio-synthetic hydraulic fluids, bio-synthetic gear oils, bio-synthetic compressor oils, bio-synthetic chain oils, and bio-synthetic turbine oils are formulated to provide exceptional oxidation stability, thermal conductivity, and foam resistance, ensuring reliable performance and extended equipment life [13].

4.2.3 Water-based lubricants

Water-based lubricants leverage water as the primary carrier fluid, complemented by additives to enhance lubricating properties. Their outstanding biodegradability and absence of harmful solvents, volatile organic compounds (VOCs), and toxic substances ensure environmental safety. Primarily employed in metal-working operations such as cutting, milling, and grinding, where cooling and lubrication are essential, they contribute to a cleaner and safer working environment. These lubricants mitigate friction, heat generation, and tool wear while facilitating easy cleanup and disposal. Coolants and cutting fluids, hydraulic fluids, industrial cleaners and degreasers, concrete form release agents, textile lubricants, and food-grade lubricants are popular water-based lubricants [14]. Water-based lubricants can incorporate various additives, as illustrated in Figure 4.1, to meet specific application needs.

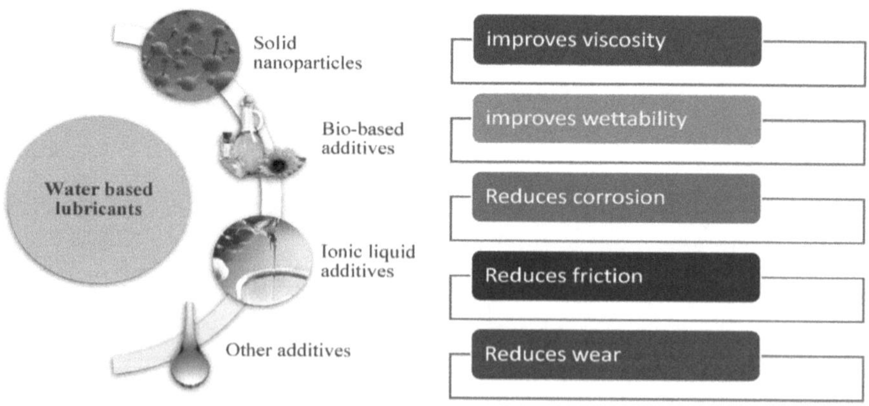

Figure 4.1 Additives for water-based lubricants offer several benefits [14].

4.2.4 Polyalkylene glycol (PAG) lubricants

PAG lubricants, synthesized from ethylene or propylene oxide, offer superior thermal stability and lubricating properties. Biodegradable and non-toxic, they are environmentally friendly alternatives to traditional mineral oils. They resist oxidation and thermal breakdown and are widely adopted in automotive air conditioning systems, refrigeration compressors, and industrial gearboxes operating under extreme conditions. PAG lubricants ensure prolonged service life and reduced environmental impact throughout their usage. These are used primarily in automotive air conditioning systems, refrigeration compressors, industrial gearboxes, hydraulic systems, metalworking applications, and textile machinery [15–18].

4.2.5 Silicone-based lubricants

Silicone-based lubricants comprise silicone oil as the base fluid, characterized by inertness, non-toxicity, and biodegradability. With excellent thermal stability, oxidation resistance, and a broad temperature range, they suit diverse applications. Commonly found in automotive, aerospace, pharmaceutical, and food processing industries where cleanliness and environmental compliance are paramount, they offer long-lasting lubrication and corrosion protection. Exhibiting low volatility, they minimize emissions and ecological contamination, contributing to sustainable practices. Silicone oils, silicone greases, silicone-based dry lubricants, silicone sprays, silicone-based release agents, and silicone rubber lubricants are some popularly used silicon lubricants [19].

4.3 LUBRICANT RENEWABILITY

The renewability of eco-friendly lubricants is a fundamental aspect that distinguishes them from conventional petroleum-based lubricants. Eco-friendly lubricants are derived from renewable sources or synthesized using environmentally benign processes, which enables them to contribute positively to sustainability efforts. Eco-friendly lubricants are often produced from renewable feedstocks such as vegetable oils, animal fats, or bio-based synthetic compounds. Unlike finite petroleum reserves used in traditional lubricants, these feedstocks can be replenished through agricultural or industrial processes [20]. Soybean oil lubricants, canola oil lubricants, sunflower oil lubricants, corn oil lubricants, and rapeseed oil lubricants can be replenished through cultivation. Shell Naturelle HF-E 15 uses renewable base oils from natural feedstocks, such as vegetable or synthetic esters. These renewable feedstocks reduce reliance on finite petroleum resources and contribute to the sustainability of the lubricant [21]. Many eco-friendly lubricants are organic or formulated to be biodegradable, meaning they can naturally break down into simpler, environmentally benign substances over time. This characteristic allows them to minimize environmental impact and reduce long-term pollution risks compared to non-biodegradable petroleum-based lubricants [22]. Panolin® HLP Synth is readily biodegradable and can break down into harmless substances when exposed to environmental conditions such as sunlight, water, and microorganisms. This characteristic reduces the ecological impact of the lubricant and helps protect sensitive ecosystems. Some eco-friendly lubricants are manufactured using closed-loop production processes, where waste and by-products from the production process are recycled or reused. This approach minimizes resource consumption and waste generation, enhancing the overall renewability of the lubricant production cycle. The Shell Naturelle HF-E 15 manufacturing process incorporates closed-loop production practices, where waste materials and by-products are recycled or reused within the production facility. This approach minimizes waste generation and enhances the overall sustainability of the production process. Specific eco-friendly lubricants are designed to have a lower carbon footprint than conventional lubricants [23]. By utilizing renewable feedstocks and employing energy-efficient manufacturing processes, these lubricants can help mitigate greenhouse gas emissions and contribute to efforts to combat climate change. An example of a green lubricant designed to have a lower carbon footprint compared to conventional lubricants is Mobil SHC™ Gear 320 WT [24]. The lubricant's synthetic formulation and additive package help reduce friction and improve energy efficiency within wind turbine gearboxes. By minimizing internal losses, Mobil SHC™ Gear 320 WT helps optimize turbine performance and reduces energy consumption, lowering carbon emissions associated with power generation. Eco-friendly lubricants exemplify sustainable practices and principles that underscore environmental stewardship and optimal resource utilization. Engineered with biodegradability as a foundational aspect, these lubricants naturally break down into harmless substances upon environmental release, thereby mitigating ecological harm. By harnessing renewable feedstocks like vegetable oils, animal fats, and bio-based synthetics, eco-friendly lubricants diminish dependence on finite fossil fuels, championing the adoption of renewable resources. Energy efficiency is central to sustainable lubrication strategies, as adequate lubrication minimizes friction and wear in machinery, fostering energy savings and operational efficiency gains over time [25]. Comprehensive life cycle assessments are integral for evaluating the environmental footprint of eco-friendly lubricants, informing decision-making, and fostering ongoing improvements in sustainability practices. Embracing circular economy principles, eco-friendly lubricants advocate the reuse, recycling, and recovery of lubricant materials, promoting a closed-loop system that optimizes resource utilization and curbs waste generation. Demonstrating adherence to rigorous eco-friendly standards, regulatory compliance, and certification underscores a commitment to environmental integrity among green lubricant manufacturers.

4.4 BIODEGRADABILITY

Biodegradability is a fundamental feature that distinguishes eco-friendly lubricants from conventional petroleum-based counterparts. These lubricants are intentionally crafted to undergo natural processes, breaking down into simpler, environmentally benign compounds upon release into the environment by the action of microorganisms like bacteria, fungi, and algae into elemental compounds such as water, carbon dioxide, and biomass [21, 26]. This inherent degradation process mitigates environmental impact and minimizes ecosystem harm. Biodegradable eco-friendly lubricants yield significant ecological advantages. Upon environmental exposure, they naturally degrade, curbing pollution risks and reducing adverse effects on aquatic life, soil organisms, and ecosystems. Many eco-friendly lubricants derive from renewable feedstocks sourced from natural origins like vegetable oils, animal fats, or bio-based synthetic compounds. These sustainable sources inherently possess biodegradability and can be replenished through agricultural or industrial processes, amplifying the ecological sustainability of eco-friendly lubricants [27].

Numerous regulatory bodies and standards organizations establish criteria and methodologies to evaluate lubricant biodegradability. Noteworthy standards encompass OECD 301, OECD 306, ASTM D5864, and ASTM D6400 [28]. Lubricants meeting stringent biodegradability criteria often earn certifications or eco-labels, signifying their environmental compatibility. While prioritizing biodegradability, eco-friendly lubricants must strike a balance with performance prerequisites. To ensure operational effectiveness across diverse applications, they must meet industry benchmarks for lubrication efficacy, wear resistance, thermal stability, and equipment material compatibility. The extent of biodegradability and environmental persistence in eco-friendly lubricants may vary based on formulation intricacies, operational environments, and exposure conditions [29]. Applications such as marine, forestry, agriculture, and sensitive ecosystems may necessitate heightened levels of biodegradability and environmental resilience to mitigate ecological impact. Biodegradability in a lubricant is a cornerstone of eco-friendly lubricants, which foster environmental sustainability and reduce ecological risks linked with lubricant usage. By formulating lubricants with biodegradable base oils and additives, manufacturers uphold environmentally friendly solutions that harmonize performance requisites with environmental stewardship across the lubricant lifecycle. Eco-friendly lubricants play a critical role in biotribology [30–41] by minimizing environmental impact while providing the necessary lubrication and protection for biological tissues in medical devices, prosthetics, and other biomechanical systems [42, 43].

4.5 REDUCED TOXICITY

Eco-friendly lubricants are pivotal in mitigating environmental toxicity through various mechanisms inherent in their formulation and utilization. These are engineered to disintegrate into benign compounds upon environmental release; eco-friendly lubricants undergo natural degradation facilitated by microorganisms. This process transforms them into simpler substances like water, carbon dioxide, and biomass, curbing the accumulation of toxic substances in soil, water bodies, and ecosystems. Polyalkylene glycols (PAGs), vegetable oils, and esters degrade naturally into water and small natural molecules [44]. Lubricants used in marine vessels contribute 51% to vessel operational discharge, 19% to tank ship spills, 10% to lube leakages, and 7% to tanker cargo washings, all adding to marine toxicity. Unlike conventional petroleum-based counterparts, eco-friendly lubricants feature formulations with reduced toxicity levels [45]. Often incorporating biodegradable base oils sourced from renewable origins such as vegetable oils or bio-based synthetics, these lubricants pose diminished environmental risks and mitigate the potential for toxic contamination in case of spills or leaks. Popularly known as EAL (environmentally accepted lubricants), they are typically made with soybean, canola oil, sunflower extracts, other vegetable matter, synthetic esters, and polyethylene glycol. Endowed with biodegradability and low toxicity attributes, eco-friendly lubricants pose lesser risks to aquatic ecosystems. In scenarios where lubricants interface with water bodies—like marine vessels or hydraulic systems—eco-friendly lubricants shield aquatic habitats and mitigate the threat of toxic contamination to marine organisms.

Eco-friendly lubricants typically eschew harmful additives, such as heavy metals, chlorine, or sulfur compounds, in traditional lubricants. By minimizing or eliminating these toxic additives, eco-friendly lubricants forestall the release of harmful substances into the environment during the application and disposal phases [46]. Eco-friendly lubricants are pivotal in reducing harmful additives, contributing to environmental protection and sustainability. Eco-friendly lubricants signify a notable departure from traditional petroleum-based lubricants by capitalizing on natural ingredients procured from renewable reservoirs. These components, including vegetable oils, animal fats, and bio-based synthetics, serve as the cornerstone of green lubricant formulations, offering manifold environmental sustainability and performance advantages. Primarily, using natural ingredients sourced from renewable origins aligns closely with sustainability principles. Abundant vegetable oils like soybean, rapeseed, and sunflower are readily replenishable through agricultural practices. Similarly, animal fats from sustainable farming methods present an environmentally conscious alternative to petroleum-derived lubricants [47]. Moreover, bio-based synthetics derived from renewable biomass sources, such as plant sugars, present a sustainable avenue for lubricant development. By integrating these natural formulations, eco-friendly lubricants markedly diminish reliance on harmful additives commonly present in conventional petroleum-based lubricants. Unlike their traditional counterparts, which often incorporate additives derived from fossil fuels and carry potential environmental and health risks, eco-friendly lubricants leverage the inherent properties of natural ingredients to augment lubrication performance. For instance, vegetable oils inherently possess lubricating properties and boast high viscosity indices, rendering them ideal base oils for eco-friendly lubricants. Conversely, animal fats exhibit exceptional boundary lubrication characteristics, bolstering the lubricant's capacity to safeguard moving parts and mitigate friction [48]. The utilization of eco-friendly lubricants in biomedical devices and implants such as knee and hip replacements is paramount to ensure both the efficacy of medical equipment and the safety of patients, aligning with the healthcare sector's commitment to environmental sustainability [49, 50].

Compared to conventional petroleum-based lubricants, bio-based synthetics can be tailored to meet precise performance prerequisites, offering heightened thermal stability, oxidation resistance, and shear stability [51–60]. Furthermore, integrating natural ingredients into eco-friendly lubricants contributes to biodegradability and environmental compatibility. Base oils in eco-friendly lubricants often include vegetable oils such as soybean, rapeseed, sunflower, and palm oil, chosen for their sustainability and lubricating properties. Sustainable farming practices yield animal fats, another natural ingredient favored in eco-friendly lubricants. Additionally, plant sugars or agricultural wastes are renewable sources for lubricant production. Specific eco-friendly lubricants integrate plant extracts and additives to improve performance and environmental compatibility. Upon release into the environment, lubricants formulated with renewable feedstocks decompose into simpler, environmentally benign substances through natural processes [61]. This mitigates the risk of environmental contamination and curtails the ecological footprint associated with lubricant usage and disposal. Collectively, including natural ingredients sourced from renewable origins underscores the commitment of eco-friendly lubricants to sustainability, environmental stewardship, and performance excellence. Through the adoption of natural formulations, eco-friendly lubricants not only diminish the requirement for harmful additives but also chart a course toward a cleaner, more sustainable future in lubrication technology.

4.6 CUSTOM DESIGN LUBRICANTS

Many eco-friendly lubricants integrate biodegradable additives that enhance lubrication properties without posing environmental risks. These additives, sourced from renewable vegetable origins, find application in various products and industries, showcasing their versatility and eco-friendliness. Citrus, corn, or soybeans are the basis for bio-based solvents utilized across industrial sectors for cleaning, coatings, and degreasing tasks. Meanwhile, natural enzymes derived from microorganisms play a crucial role in laundry detergents, dishwashing liquids, and wastewater treatment processes by facilitating the breakdown of organic matter. In the realm of materials, plant-derived plasticizers augment the flexibility and durability of plastics and polymers, reducing the environmental impact of these products while enhancing their functionality [62]. Likewise, additives enhancing lubrication properties in oils and lubricants derived from renewable sources like vegetable oils and animal fats embody biodegradability and environmental sustainability. In the realm of food, cosmetics, and pharmaceuticals, natural antioxidants extracted from plants such as rosemary, green tea, and grape seed extend shelf life and counteract oxidation. Wood pulp is the source material for bio-based flame retardants, employed in textiles, electronics, and construction materials to diminish flammability without compromising biodegradability. In agricultural contexts, microbial biostimulants contain beneficial microorganisms that foster soil health and augment plant growth, reducing the reliance on chemical fertilizers and pesticides. Extracts and essential oils derived from plants serve as biodegradable additives in air fresheners, cleaning products, and personal care items, effectively neutralizing odors without harmful chemicals. Furthermore, plant-derived surfactants, utilized in cleaning products and personal care items, offer biodegradable alternatives to synthetic surfactants, reducing environmental impact during use and disposal. These biodegradable additives, sourced from renewable origins, deliver adequate performance and embody a commitment to sustainable practices across diverse industries. Eco-friendly lubricants are formulated to be free from harmful substances like sulfur and chlorine, which can contribute to environmental pollution and health hazards. Instead, they utilize alternative additives or natural compounds that provide lubrication properties without compromising ecological integrity [63]. Eco-friendly lubricants, formulated to exclude harmful substances such as sulfur and chlorine, are pivotal solutions for ensuring environmental compatibility and diminishing health risks. Prioritizing sustainability and eco-friendliness, these lubricants deliver effective performance across diverse applications. Vegetable-based lubricants are crafted from renewable vegetable oils like soybean, rapeseed, and sunflower oil; these lubricants eschew sulfur and chlorine additives. They provide superior lubrication properties while minimizing ecological impact. Bio-based synthetic lubricants, leveraging renewable feedstocks, offer heightened performance attributes without including sulfur and chlorine compounds. They emerge as sustainable alternatives to conventional lubricants. Devoid of sulfur and chlorine additives, water-based lubricants employ water as their primary carrier fluid. Widely used in metal-working and industrial contexts, they ensure environmental compatibility. Lubricants used for vacuum applications often include specialized synthetic oils, such as perfluoropolyether (PFPE) and silicone-based lubricants, due to their low vapor pressure and chemical stability in low-pressure environments. Eco-friendly lubricants used in vacuum applications like vacuum fryers are designed to withstand extreme conditions while minimizing contamination risks, ensuring both equipment longevity and environmental safety [64, 65].

4.6.1 Silicone-based lubricants

Silicone-based lubricants, excluding sulfur and chlorine compounds, boast remarkable thermal stability and oxidation resistance. Their application spans automotive, aerospace, and high-performance sectors. Silicone-based lubricants find extensive usage across industries due to their exceptional thermal stability, resistance to oxidation, and ability to withstand a wide range of temperatures [66]. Silicone oil lubricants are versatile and have various applications. Renowned for their superior lubricating properties, they are often employed in mechanical components, plastic parts, and medical devices. Formulated with multiple thickeners, silicone greases offer enhanced lubrication and protective qualities. Due to their water-repelling and corrosion-resistant attributes, they are commonly utilized in automotive settings, electrical connectors, and O-ring applications. Silicone spray lubricants are user-friendly and provide adequate lubrication for various surfaces. They are frequently employed in household scenarios such as lubricating door hinges, locks, drawer slides, and equipment maintenance in industrial environments. These thick, viscous formulations deliver long-lasting lubrication and protection. They are widely used in automotive, marine, and aerospace industries where durability and performance are paramount—specifically designed to insulate and safeguard electrical components from moisture, corrosion, and arcing. It is commonly found in electrical connectors, spark plugs, and automotive ignition systems. Tailored to meet stringent regulatory standards, silicone lubricants for medical devices ensure compatibility with sensitive materials. They are extensively used in medical equipment, surgical instruments, and catheters to facilitate smooth operation and patient comfort [67].

4.6.2 Polyalkylene glyco (PAG) lubricants

PAG lubricants, void of sulfur and chlorine additives, offer exceptional thermal stability and lubricating properties. Commonly found in automotive air conditioning systems, they underscore environmental compatibility in industrial applications. These eco-friendly lubricants respond to the escalating demand for environmentally conscious lubrication options. By eradicating substances like sulfur and chlorine, they actively participate in curbing environmental pollution and advancing sustainability across industries. Eco-friendly lubricants commonly employ non-toxic anti-foaming agents derived from natural sources to prevent foam formation during operation [68]. These additives help maintain lubricant stability and performance without introducing harmful chemicals into the environment. Eco-friendly lubricants incorporate viscosity modifiers that are environmentally friendly and biodegradable. These modifiers help maintain lubricant viscosity across various operating conditions without relying on toxic or non-biodegradable additives. Some eco-friendly lubricants contain corrosion inhibitors derived from plant-based compounds, which effectively protect metal surfaces against corrosion without introducing harmful chemicals into the environment. These inhibitors help extend equipment lifespan while promoting environmental sustainability. Many eco-friendly lubricants adhere to stringent environmental regulations and standards mandated by regulatory bodies like the Environmental Protection Agency (EPA), the European Chemicals Agency (ECHA), OSHA, ISO, etc. Manufacturers ensure compliance with criteria about biodegradability, eco-toxicity, and overall environmental performance, mitigating the likelihood of ecological toxicity and upholding regulatory mandates [69–76].

Embracing eco-friendly lubricants fosters the cultivation of sustainable practices within industries, emphasizing environmental stewardship. By opting for eco-friendly lubricants, enterprises underscore their commitment to mitigating environmental toxicity, conserving natural resources, and minimizing ecological harm. Promoting sustainable practices through eco-friendly lubricants involves employing various strategies to increase awareness, encourage adoption, and foster innovation within the lubricant industry. These strategies include launching educational campaigns directed at businesses, industries, and consumers to underscore the environmental benefits of eco-friendly lubricants, emphasizing their reduced environmental impact and improved resource efficiency compared to conventional petroleum-based alternatives. Additionally, advocating for third-party certifications and labels, such as the USDA Certified Biobased Product Label or the European Union Ecolabel, validates Eco-friendly lubricants' environmental credentials, assisting consumers in making informed purchasing decisions. Offering incentives and recognition to businesses that integrate eco-friendly lubricants into their sustainability initiatives, fostering partnerships and collaborations among stakeholders to collectively promote the use of eco-friendly lubricants collectively, investing in research and development to enhance the performance and sustainability of these lubricants, advocating for supportive regulatory frameworks, showcasing successful implementations through demonstration projects, and providing training programs to equip stakeholders with the necessary knowledge and skills all contribute to the promotion of eco-friendly lubricants. Through these concerted efforts, stakeholders can effectively advocate for the widespread adoption of Eco-friendly lubricants, thereby driving sustainable practices and making meaningful contributions to global environmental preservation endeavors [77].

4.6.3 Energy-efficient production

Eco-friendly lubricants are instrumental in advancing energy-efficient production practices across diverse industries. Engineered to diminish friction and wear amidst machinery and equipment components, Eco-friendly lubricants minimize energy dissipation from friction, thereby enhancing overall production efficiency. Eco-friendly lubricants reduce friction and wear within machinery and equipment, improve operational efficiency, and extend component lifespan [77]. Utilizing advanced formulations and sustainable ingredients, eco-friendly lubricants provide improved lubrication properties to counteract the adverse effects of friction and wear. For example, consider the use of biodegradable vegetable-based lubricants in industrial gearboxes. Derived from renewable sources like soybean or rapeseed oil, these lubricants boast high viscosity indices and exceptional lubricity, ensuring smooth operation and minimal friction within gear assemblies. As a result, components experience reduced wear and tear over time, leading to prolonged service life and increased machinery reliability. Additionally, eco-friendly lubricants exhibit excellent thermal stability and oxidation resistance, which are essential for mitigating heat generation due to friction and preventing component degradation. Maintaining consistent lubricating film thicknesses across varying operating conditions, eco-friendly lubricants effectively minimize metal-to-metal contact and associated wear, even in high-temperature environments. Eco-friendly lubricants also offer advantages in automotive engines by enhancing fuel efficiency and promoting engine longevity. Bio-based engine oils formulated with renewable base oils and additives provide superior lubrication properties, reducing frictional losses within engine components like pistons, cylinders, and bearings. This results in improved fuel economy, reduced emissions, and prolonged durability of critical engine parts. Eco-friendly lubricants optimize equipment performance and align with sustainability objectives with their environmentally friendly compositions and superior lubricating capabilities. They are essential in modern mechanical systems, contributing to energy efficiency and environmental preservation across various industries.

Eco-friendly lubricants often outperform conventional counterparts, sustaining optimal machinery conditions that translate into heightened energy efficiency and diminished energy consumption during production cycles [78]. Eco-friendly lubricants are vital in elevating performance across diverse industrial and automotive sectors through their advanced lubrication properties and commitment to environmental sustainability. Eco-friendly lubricants are meticulously crafted to deliver superior lubrication efficiency compared to conventional counterparts. For instance, bio-based lubricants engineered with renewable base oils, like vegetable oils, showcase exceptional lubricity and film strength, ensuring seamless machinery and equipment operation. This heightened lubrication efficiency minimizes frictional losses, mitigates component wear, and optimizes overall system performance. Eco-friendly lubricants significantly bolster energy efficiency by curbing frictional losses and optimizing equipment performance. Consider bio-based engine oils fortified with friction modifiers and anti-wear additives. Eco-friendly lubricants actively extend the lifespan of machinery and mechanical systems by mitigating friction and wear [78]. They effectively forestall premature component failure and deterioration. For instance, bio-based hydraulic fluids in hydraulic systems boast enhanced anti-wear properties, corrosion resistance, and oxidative stability. Consequently, they prolong the operational lifespan of hydraulic equipment and reduce downtime. Eco-friendly lubricants serve as environmentally conscious alternatives to conventional petroleum-based lubricants. Take, for instance, synthetic ester-based lubricants commonly utilized in marine settings. Compared to mineral oil-based alternatives, these lubricants offer biodegradability, non-toxicity, and diminished environmental impact. This heightened environmental compatibility ensures adherence to regulatory standards and fosters sustainable practices in various industries. By furnishing adequate lubrication and shielding against wear and corrosion, Eco-friendly lubricants extend the operational lifespan of machinery and equipment. This longevity reduces the need for frequent maintenance or replacements, culminating in energy savings over time. Eco-friendly lubricants are pivotal in elongating equipment lifespan across various industries, owing to their environmentally friendly nature and exceptional performance attributes. Engineered to minimize friction and wear between machinery components, eco-friendly lubricants establish a protective layer that diminishes metal-to-metal contact. This proactive measure mitigates wear and tear, safeguarding critical parts and extending equipment lifespan. Specific eco-friendly lubricants boast properties enabling them to function adeptly at lower temperatures than traditional variants [78]. Even in conventional manufacturing processes and computerized numerical control (CNC) operations, lubrication plays an important role [79].

4.7 PERFORMANCE AND COMPATIBILITY

Eco-friendly lubricants are pivotal in improving performance and compatibility across various applications. Eco-friendly lubricants often outperform traditional ones by offering superior lubrication properties. They exhibit better film strength, lower friction, and enhanced wear protection, resulting in smoother machinery operation and improved performance.

Eco-friendly lubricants optimize machinery operation by reducing frictional losses, minimizing component wear, and facilitating efficient energy transfer. This leads to increased reliability, decreased downtime, and heightened productivity. By mitigating friction and wear, eco-friendly lubricants prolong the lifespan of the equipment. Components endure less mechanical stress and degradation, translating to longer service intervals and reduced maintenance expenses [80]. Formulated to align with a wide range of seals, gaskets, and materials used in machinery, eco-friendly lubricants maintain seal integrity and prevent material degradation. This ensures smooth operation and prevents leaks or malfunctions. Many eco-friendly lubricants are eco-friendly and biodegradable, making them suitable for sensitive ecosystems and regulatory standards. They minimize contamination risk and ensure compliance with environmental regulations. Eco-friendly lubricants are designed to adapt to various operating conditions and equipment types. Whether in automotive, industrial, or marine settings, they offer compatibility with diverse systems and components, ensuring optimal performance in different environments. Derived from renewable sources and containing biodegradable components, eco-friendly lubricants have a lower environmental impact than petroleum-based counterparts. They naturally break down over time, reducing soil and water pollution and supporting sustainable practices. Eco-friendly lubricants reduce emissions and improve air quality by promoting efficient energy transfer and minimizing frictional losses. They help mitigate machinery and equipment's environmental footprint, aligning with sustainability goals and regulatory requirements [81].

4.8 REGULATORY COMPLIANCE

Regulatory authorities worldwide establish environmental guidelines and standards that mandate adherence to eco-friendly lubricants. Regulatory authorities worldwide establish ecological guidelines and standards to encourage using eco-friendly lubricants. These standards aim to promote sustainability and reduce the environmental impact of lubricants in various industries, from automotive to manufacturing [82]. The objective is to ensure that these lubricants minimize pollution, are safe for ecosystems, and have reduced toxicity for human health. Table 4.2 provides a summary of the authorities responsible for eco-friendly lubricants. The Environmental Protection Agency (EPA) is responsible for setting national environmental standards in the United States. The EPA focuses on minimizing pollution and ensuring that lubricants used in industry and transportation do not contribute to environmental degradation [21]. This involves promoting biodegradable lubricants and reducing harmful emissions associated with lubricant production and use. Similarly, in the European Union, the European Chemicals Agency (ECHA) plays a critical

Table 4.2 A brief summary of the regulatory authorities for eco-friendly lubricants

Region	*Regulatory authority*	*Possible standards*	*Focus*
Europe	European Chemicals Agency (ECHA) European Parliament (EP)	Biodegradability (e.g., OECD 301B test) Toxicity (e.g., OECD 201 and 202 tests for aquatic toxicity) Labeling (e.g., EU Ecolabel criteria)	Reduced environmental impact throughout the cycle
North America	United States Environmental Protection Agency (EPA) Canadian Environmental Protection Agency (CEPA)	Minimum biodegradability levels (e.g., EPA's Vessel General Permit) Restrictions on harmful additives Recycling and disposal requirements	Reduced environmental impact during use and disposal
Asia	Varies by country (e.g. China's Ministry of Ecology and Environment (MEE), Japan's Environmental Ministry (MOE)), International Organization for Standardization (ISO)	Emerging standards focusing on biodegradability (e.g., China's GB 17479 standards) and recyclability ISO 14001 (Environmental Management Systems) and ISO 14040 (Life Cycle Assessment)	Alignment with broader environmental goals

role in regulating chemicals, including those used in lubricants. Through regulations like REACH (Registration, Evaluation, Authorization, and Restriction of Chemicals), the EU encourages using eco-friendly lubricants and mandates rigorous testing to ensure safety and sustainability. The European Environment Agency (EEA) oversees environmental sustainability across the EU. In other regions, such as Canada, Asia-Pacific, and Australia, regulatory bodies like Environment and Climate Change Canada (ECCC) and their counterparts in Japan and South Korea also set environmental standards promoting eco-friendly lubricants. These regulations often focus on biodegradability, reduced toxicity, and using renewable sources for lubricant production [83].

Specific guidelines for eco-friendly lubricants include requirements for biodegradability, reduced toxicity, and using renewable or bio-based materials. This can also extend to waste management practices, emphasizing the proper disposal or recycling of used lubricants to minimize their environmental footprint. Industries must comply with these regulations and may face fines or penalties for non-compliance. Beyond regulatory adherence, companies often adopt eco-friendly lubricants as part of their corporate social responsibility efforts, aligning with consumer demands for environmentally sustainable products.

Here is an overview of some key regulatory bodies and standards associated with various lubricants, including eco-friendly lubricants.

4.8.1 Environmental Protection Agency (EPA)

The EPA, responsible for environmental regulations in the United States, oversees various aspects of environmental protection, including managing pollutants and hazardous substances. The Environmental Protection Agency (EPA) in the United States assumes a central role in regulating environmental concerns, including those related to lubricants. Despite the absence of dedicated regulations for eco-friendly lubricants, the EPA's overarching directives encompass aspects relevant to lubricant usage and disposal. Through initiatives like the Resource Conservation and Recovery Act (RCRA), the EPA oversees the management and disposal of waste, including lubricants, fostering the adoption of eco-friendlier alternatives like Eco-friendly lubricants to mitigate environmental impact. The Toxic Substances Control Act (TSCA) empowers the EPA to regulate chemicals in lubricants, ensuring assessments for potential environmental and health hazards. Furthermore, the EPA's Safer Choice Program advocates for safer chemicals in consumer products, allowing manufacturers to certify lubricants meeting environmental and human health safety standards. In tandem, the Pollution Prevention (P2) program targets pollutant reduction associated with lubricants, aligning with eco-friendly lubricants' design to minimize environmental harm. The EPA's encouragement of environmentally responsible practices underscores the adoption of products like eco-friendly lubricants, reflecting a broader commitment to sustainability. Although specific regulations for eco-friendly lubricants are absent, the EPA's regulatory landscape promotes environmental responsibility, providing stakeholders with the impetus to choose eco-friendly lubricants voluntarily. This proactive approach reduces environmental footprint and ensures compliance with broader environmental regulations administered by the EPA [84–93].

4.8.2 European Chemicals Agency (ECHA)

Within the European Union (EU), the ECHA regulates chemical usage, implementing regulations like REACH (Registration, Evaluation, Authorization, and Restriction of Chemicals) to ensure safe chemical usage and environmental sustainability. The European Chemicals Agency (ECHA) plays a pivotal regulatory role in the European Union (EU), overseeing chemicals utilized across various industries, including lubricants. Although ECHA's directives do not explicitly focus on eco-friendly lubricants, its regulatory framework encompasses environmental considerations relevant to lubricant formulation, usage, and disposal. Major regulations like the Registration, Evaluation, Authorization, and Restriction of Chemicals (REACH) mandate thorough assessments of product chemicals, ensuring their environmental and health impacts are thoroughly evaluated. The Classification, Labeling, and Packaging (CLP) Regulation ensures accurate classification and labeling of lubricants, including green variants, based on their environmental and health risks. Furthermore, the Biocidal Products Regulation (BPR) assesses biocidal substances in lubricants for their environmental impact and effectiveness. The Restriction of Hazardous Substances (RoHS) Directive indirectly influences lubricant formulations by limiting hazardous substances in electrical and electronic equipment, encouraging the adoption of environmentally friendly options like Eco-friendly lubricants. Moreover, the Waste Framework Directive (WFD) establishes guidelines for waste management, advocating for environmentally sound practices, including adopting eco-friendly lubricants to mitigate environmental harm. While eco-friendly lubricants are not explicitly targeted, ECHA's regulations promote safer and eco-friendly chemicals, prompting manufacturers and users to prioritize sustainability and environmental protection in lubricant selection and usage throughout the EU [94, 95].

4.8.3 International Organization for Standardization (ISO)

ISO develops global standards across industries, including lubricants. ISO 14001 (Environmental Management Systems) and ISO 9001 (Quality Management Systems) provide environmental management and quality assurance frameworks. The International Organization for Standardization (ISO) is a global institution responsible for formulating and disseminating international standards across various sectors, including the lubricant industry. Although ISO does not provide specific standards exclusively tailored to "green" lubricants, its existing protocols encompass sustainability and environmental compatibility elements pertinent to lubricant development, manufacture, and application.

ISO standards concerning lubricants typically address performance, quality, safety, and environmental impact. While these standards do not overtly target "green" lubricants, they integrate criteria and considerations aligned with sustainability principles and environmental responsibility. An example of a significant ISO standard applicable to lubricants is ISO 14001, which pertains to Environmental Management Systems (EMS). ISO 14001 furnishes guidelines for organizations to establish, implement, maintain, and continuously improve Environmental Management Systems. Although not directly addressing lubricants, entities involved in lubricant-related activities can adhere to ISO 14001 to enhance environmental performance and diminish ecological footprints. ISO 9001 represents another notable standard developed by ISO, focusing on Quality Management Systems (QMS). While primarily concentrating on quality management principles, ISO 9001 enables organizations within the lubricant sector to ensure consistent product quality, customer satisfaction, and regulatory compliance. Adherence to ISO 9001 allows lubricant manufacturers to refine processes and procedures, potentially encompassing sustainability and environmental considerations [95].

4.8.4 Society of Automotive Engineers (SAE)

SAE publishes standards for automotive and aerospace sectors, such as SAE J2360, outlining lubricant performance requirements and testing protocols. The Society of Automotive Engineers (SAE) is a highly respected professional organization tasked with developing standards and guidelines for the automotive and aerospace sectors worldwide. While the SAE does not create specific standards exclusively for "green" lubricants, its current standards and practices indirectly encompass environmental considerations and sustainability principles relevant to lubricants used in automotive and aerospace applications. SAE lubricant standards primarily focus on establishing performance requirements, testing methodologies, and specifications across various lubricant applications. For example, SAE J2360 outlines performance requirements and testing protocols for industrial gear lubricants, while SAE J300 specifies viscosity grades for engine oils used in automotive engines. Despite the absence of a direct emphasis on environmental sustainability, these standards implicitly encourage it by emphasizing lubricant efficiency, durability, and compatibility with engine and equipment specifications. Furthermore, the SAE routinely updates its standards to reflect advancements in lubricant technology, evolving regulatory requirements, and industry best practices. With the increasing prominence of environmental concerns, these revisions may incorporate considerations aimed at reducing harmful additives, enhancing biodegradability, and improving the overall ecological compatibility of lubricants. While these updates may not explicitly target "green" lubricants, they inspire lubricant manufacturers to innovate and adopt environmentally responsible practices in their formulations and processes. Additionally, the SAE actively engages in research, development, and knowledge-sharing initiatives within the automotive and aerospace sectors [96].

4.8.5 National Sanitation Foundation (NSF International)

NSF International certifies public health and safety products, including lubricants used in food processing (NSF H1 standard), ensuring they meet safety and environmental standards. NSF International, commonly called the National Sanitation Foundation, is a widely recognized organization responsible for establishing standards and certification programs for public health and safety products, including lubricants. While NSF International does not develop specific standards exclusively dedicated to "green" lubricants, its current standards and certifications incorporate environmental sustainability and safety considerations in lubricant formulations. NSF International's prominent certification is the NSF H1 certification, designed for lubricants in food processing facilities. Although its primary focus is to ensure lubricants meet stringent safety standards and do not pose risks to food products, the certification indirectly promotes environmental sustainability by encouraging using lubricants for human health and the ecosystem. Lubricants certified under NSF H1 standards must adhere to specific criteria regarding ingredients, formulation, and manufacturing processes to ensure compliance with food safety regulations while minimizing environmental impact. Moreover, through diverse programs, NSF International extends certifications to lubricants in other sectors like automotive and manufacturing. While these certifications may not explicitly prioritize environmental sustainability, they often encompass requirements concerning biodegradability, toxicity, and eco-friendliness, indirectly fostering the promotion of eco-friendly lubricants. Lubricants certified by NSF International undergo thorough testing and assessment to ensure they meet performance, safety, and environmental standards benchmarks, reassuring consumers, businesses, and regulatory bodies. Beyond certifications, NSF International collaborates with industry stakeholders, lubricant manufacturers, and regulatory agencies to tackle emerging trends and challenges in lubricant technology and sustainability [97].

4.9 WASTE REDUCTION AND RECYCLING

Eco-friendly lubricants are pivotal in waste reduction and recycling initiatives across diverse industries. They often comprise biodegradable compounds from renewable materials like soybean oil, canola oil, sunflower oil, corn oil, and rapeseed oil. Unlike traditional petroleum-based alternatives, they naturally decompose over time, curbing the accumulation of non-biodegradable waste. For instance, bio-based hydraulic fluids derived from renewable vegetable oils undergo natural degradation, lessening environmental harm and fostering waste reduction. Eco-friendly lubricants mitigate the risk of contamination in industrial settings and ecosystems. With fewer harmful additives compared to conventional oils, they minimize the likelihood of soil, water, and air contamination. Using environmentally friendly lubricants in machinery mitigates soil pollution, particularly in agricultural environments, promoting ecosystem health and minimizing the need for waste cleanup endeavors [83]. Designed to align with recycling and reclamation processes, eco-friendly lubricants feature biodegradable properties and reduced toxicity, simplifying separation from other waste materials for recycling purposes. For example, bio-based lubricants employed in metal-working applications can be segregated from metal shavings and repurposed, fostering sustainable manufacturing practices and reducing landfill waste. Eco-friendly lubricants advocate for closed-loop systems, where used lubricants are collected, treated, and reintroduced into the production cycle. Unlike conventional counterparts that may contain harmful substances hindering recycling efforts, eco-friendly lubricants pose less environmental risk during the recycling process. This encourages industries to embrace closed-loop systems, minimizing the necessity for new lubricant production and reducing waste generation [97]. By adopting eco-friendly lubricants, industries adhere to stringent environmental regulations concerning waste management and disposal. This mitigates the risk of fines and penalties associated with improper waste-handling practices. For instance, industries utilizing biodegradable lubricants uphold rules governing hazardous waste disposal, furthering waste reduction endeavors. Eco-friendly lubricants endorse sustainable practices centered on waste reduction and resource preservation. By opting for environmentally friendly lubricants, industries showcase their commitment to environmental stewardship and sustainable growth. For instance, the utilization of bio-based lubricants in marine applications minimizes the probability of oil spills, safeguards ecosystems, and promotes responsible waste management practices [1].

4.10 CONCLUSION

In conclusion, the emergence of eco-friendly lubricants represents a significant advancement in pursuing environmental sustainability within various industries. These lubricants, derived from renewable resources or synthesized through green processes, offer numerous advantages over conventional petroleum-based counterparts, including biodegradability, reduced toxicity, and lower emissions. Their applications span various sectors, from automotive and manufacturing to agriculture and marine, where they contribute to enhanced operational efficiency and a reduced environmental footprint. The classification of eco-friendly lubricants into vegetable-based, bio-synthetic, water-based, polyalkylene glycol (PAG), and silicone-based categories demonstrates the diversity and versatility of these products. This adaptability allows industries to select lubricants that align with specific performance and compatibility requirements while maintaining compliance with environmental regulations. Eco-friendly lubricants contribute to broader sustainability goals through their renewable nature and lower ecological impact. The emphasis on reduced toxicity and biodegradability promotes safer working environments and decreases long-term environmental risks. Furthermore, these lubricants support regulatory compliance, ensuring adherence to standards set by agencies like the Environmental Protection Agency (EPA), the European Chemicals Agency (ECHA), and the International Organization for Standardization (ISO). Ultimately, adopting eco-friendly lubricants signifies a shift toward more sustainable industrial practices. By integrating these lubricants into operational processes, industries can achieve performance goals while demonstrating a commitment to environmental stewardship. This transition fosters eco-conscious behaviors and sets the stage for a cleaner and more sustainable future.

REFERENCES:

1. Boyde, S., *Green Chemistry*, 2002. **4**(4): p. 293–307.
2. Johnson, M. and M. Miller, *Tribology & Lubrication Technology*, 2010. **66**(10): p. 28–34.
3. Bisht, R.P.S., et al., *Journal of Synthetic Lubrication*, 2002. **19**(3): p. 243–248.
4. Covert, T., et al., *Journal of Economic Perspectives*, 2016. **30**(1): p. 117–138.
5. Khare, P., et al., *Atmospheric Chemistry and Physics*, 2018. **18**(8): p. 5391–5413.
6. He, X., et al., *ACS Sustainable Chemistry & Engineering*, 2024. **12**(32): p. 12, 19, 7226-7234.
7. Karthikeyan, K.M.B., et al., *Materials Today: Proceedings*, 2021. **39**: p. 285–291.
8. Chakraborty, S., et al., *Materials Letters*, 2020. **274**: p. 128004.
9. Kumar, A., et al., *Tribology Transactions*, 2018. **61**(5): p. 951–959.
10. Kumar, A., et al., Recent advancements and future trends in next-generation materials for biomedical applications, in: *Advanced Materials for Biomedical Applications*. CRC Press, 2022, p. 1–19.
11. Hörner, D., *Journal of Synthetic Lubrication*, 2002. **18**(4): p. 327–347.
12. Maleque, M.A., et al., *Industrial lubrication and Tribology*, 2003. **55**(3): p. 137–143.
13. Feiferytė, A., et al., *Journal of Cleaner Production*, 2015. **95**: p. 281–290.
14. Rahman, M.H., et al., *Lubricants*, 2021. **9**(8): p. 73.
15. Paredes, X., et al., *Tribology Letters*, 2012. **45**: p. 89–100.
16. Chakraborty, S., et al., *Materials Letters*, 2019. **257**: p. 126739.
17. Roy, D., et al., *Journal of Alloys and Compounds*, 2021. **854**: p. 157174.
18. Roy, D., et al., *Materials Letters*, 2020. **271**: p. 127780.
19. Ngaile, G. and F. Botz, *Journal of Tribology*, 2008. **130**: p. 021801.
20. Gnanasekaran, D. and V.P. Chavidi, *Vegetable Oil Based Bio-Lubricants and Transformer Fluids: Applications in Power Plants*. Springer Nature, 2018, p. 29–47.
21. Salih, N. and J. Salimon, *Biointerface Research in Applied Chemistry*, 2021. **11**(5): p. 13303–13327.
22. Lathi, P.S., et al., *Applied Catalysis B: Environmental*, 2007. **69**(3–4): p. 207–212.
23. Omair, M., et al., *Sustainability*, 2017. **9**(5): p. 714.
24. Cigno, E., *Mechanical Engineering*. Rochester Institute of Technology, 2016.
25. Shah, R., et al., *Lubricants*, 2021. **9**(2): p. 21.
26. Pichler, J., et al., *Green Chemistry Letters and Reviews*, 2023. **16**(1): p. 2185547.
27. Goyan, R.L., et al., *Tribology & Lubrication Technology*, 1998. **54**(7): p. 10.

28. Luna, F.M.T., et al., *Tribology International*, 2015. **92**: p. 301–306.
29. Hooper, D.L. and D.I. Hoel, *SAE Transactions*, 1996. **105**: p. 1149–1157.
30. Kumar, A., et al., Introduction to biotribology: A science of surface interaction, in: *Application of Biotribology in Biomedical System*. Springer, 2024.
31. Datta, A., et al., Advanced materials in biological implants and surgical tools, in: *Advanced Materials for Biomedical Applications*. CRC Press, 2022, p. 21–43.
32. Kumar, A., et al., Recent advancements and future trends in next-generation materials for biomedical applications, in: *Advanced Materials for Biomedical Applications*. CRC Press, 2022, p. 1–19.
33. Kumar, A., et al., Liquid slippage in confined flows: Effect of periodic micropatterns of arbitrary pitch and amplitude. ASME 2016 5th International Conference on Micro/Nanoscale Heat and Mass Transfer, 2016.
34. Kumar, A., et al., *Physics of Fluids*, 2016, p. 28.
35. Kumar, A., et al., Reduction of hydraulic friction in confined flows by laser texturing: Experiments and theoretical validation. ASME 2018 16th International Conference on Nanochannels, Microchannels, and Minichannels. ICNMM, 2018.
36. Kumar, A., et al., *Journal of Heat Transfer*, 2018: p. 140.
37. Kumar, A., Liquid slippage in con fined flows: Effect of periodic micropatterns of arbitrary pitch and amplitude. ASME 2016 5th Micro/Nanoscale Heat and Mass Transfer International Conference (MNHMT2016), Biopolis, 2016.
38. Kumar, A., et al., Biomedical study of femur bone fracture and healing, in: *Advanced Materials for Biomedical Applications*. CRC Press, 2022, p. 235–250.
39. Gupta, A. K., et al., Composites for drug eluting devices: Emerging biomedical applications, in *Applications of Biotribology in Biomedical Systems*, Springer Press, Springer Nature Group, 2024.
40. Choudhari, A. et al., Wear and friction mechanism study in knee and hip rehabilitation: A comprehensive review, in: *Applications of Biotribology in Biomedical Systems*, Springer Press, Springer Nature Group.
41. Kumar, A., et al., Biotribology, in *Application of Biotribology in Biomedical System*. Springer, 2024.
42. Chowdhury, N., et al., Characteristics of hydrogel properties in the advancement of biotribology, in: *Application of Biotribology in Biomedical System*. Springer, 2024.
43. Avinash et al., Tribological hurdles in biomedical manufacturing: A comprehensive examination, in: *Application of Biotribology in Biomedical System*. Springer, 2024.
44. Benedicto, E., et al., *Procedia Engineering*, 2017. **184**: p. 99–116.
45. Kass, M., et al., *Oak Ridge National Lab.(ORNL)*. Oak Ridge National Laboratory, 2019.
46. Falleh, H., et al., *Food Chemistry*, 2020. **330**: p. 127268.
47. Sharma, A., *Asian Journal of Home Science*, 2013. **8**(2): p. 768–771.
48. Hamdan, S.H., et al., *Tribology International*, 2017. **113**: p. 76–82.
49. Gupta, A.K., et al., Composites for drug eluting devices: Emerging biomedical applications, in: *Application of Biotribology in Biomedical System*. Springer, 2024.

50. Choudhari, A., et al., Wear and friction mechanism study in knee and hip rehabilitation: A comprehensive review, in: *Application of Biotribology in Biomedical System*. Springer, 2024.
51. Kumar A, et al., *Tribology Transactions*, 2018. **61**: p. 951–959.
52. Kumar A, and U. Panda (2022) Microfluidics-based devices and their role on point-of-care testing, in: *Biosensor Based Advanced Cancer Diagnostics*. Elsevier, p. 197–224.
53. Kumar, A., et al., Laser-assisted fabrication of polymers by pushing down the limit of resolution, in: *Trends in Fabrication of Polymers and Polymer Composites*. AIP Publishing Books, 2022, p. 1–3.
54. Kumar, A., et al., Clinically available/under trial drugs and vaccines for treatment of SARS-COV-2, in: *Computational Approaches for Novel Therapeutic and Diagnostic Designing to Mitigate SARS-CoV2 Infection*. Elsevier, 2022, p. 451–488.
55. Kumar, A., et al., *ACS Applied Bio Materials*, 2022. **5**: p. 2046–2068.
56. Kumar, A., et al., Physics of laser—matter interaction in laser-based manufacturing, in: *Laser-based Technologies for Sustainable Manufacturing*. CRC Press, 2023, p. 45–54.
57. Kumar, A., et al., *Lubricants*, 2023. **11**: p. 64.
58. Heidari, M., et al., Navigating the landscape: Cutting-edge biomedical manufacturing techniques, in: *Applications of Biotribology in Biomedical Systems*. Springer Press, Springer Nature Group, 2024.
59. Gour, R. et al., Animal tribology, in: *Applications of Biotribology in Biomedical Systems*, Springer Press, Springer Nature Group, 2024.
60. Chowdary, K., et al., *Journal of Molecular Liquids*, 2021. **339**: p. 116717.
61. Hamnas, A. and G. *Unnikrishnan, Renewable and Sustainable Energy Reviews*, 2023. **182**: p. 113413.
62. Solvent-Based, N. and A. Spray, *Tribology & Lubrication Technology*, 2014. **6**: p. 3.
63. Rohr, O., *Industrial Lubrication and Tribology*, 2002. **54**(4): p. 153–164.
64. Gupta, A., et al. Design and analysis of vertical vacuum fryer, in: *Proceedings of International Conference on Intelligent Manufacturing and Automation. Lecture Notes in Mechanical Engineering*. Springer, 2019. DOI: 10.1007/978-981-13-2490-1_13
65. Choudhari, A., et al., Design and development of vacuum frying machine for the production of high-quality fried products, in: *Proceedings of International Conference on Intelligent Manufacturing and Automation: ICIMA 2022*. Springer, 2023, DOI: 10.1007/978-981-19-7971-2_50.
66. Baumgarten, B., et al., *Forensic Chemistry*, 2018. **8**: p. 28–39.
67. Miyamoto, T., et al., *Langmuir*, 2019. **35**(48): p. 15784–15794.
68. Jain, V.K. and D.S. Shukla, *Wear*, 1997. **208**(1–2): p. 11–16.
69. Greaves, M.R., Polyalkylene glycols, in: *Synthetics, Mineral Oils, and Bio-Based Lubricants*. CRC Press, 2020, p. 119–146.
70. Kumar S, et al., *Trends in Fabrication of Polymers and Polymer Composites*. AIP Publishing Books, 2022: p. 10–11.
71. Leventini S.D., et al., *Tribology Letter*, 2022: **70**: p. 127.
72. Mehra, A.K., et al., *Composites Communications*, 2021: **25**: p. 100732.
73. Parihar, A., et al., *Advanced Biology*, 2023. **7**: p. 2200285.

74. Parmar, V., et al., *Journal of Laser Applications*, 2018: p. 30. https://pubs.aip.org/lia/jla/article-abstract/30/3/032001/700091/Oxidation-facilitated-anti-microbial-ability-of?redirectedFrom=fulltext.
75. Pal, A. et al., Tribological measurements of human skin, in: *Applications of Biotribology in Biomedical Systems*. Springer Press, Springer Nature Group, 2024.
76. Kumar, A., et al., Tribological hurdles in biomedical manufacturing: A comprehensive examination, in: *Applications of Biotribology in Biomedical Systems*. Springer Press, 2024.
77. Guo, Z., et al., *International Journal of Environmental Research and Public Health*, 2019. **16**(7): p. 1282.
78. Anand, A., et al., *Materials Today: Proceedings*, 2017. **4**(2): p. 3659–3665.
79. Choudhari, A., et al. Design and manufacturing of compact and portable smart CNC machine, in: *Proceedings of International Conference on Intelligent Manufacturing and Automation. Lecture Notes in Mechanical Engineering*. Springer, 2020. DOI: 10.1007/978-981-15-4485-9_21.
80. Menezes, P.L., et al., *Green Tribology: Biomimetics, Energy Conservation and Sustainability*. Springer, 2012: p. 265–286.
81. Mannekote, J.K., et al., *Renewable and Sustainable Energy Reviews*, 2018. **81**: p. 1787–1801.
82. Environmental Protection Agency, *Environmentally Acceptable Lubricants*. Environmental Protection Agency, 2011.
83. Pinheiro, C.T., et al., *Critical Reviews in Environmental Science and Technology*, 2021. **51**(18): p. 2015–2050.
84. Pollard, J.E., et al., *Environmental Monitoring Systems Laboratory Office of Research and Development us Environmental Protection Agency Cincinnati*. The Online Books Page, 1990, p. 45268.
85. Pathak, A., et al., Application of laser technology in the mechanical and machine manufacturing industry, in: *Laser-based Technologies for Sustainable Manufacturing*. CRC Press, 2023, p. 107–155.
86. Gupta, S. et al., *International Journal of Recent Technology and Engineering*, 2020. **8**: p. 4569–4573.
87. Shrivastava Prateek, S.R., et al. Investigation of torsional rod to minimize vibration in automobile using ANSYS. 2020 1st International Conference on Innovative Research in Applied Science, Engineering and Technology (IRASET), 2020, p. 1–6.
88. Singh, M., et al., *Physical Review Fluids*, 2020. **5**: p. 102101.
89. Singh, R.K., et al., *Microsystem Technologies*, 2014. **20**: p. 101–111.
90. Vats, P., et al., Laser-based additive manufacturing, in: *Laser-based Technologies for Sustainable Manufacturing*. CRC Press, 2023, p. 67–83.
91. Dias, D., et al., *Lubricants*. 2024. **12**(4): p. 114.
92. Kumar, A. et al., Introduction to biotribology: A science of surface interaction, in: *Applications of Biotribology in Biomedical Systems*, Springer Press, Springer Nature Group, 2024.
93. N. Chowdhury, et al., Characterization of hydrogel properties in the advancement of biotribology, in: *Applications of Biotribology in Biomedical Systems*, Springer Press, Springer Nature Group, 2024.
94. Torbacke, M. and E. Kassfeldt, *Lubricants: Introduction to Properties and Performance*. John Wiley & Sons, 2014.

95. Caines, A.J., et al., *Automotive Lubricants Reference Book*, Vol. 354. John Wiley & Sons, 2004.
96. Lawate, S., Lubricants and fluids for the food industry, in: *Lubricant Additives*. CRC Press, 2017, p. 485–502.
97. Belinskaia, I. and A. Kartoshkin, *Engineering for Rural Development*, 2017. **16**: p. 378–382.

Chapter 5

Biolubricants from non-edible vegetable oils

Amit J. Agrawal, Pratibha Agrawal, and Vijay Karadbhajane

5.1 INTRODUCTION

"Lubricants" are substances that are utilized to limit the grinding, wear, and warmth, typically, as a liquid film keeps the contradicting surfaces separated. Oils increase the toughness of gears by controlling bothersome discharges and keeping them away from breakdowns or separation, accordingly bringing about energy and cost reserve funds. Lubricants are essentially utilized in modern and transportation areas and are generally classified depending on their applications as water-powered liquids, electrical transformer liquids, coolants, sealants, separators, heat move liquids, and so forth. Business greases are ordinarily made out of an oil base stock planned with added substances as needed for explicit purposes. Because of their actual state, lubricants can be of four sorts: fluids (oils), semi-fluids (greases), solids (dry greases), and gas (packed air or different gases). Fluid and semi-fluid lubricants are all the more regularly utilized and are overwhelmingly delivered from mineral oils. Contingent upon its beginning, lubricants are additionally recognized as mineral, synthetic, and bio-based.

Soni and Agarwal (2014) presented an exploration that gives an outline of the combination, tribological conduct; the impact of structure on erosion/wear, load-bearing limit, opposition to increase in the sample temperature, and differing response of hostility to wear/outrageous pressure additives within the sight of vegetable oil/subsidiary structures has additionally been reviewed. According to the authors, an appropriate choice of base oil and additives is accordingly fundamental for a proficient synthesis of lubricating oil (Soni & Agarwal, 2014).

Experimental work was completed by researchers such as Sanni et al. (2017) wherein palm kernel oil (PKO) and soybean oil were utilized as crude materials to synthesize lubricating oils. It was found that the PKO gave a higher biolubricant oil yield in contrast to soybean oil. The addition

DOI: 10.1201/9781003535447-5

of ethylene glycol combined with subsequent mixing of the oils with mineral oils assisted with adjusting the results of the second transesterification which brought about the ideal lube oils (Sanni et al., 2017).

It is accounted that over 95% of regular lubricants are straightforwardly mineral oil based (Schneider, 2006), and half of these burned-through oils are figured to wind up in the climate because of spillages, spills, or spent release (Horner, 2002). It portrays an exceptionally weak circumstance of our current circumstances and its creatures. Such issues can't be controlled simply by effective waste-handling innovations, thus variation in lubricant base stock choice is strongly suggested.

Based on the physical state, lubricants may be classified into four categories (Figure 5.1).

Of all the four types, liquid lubricating oils and semisolid lubricants also termed greases are more popular and have wide industrial applications. But at the same time, both are composed of petroleum-based mineral oil, which is limited as well as non-biodegradable. The present research focuses on the synthesis of these two types of lubricants from renewable sources.

5.2 LUBRICATING OILS

Current lubricants are for the most part made out of over 80% base oil and a more modest measure of utilitarian added substances (Dmytryshyn et al., 2004; Heikal et al., 2017). Accordingly, the base oil for the most part decides lubricant properties such as oxidative dependability stream and lubricity properties.

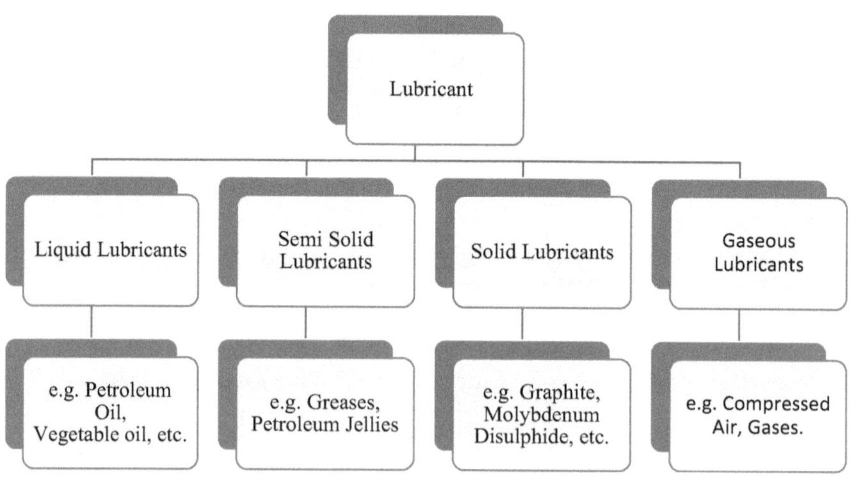

Figure 5.1 Types of lubricants.

There are three classifications in base oils: mineral oils, synthetic oils, and vegetable oils.

5.2.1 Mineral oils

The early modern lubricating oils were generally refined from unrefined petrol and were comprised of divisions bubbling at temperatures somewhere in the range of 300–370°C and normally included somewhere in the range of 20–40 carbon particles in every atom. Atoms present in these divisions incorporate long-chain alkanes, cycloalkanes, and fragrant mixtures (Ghadge & Rahman, 2005).

5.2.2 Synthetic oils

The always expanding request made by a quick industry provoked the advancement of another sort of modern oils, and hence synthetic lubricating oils were created. Synthetic lubricating oils are logically designed unpredictable options in contrast to refined oils and give a more noteworthy presentation and unwavering quality (C. Liu et al., 2014). Synthetic oil is a combination of mixtures that are made by synthetic adjustment of petrol. Synthetic liquids are classified by their arrangement or their substance structure. Normally manufactured liquids incorporate engineered hydrocarbons. Synthetic liquids have numerous benefits when compared to crude oil; in any case, the creation cost for synthetic oil is by and large four to multiple times higher than mineral oil. The greater expense restricts the market for manufactured oils (Forbus, 2006).

5.2.3 Vegetable oils

Vegetable oils, nowadays, are acquiring prominence as a sustainable wellspring of lubricants. Vegetable oils can be straightforwardly utilized as lubricants because of their slickness; however, their helpless oxidation dependability confines their immediate use as lubricants. Then again, the change of these oils as bio-greases adds to the objectives of energy freedom and a cleaner climate, which is a significant driver for the developing interest in bio-based lubricants. The utilization of vegetable oils as lubricants decreases the natural impact, improves the exhibition of apparatus just, and ensures operational well-being. In addition, the biodegradability of vegetable oils is more when compared to mineral oil (Campanella et al., 2010; Fox & Stachowiak, 2007; Karmakar et al., 2017; Sharma et al., 2006). As mechanical clients understand the advantages of biodegradable oil, its appropriation will increase and its applications will develop across businesses where ecological and security concerns are high (Honary & Richter, 2011).

5.3 ROLE AND FUNCTIONS OF LUBRICANTS

The essential job of lubricants is to decrease contact between moving parts, which are in relative movement with one another. The fundamental purposes behind the utilization of lubricants are as follows:

1. To lessen wear and prevent heat misfortune that outcomes because of contact between moving surfaces.
2. To shield it from erosion and lessen oxidation.
3. A separator in transformer applications.
4. A fixing specialist against soil, residue, and water.

The role of lubricants along with their applications are summarized in Table 5.1 (Hafizah & Salmon, 2010).

In addition to the primary purposes, Figure 5.2 summarizes the key functions a lubricant performs.

Lubricating oils should have explicit physical and substance attributes. Maybe the necessity of oils is that the oil ought to stay fluid over a wide scope of temperatures. Practically speaking, the usable fluid reach is restricted by the pour point (PP) at low temperatures and the glimmer point at high temperatures. The pour point ought to be low to guarantee that the oil is siphon-capable when the supplies begin from amazingly low temperatures. The flash point should be high to permit the protected activity and the least volatilization at the greatest working temperature. Current oils are by and large made out of over 80% base oil and a more modest measure of useful additives (Heikal et al., 2017). An excellent lubricant base fluid is needed to have the least volatility, excellent heat and oxidative stability, good cold fluidity, and a high viscosity index.

5.4 PROPERTIES OF LUBRICANTS

A good lubricant needs to meet different systems of tribology separated from diminishing erosion between the moving substrates. An ideal oil ought to have the following qualities.

1. It should have low variation in viscosity with an increase in temperature.
2. It ought to have flash and fire points higher than the working temperature of the machine.
3. It ought to have high sleekness (oiliness).
4. The cloud and pour points of a decent oil must be consistently lower than the working temperature of the machine.
5. The oil should be very less volatile.

Table 5.1 Role of lubricants

Sr. No.	*Role of lubricants*	*Application*
1.	Allow ease of starting	The majority of the motor wear happens during the engine startup, principally because of a lack of lubrication.
2.	Should have proper viscosity at higher working temperatures	The oil should have a viscosity in a proper range to ensure the formation of a sufficiently thick lubricating film.
3.	Lubricate and prevent wear	The oil should be able to maintain the appropriate thickness of lubricating film to avoid the metal surfaces from coming in contact with each other.
4.	Reduce friction	The legitimate thickness of the oil film will lessen rubbing and wear. It diminishes the frictional coefficient of the motor parts during the beginning when the oil is adequate and the frictional losses may occur. Thus, controlling the contact between the engine parts will minimize the energy utilization consequently improving the efficiency.
5.	Protection against rust and corrosion	The presence of components like water and oxygen in acidic conditions initiates corrosion of engine parts. If lubricating film is present, water and oxygen are not allowed inside the engine.
6.	Keep engine parts clean	Motor oils are intended to restrict the entry of combustion products like free radicals, soot, sulfur, and nitrogen oxides inside the crankcase and associate with the oil to generate polar deposit precursors and destructive materials by suspending them in the volume of lubricant.
7.	Seal combustion pressures	The surfaces of cylinder rings, ring notches, and chamber dividers don't have an ideal fit because of the machining restrictions. Lubricating oils consequently improve the seal by occupying spaces with a film thickness of 0.025 mm. This is to prevent the deficiency of the high burning and compression pressures in the long run, which results in a decrease in the motor force and productivity.
8.	Cool engine parts	Lubricants also act as a coolant when the parts can't be successfully cooled by the coolant, either given their area or the part temperature is amazingly high, which prompts the fast vaporization of water.

6. It should deposit a minimum amount of carbon during use.
7. It must have a higher obstruction toward oxidation and consumption.
8. It ought to have great cleanser quality.

5.5 DEMAND FOR BIOLUBRICANTS

According to the energy insights of the Ministry of Statistics and Program Implementation Government of India (2018), India devours 3.5 million

Figure 5.2 Key functions of a lubricant.

tons of lubricating oils every annum and India holds the sixth largest lubricant market on the planet (Central Statistics Office, Ministry of Statistics and Programme Implementation, 2018).

The usage levels of lubricating oils in India have reliably been ludicrous and thus India is getting crude petroleum and oil-based merchandise from Gulf countries. As of now, petroleum-based resources are confined while the energy demand continues to rise and ozone-hunging substance releases are transforming into a more important concern.

As demonstrated by a report by the International Energy Agency (2018), China and India together will contribute practically 50% of overall oil revenue soon. Along these lines, the need of the day is to find substitutes for petroleum-based things.

For the attainable monetary and current turn of events, secured and reasonable resources of energy have to be explored, and new techniques in imaginative work, creation, and economy must be practiced (Naik et al., 2010). India is a tropical nation having rich vegetable assets, the transformation of such vegetable assets into powers and other valuable compound items would essentially diminish reliance on exhaustible and non-renewable energy sources. Besides, accomplishing a contamination-free climate is a need of great importance, and the inclination to protect the environment has acquired critical significance in generating innovations to combine eco-accommodating and bio-based items.

Bio-based oils are costly in contrast to manufactured oils or mineral oils. In any case, ecological guidelines are required to give huge development freedoms to the bio-based grease market. The market of biolubricants is relied upon to develop a guideline affirmed by the International Maritime Organization on the utilization of EALs (environmentally acceptable oils). Research and development exercises are being completed to create

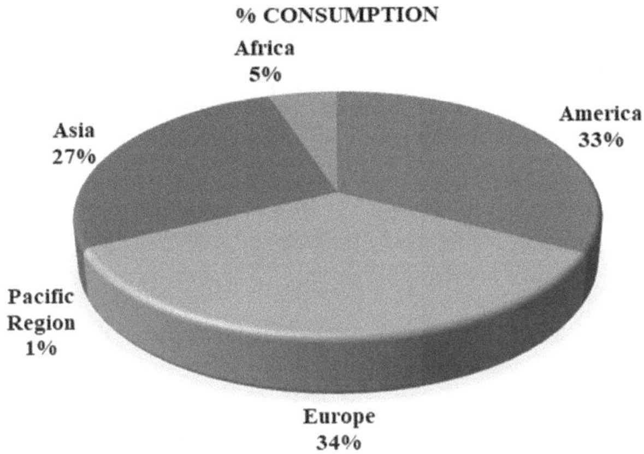

Figure 5.3 Worldwide consumption of lubricants (Legrand J., 1998).

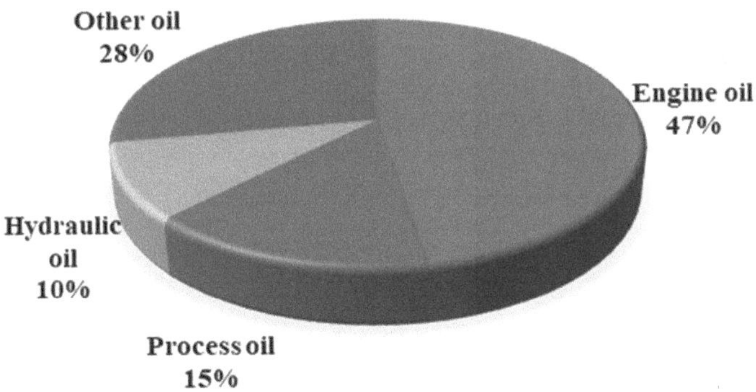

Figure 5.4 Market of various types of lubricating oils in India (2018).

financially savvy and superior bio-based oils that hold the possibility to replace conventional lubricants. Figure 5.3 uncovers the volume of the overall lubricant market devoured by various nations.

Figure 5.4 shows the use of lubricating oils in various forms in India. Only 13–32% of all pre-owned oils get back to the climate with adjusted actual properties and appearances (Wagner et al., 2001). Remaining oil adds to the pollution in the environment.

5.6 BIOLUBRICANTS: BENEFITS AND PROPERTIES

Biolubricants are esters of higher alcohols synthesized from vegetable oil-based feedstock and have greasing-up properties like those of mineral oil-based lubricants (C. Liu et al., 2014).

The benefits or advantages of lubricants can be summarized as follows:

- Lower emissions due to the higher boiling points of esters.
- Total absence of aromatic compounds, also vegetable oils are highly biodegradable and non-water polluting.
- Biolubricants are skin-friendly as compared to conventional lubricants.
- Biolubricants have a higher viscosity index, i.e., viscosity–temperature relationship is better than mineral oil.
- Higher safety on the shop floor due to high fire points.

5.7 PLANT OILS AS SOURCES OF BIOLUBRICANTS

Plant oils can be classified into various ways like edible or non-edible.

5.7.1 Edible vegetable oils

A fluid fat that is equipped for being eaten as a food or food access. For the steadily expanding population of a nation like India, edible oil seeds utilization for lubricant needs will be unable to meet homegrown prerequisites. As an option, non-edible vegetable oil and tree-borne seeds can end up being advantageous (Dwivedi & Sapre, 2002; Z. Liu et al., 2018).

5.7.2 Non-edible vegetable oils

These items are from non-consumable vegetable oils, or if nothing else, the same execution as oil-based items other than being more affordable. Non-consumable vegetable oils are environment-friendly and effectively accessible sources of biolubricants (M.T.S. Syaima et al., 2014; Mobarak, 2013). Contingent upon environmental and soil conditions, various countries are investigating diverse oils that are being viewed as better alternatives for renewable fuels and lubricants (Karmakar et al., 2017; Z. Liu et al., 2018). Since consumable oil demand is higher than its production, and being a tropical country, India is wealthy in woodland assets having a wide scope of trees, which yield a critical amount of oilseeds (Joshi et al., 2015).

5.8 EXPERIMENTAL

This section summarizes the synthesis of lubricants via chemical modifications, namely transesterification using various polyols. Afterwards, a detailed comparative analysis of the synthesized biolubricants was carried out. The synthesis of various types of biolubricants was carried out through the following steps and procedures:

1. Preliminary analysis of raw vegetable oils.
2. Synthesis of methyl esters of the respective oils using esterification and trans-esterification procedures.
3. Synthesis of trimethylolpropane (TMP)-based lubricants from methyl esters.
4. Synthesis of diethylene glycol (DEG)-based lubricants from oil methyl esters.
5. Synthesis of pentaerythritol (PE)-based lubricants from raw oils.

5.8.1 Synthesis of fatty acid methyl esters (FAMEs) of vegetable oils

Oils that were used in the experimental work initially had an extremely high acid value corresponding to high free fatty acid (FFA) levels of about 30%. Efforts were made to convert these high FFA-content oils into fatty acid methyl esters (FAMEs). Thus, the high FFA level of both oils was reduced to less than 1% before the transesterification process (Ghadge & Raheman, 2005; Issariyakul & Dalai, 2014). Experiments were conducted in a laboratory-scale setup as shown in Figure 5.1, which consisted of a 1000 ml two-neck round-bottom flask. A stirrer was introduced through one neck and the other neck consisted of a water condenser to avoid the losses of methanol. The flask was kept in a water bath that was maintained at a temperature of 60–65°C, just below the boiling point of methanol.

5.8.1.1 Stage 1: pretreatment process

It is a two-step process. In the first step, a 0.35 v/v ratio of methanol-to-oil was used. 1% of H_2SO_4 as an acid catalyst was added to the reaction mixture with constant stirring. The reaction was carried out for 1 h. Then, the layer of mixture of water and methanol was separated from the layer of esterified oil. The acid value of the separated oil was determined. The acid value was 12.5 mg KOH/g sample, which was much higher than the desired acid value of 2.0 mg KOH/g sample, hence a second step of esterification was carried out (Issariyakul & Dalai, 2014; Karmee & Chadha, 2005) as shown in Figure 5.5. The final product was used further in the transesterification reaction.

5.8.1.2 Stage 2: transesterification

The transesterification reaction was carried out with 0.25 v/v methanol and 0.7% w/v KOH as an alkaline catalyst. The reaction was carried out at 60°C for 1 h. The mixture was allowed to settle to get the ester layer on the top, separated as fatty acid methyl esters.

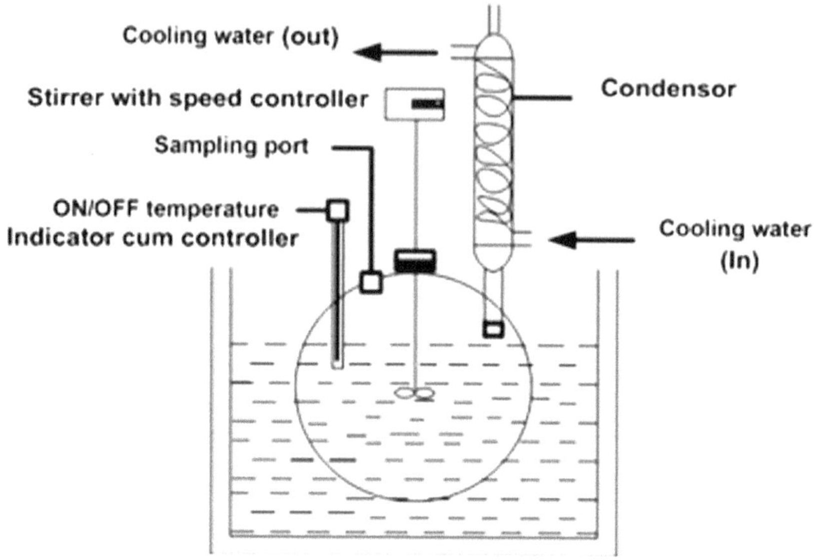

Figure 5.5 Experimental setup for fatty acid methyl ester synthesis.

5.8.2 Synthesis of TMP triesters (lubricants)

In this part, trimethylolpropane (TMP) was synthesized from FAMEs and TMP. TAG (triacylglyceride) molecules of oils were modified to produce respective FAMEs by a two-stage base-catalyzed transesterification and then linked to TMP. In the first stage, vegetable oils were transesterified with excess methanol in the presence of potassium hydroxide (KOH) to produce crude FAMEs. After the initial step, the FAMEs of mahua and karanja oils were refined and further transesterified with TMP upon heating under vacuum in a reaction catalyzed by sodium methoxide. The conversion of most of the TMP present was successfully achieved by adding an excess of FAME to the reaction mixture in the second step. The reaction chemistry for TMP transesterification may be shown in Figure 5.6 for which the reactor can be set as shown in Figure 5.7.

As illustrated in Figure 5.7, the synthesis of TMP triesters for methyl esters of both oils was carried out by reaction of respective FAMEs with trimethylolpropane in 200 batches, where 200 g of FAME was weighed in a flask, connected to a water condenser and equipped with a vacuum pump. The flask was placed in a paraffin bath over a magnetic stirrer with temperature control and stirring speed adjustment along with a time indicator. Initially, trimethylolpropane was added at the molar ratio of 4:1 (oil:methyl ester) along with 1 wt% sodium methoxide as a catalyst. The heating and stirring were started and the reaction was carried out at 130°C and 185 rpm. A vacuum of 60 mmHg was maintained throughout the reaction time.

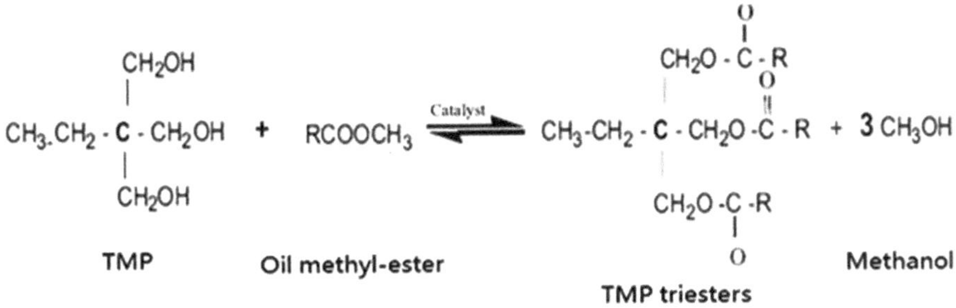

Figure 5.6 TMP transesterification of FAME to lubricants.

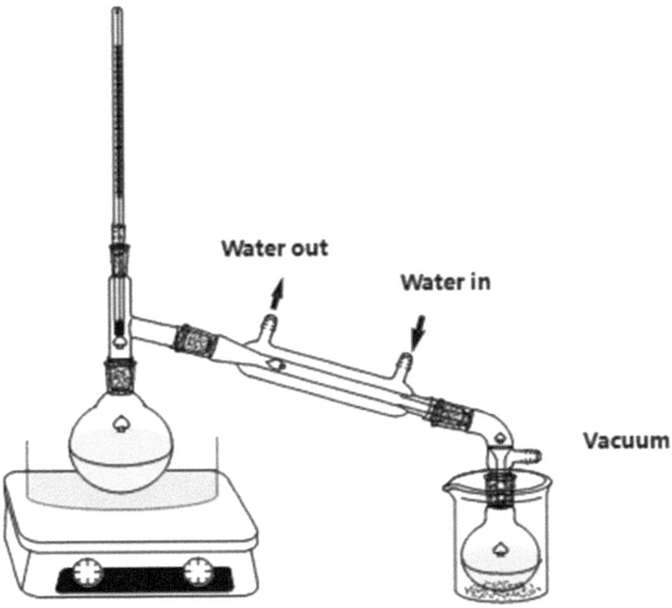

Figure 5.7 Schematics of synthesis of TMP triesters in one-pot reaction.

The same procedure was followed for lubricant synthesis of mahua and karanja oil (Hafizah & Salmon, 2010; Mobarak, 2013; Salih, Salimon, Yousif et al., 2013).

A vacuum of 60 mmHg was maintained for 180 min of reaction time. After completion of the reaction, the product was cooled to room temperature and tested for its various physicochemical properties. The molar ratio of FAME:TMP was varied and the uniform product formation was obtained at the molar ratio of 6:1. The analysis of the products is shown in Table 5.2.

From Table 5.2, it is clear that biolubricants of methyl esters of mahua oil and karanja oil gave better viscosity as compared to the conventional

Table 5.2 Analysis of TMP triesters of mahua oil methyl ester and karanja oil methyl ester

Sr. No.	*Properties*	*Analytical method*	*Mahua rimester (biolubricant)*	*Karanja rimester (biolubricant)*	*Commercial 4T engine oil*
1.	Density	ASTM-D1298	0.879 g/ml	0.88 g/ml	0.97 g/ml
2.	Viscosity @40°C	ASTM-D445	69.3 cSt	43 cSt	42 cSt
3.	Viscosity @100°C	ASTM-D445	9.53 cSt	6.5 cSt	6.4 cSt
4.	Viscosity index	ASTM-D2270	116.529	100.72	100.58
5.	Pour point	ASTM-D97	0°C	0°C	−15°C
6.	Acid value	ASTM-D664	1.65 mg KOH/g	1.71 mg KOH/g	0.49 mg KOH/g

4T oil. This resulted in higher viscosity indices also as compared to the conventional 4T engine oil. Mahua oil biolubricant showed the maximum viscosity index, which is very good in terms of biolubricant properties as the viscosity index is one of the major requirements for lubricating oils. Here, it is to be mentioned that there is no additive in the lubricant, whereas the conventional oil procured from the market already has viscosity index improvers added to it.

Also, the pour point values and acid values obtained for the lubricants are a bit higher than the specified range and in comparison to the conventional 4T oil. But the values are not too high and the standard values can be met easily by the addition of some additives.

5.8.3 Synthesis of diethylene glycol-based biolubricants

Synthesis of diethylene glycol-based lubricants was also carried out by the two-stage base-catalyzed transesterification process. In the first stage, vegetable oils were transesterified with excess methanol in the presence of potassium hydroxide to produce crude FAME, which was transesterified with diethylene glycol after purification.

The reaction was carried out in 200 ml batches in the presence of sodium methoxide as a catalyst (Bilal S, 2013). After completion of the reaction, the product that was obtained was cooled and tested for its various physical properties. The weight ratio of FAME:diethylene glycol was varied, and the optimum ratio was found to be 3.5:1 for FAME to DEG. From Table 5.3, it can be said that diethylene glycol can become a convenient source for the production of lubricants because it can be obtained from glycerol, which is one of the byproducts of methyl ester synthesis; moreover, it is easy to carry out biolubricant synthesis as the reactants are measured on a weight basis

Table 5.3 Analysis of DEG-based biolubricants of mahua and karanja oil methyl esters

Sr. No.	*Properties*	*Analytical method*	*Mahua DEG ester (biolubricant)*	*Karanja DEG ester (biolubricant)*	*Commercial 4T engine oil*
1.	Density	ASTM-D1298	0.95 g/ml	0.88 gm/ml	0.97 g/ml
2.	Viscosity @40°C	ASTM-D445	61.32 cSt	52.87 cSt	42 cSt
3.	Viscosity @100°C	ASTM-D445	9.41 cSt	8.04 cSt	6.4 cSt
4.	Viscosity index	ASTM-D2270	134.137	121.143	100.58
5.	Pour point	ASTM-D97	$5^{\circ C}$	$3^{\circ C}$	−15°C
6.	Acid value	ASTM-D664	1.12 mg KOH/g	1.43 mg KOH/g	0.49 mg KOH/g

and not on a molar basis. Here also the requirement of methyl ester is in excess to avoid the backward reaction.

From the readings provided in Table 5.3, it can be claimed that diethylene glycol also produces biolubricants with higher viscosities and viscosity indices. Lubricants with higher viscosity index (VI) (>130) find a wide variety of applications. The density of the final biolubricants obtained is also a bit higher as compared to that of TMP lubricants. Also, the acid value is reduced in the permissible range according to the commercial lubricants. But the viscosities at lower temperatures in this case also are quite high that they may alter the flowability of the biolubricant during the startup of the engine. Similarly, the pour points obtained are also a bit on the higher side, but the values may be minimized by the use of pour point depressants.

5.8.4 Synthesis of pentaerythritol (PE)-based biolubricants

The replacement of the glycerol backbone of the TAG molecule with a stable polyol results in the formation of esters where the thermal stability of vegetable oils can be improved by replacing the glyceride moiety in the original molecule with polyhydric alcohols (such as trimethylolpropane or pentaerythritol) (C. Liu et al., 2014).

The reaction chemistry can be illustrated in Figure 5.8.

Polyol esters were synthesized by the esterification reaction of the oils with pentaerythritol (PE). The molar ratio of oil:PE was varied and the optimum molar ratio for uniform product formation was found to be 4.5:1. The synthesized biolubricants were then tested for their physicochemical properties and the results are mentioned in Table 5.4.

From the readings provided in Table 5.4, it can be claimed that pentaerythritol produces biolubricants with higher viscosities and viscosity indices. Also, the acid value is reduced in the permissible range according to the

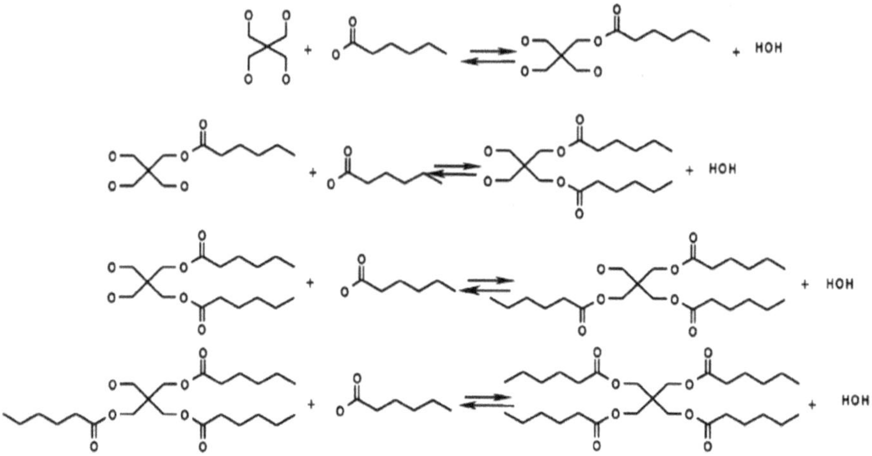

Figure 5.8 Reaction mechanism for synthesis of pentaerythritol esters from high FFA oils.

Table 5.4 Analysis of PE esters of mahua and karanja oils

Sr. No.	*Properties*	*Analytical method*	*Mahua PE ester (biolubricant)*	*Karanja PE ester (biolubricant)*	*Commercial 4T engine oil*
1.	Density	ASTM-D1298	0.95 g/ml	0.88 g/ml	0.97 g/ml
2.	Viscosity @40°C	ASTM-D445	67.517 cSt	63.14 cSt	42 cSt
3.	Viscosity @100°C	ASTM-D445	11.69 cSt	10.15 cSt	6.4 cSt
4.	Viscosity index	ASTM-D2270	169.67	147.38	100.58
5.	Pour point	ASTM-D97	10°C	11°C	−15°C
6	Acid value	ASTM-D664	0.92 mg KOH/g	1.31 mg KOH/g	0.49 mg KOH/g

commercial lubricants. However, the viscosities at lower temperatures are so high that they may alter the flowability of the biolubricant during the startup of the engine. One of the reasons behind higher viscosities may be the presence of the triglyceride chain (Aziz et al., 2014; C. Liu et al., 2014).

5.9 RESULTS

5.9.1 Oxidation stability

Oxidation of all the synthesized biolubricant samples of mahua and karanja oils was carried out at 200°C for 6 h and the variation in density and viscosity at 40°C was measured before and after the oxidation process. The values of change in density and viscosity at 40°C are reported in Table 5.5. From the table, it is clear that mahua oil has the highest viscosity ratio. The higher the viscosity ratio, the poorer the oxidation stability of the sample. It means that the oxidation stability of mahua oil is the poorest among all the samples. The lubricant with a viscosity ratio close to commercial oil is found to be a biolubricant synthesized using TMP and PE. Its viscosity ratio was found to be 1.26, whereas for a commercial lubricant, the value was found to be 1.16.

5.9.2 FTIR analysis

The structure of the TMP triester produced from the esterification of MOME with trimethylolpropane was confirmed from the infrared spectra by the appearance of the C=O stretching vibration for the ester peak at 1739 cm^{-1} (Salih, Salimon, & Jantan, 2013). In addition, there is a shift in the wavelength of 1709 cm^{-1} for the spectrum of fatty acids to the wavelength of 1739 cm^{-1} for the spectrum of ester TMP (Hafizah & Salmon, 2010). TMP shows a distinct broad peak from the 3500–3000 cm^{-1} region that refers to the stretching of the alcohol O–H group. The absence of the peak in this region confirms that the O–H bond owned by TMP has been replaced with methyl ester (Aziz et al., 2020). It is shown in Figure 5.9.

The Fourier-transform infrared spectroscopy (FTIR) spectrum for karanja TMP triester shows that the peak is obtained at 1743 cm^{-1}. It confirms the presence of oxygen in the molecule and the presence of ester functional groups in the system. The reducing intensity of primary alcohol of trimethylolpropane peak at 1090–1000 cm^{-1} and the appearance of C–O–C stretching vibration peak of trimethylolpropane ester at 1160–1050 cm^{-1} (Salih, Salimon, & Jantan, 2013) can also be seen. It is shown in Figure 5.10.

The FTIR spectrum for MOME DEG lubricant is shown in Figure 5.11. The peak is obtained at 1740 cm^{-1}. It confirms the presence of oxygen in the molecule and the presence of ester functional groups in the system (Shaba et al., 2018). The stretching of the O–CH_3 group is represented by the absorbance at 1196 cm^{-1} and it is typically for methyl esters, i.e., biodiesel (Rabelo et al., 2015).

From Figure 5.12, it is clear that the peak is obtained at 1744 cm^{-1}, again confirming the presence of an ester functional group. The absence of the spectrum of the carboxylic acid functional group also confirms the presence of lubricants (Shaba et al., 2018). The stretching of the O–CH_3 group is represented by the absorbance at 1196 cm^{-1}, and it is typically for methyl esters, i.e., biodiesel (Rabelo et al., 2015). The absence of this peak in Figure 5.12 confirms that the methyl esters reacted with the diethylene glycol.

Table 5.5 Oxidation stability for all types of lubricants of mahua and karanja oils

Sr. No.	*Properties*	*Mahua oil*	*TMP biolubricant*	*PE biolubricant*	*DEG biolubricant*	*Commercial 4T oil*	*Karanja oil*	*TMP biolubricant*	*PE biolubricant*	*DEG biolubricant*	*Commercial 4T oil*
1.	Density, g/ml (before oxidation)	0.907	0.879	0.95	0.95	0.97	0.936	0.88	0.88	0.88	0.97
2.	Density, g/ml (after oxidation @200°C for 6 h)	0.911	0.89	0.97	0.971	0.986	0.949	0.889	0.883	0.883	0.986
3.	Viscosity @40°C, cSt (before oxidation)	37.18	69.3	67.517	61.32	45	54.54	43	63.14	58.37	45
4.	Viscosity @40°C, cSt (after oxidation @200°C for 6 h)	66.18	87.318	85.071	78.49	52.2	93.26	53.75	80.18	73.55	52.2
5.	Viscosity ratio (after oxidation/ before oxidation)	1.78	1.26	1.26	1.28	1.16	1.71	1.25	1.27	1.26	1.16

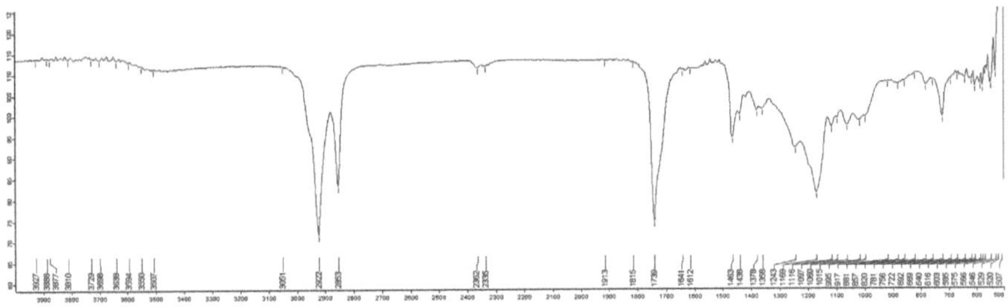

Figure 5.9 FTIR of MOME TMP triester.

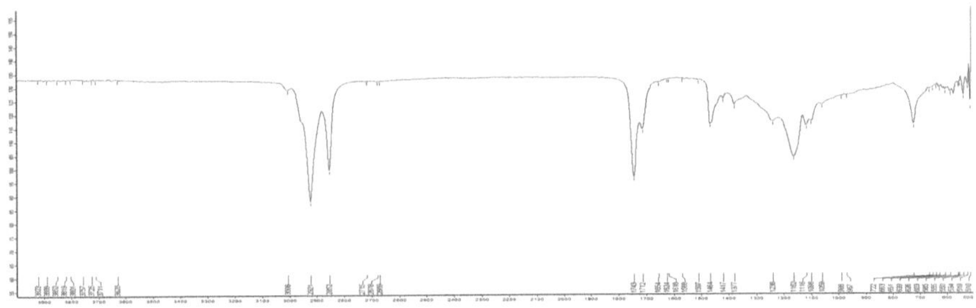

Figure 5.10 FTIR spectra of KOME TMP triester.

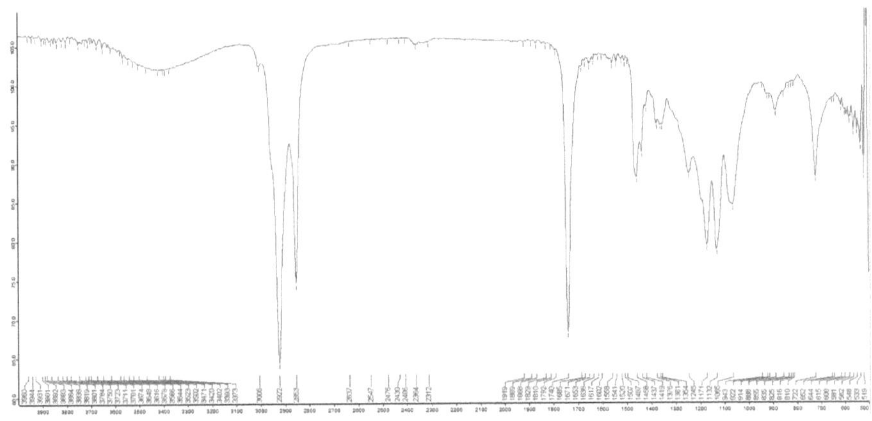

Figure 5.11 FTIR spectra of MOME DEG triester.

5.10 CONCLUSIONS

Biolubricants from mahua and karanja oils were synthesized through various routes, namely esterification and transesterification, using TMP, DEG, and PE. Detailed analysis of the synthesized biolubricants was carried out

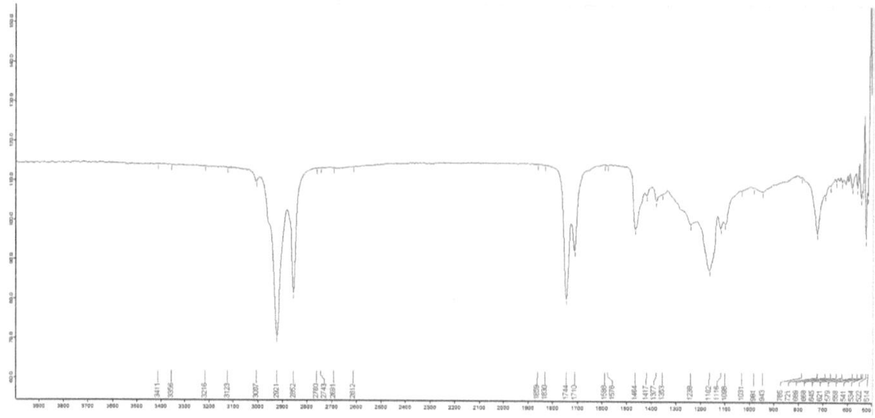

Figure 5.12 FTIR spectra of KOME DEG biolubricant.

as per the ASTM standards, and the suitability of the biolubricants was decided. A comparison of all these biolubricants was made with commercial 4T engine oil. Based on the experimental work and detailed analysis, major conclusions drawn are presented here.

- Analysis of raw mahua and karanja oils shows that they have good viscosity and viscosity indices. However, their higher pour point, higher acid values, and poor thermo-oxidative stability make them unsuitable for their direct application as biolubricants.
- Both mahua and karanja oils can be converted into their fatty acid methyl esters (FAMEs), which are one of the major starting materials for conversion into biolubricants through the esterification and transesterification processes. In this process, the acid values and pour points of both oils are reduced by a huge amount.
- FAMEs of both oils can be effectively converted into TMP-based biolubricants. The molar ratio of 6:1 for FAME to TMP is required for efficient conversion and assurance of forward reaction into biolubricants.
- Comparison of TMP biolubricants with commercial lubricants proved that the biolubricants have higher viscosity and viscosity indices as compared to commercial lubricants with mahua biolubricants yielding the highest values. This may be due to the presence of saturated fatty acids at higher proportions in mahua oil. In applications where a bit lower viscosity is required, the use of mahua biolubricant may not be beneficial.
- Synthesis of biolubricants from diethylene glycol was also achieved successfully in the weight ratio of 3.5:1 for oil to glycol.
- The viscosity indices of DEG biolubricant were found to be even higher than that of TMP biolubricants. But, in this case, the pour

point was also increased by an average value of 4°C. Such types of biolubricants need the addition of pour point depressants before their final use.

- When TMP is replaced by pentaerythritol (PE), biolubricants can be directly synthesized from raw oils instead of their methyl esters. The optimum molar ratio arrived at 4.5:1 for oil to PE for efficient biolubricant synthesis.
- When pentaerythritol was utilized, the pour point of the final products increased further. This may be due to the higher polyol alcohol utilized for the synthesis.
- From the abovementioned all three cases, it is clear that the synthesis of biolubricants using polyol alcohols is a reversible reaction. Hence, higher molar ratios of oil to polyols are required to restrict the reversible reactions.

5.11 FUTURE SCOPE

There is a vast scope of the present work. In connection with this, further work may be undertaken to utilize the data generated for designing and fabrication of large-scale industrial flow reactors to produce biolubricants. Besides, the product of the current work requires a more meticulous study of its tribological properties to identify its specific field of application.

Some prospects of the work are as follows:

- A good formulation of biolubricant can be made by incorporating an optimum number of additives.
- Exploration of other vegetable oils and chemically modifying them could lead to a variety of bio-based lubricants.
- Other low-cost feedstocks may be explored especially with high FFA.
- Some more novel catalysts can be explored for the synthesis.
- A formulation study can be done for the enhancement of low-temperature properties.

REFERENCES

Aziz, N. A. M., Yunus, R., Hamid, H. A., Ghassan, A. A. K., Omar, R., Rashid, U., & Abbas, Z. (2020). An acceleration of microwave-assisted transesterification of palm oil-based methyl ester into trimethylolpropane ester. *Scientific Reports*. https://doi.org/10.1038/s41598-020-76775-y

Aziz, N. A. M., Yunus, R., Rashid, U., & Syam, A. M. (2014). Application of response surface methodology (RSM) for optimizing the palm-based pentaerythritol ester synthesis. *Industrial Crops and Products*, 62, 305–312. https://doi.org/10.1016/j.indcrop.2014.08.040

Bilal, S. (2013). Production of biolubricant from Jatropha curcas seed oil. *Journal of Chemical Engineering and Materials Science*, *4*(6), 72–79. https://doi.org/10.5897/jcems2013.0164

Campanella, A., Rustoy, E., Baldessari, A., & Baltanás, M. A. (2010). Lubricants from chemically modified vegetable oils. *Bioresource Technology*, *101*(1), 245–254. https://doi.org/10.1016/j.biortech.2009.08.035

Central Statistics Office, Ministry of Statistics and Programme Implementation, G. of I. (2018). *Energy statistics-2018 energy statistics 2018 STATISTICS ENERGY (Twenty Fifth Issue). 2018* (pp. 1–110). Central Statistics Office, Ministry of Statistics and Programme Implementation.

Dmytryshyn, S. L., Dalai, A. K., Chaudhari, S. T., Mishra, H. K., & Reaney, M. J. (2004). Synthesis and characterization of vegetable oil derived esters: Evaluation for their diesel additive properties. *Bioresource Technology*, *92*(1), 55–64. https://doi.org/10.1016/j.biortech.2003.07.009

Dwivedi, M. C., & Sapre, S. (2002). Total vegetable-oil based greases prepared from castor oil. *Journal of Synthetic Lubrication*, *19*(3), 229–241. https://doi.org/10.1002/jsl.3000190305

Forbus, R. (2006). Practical advances in petroleum processing. *Practical Advances in Petroleum Processing, November.* https://doi.org/10.1007/978-0-387-25789-1

Fox, N. J., & Stachowiak, G. W. (2007). Vegetable oil-based lubricants: A review of oxidation. *Tribology International*, *40*(7), 1035–1046. https://doi.org/10.1016/j.triboint.2006.10.001

Ghadge, S. V., & Raheman, H. (2005). Biodiesel production from mahua (*Madhuca indica*) oil having high free fatty acids. *Biomass and Bioenergy*. https://doi.org/10.1016/j.biombioe.2004.11.009

Hafizah, A. N., & Salmon, J. (2010). Synthesis and characterization of ester trimethylolpropane based *Jatropha curcas* oil as. *Journal of Science and Technology*, *2*, 47–58.

Heikal, E. K., Elmelawy, M. S., Khalil, S. A., & Elbasuny, N. M. (2017). Manufacturing of environment friendly biolubricants from vegetable oils. *Egyptian Journal of Petroleum*. https://doi.org/10.1016/j.ejpe.2016.03.003

Honary, L. A. T., & Richter, E. (2011). Biobased lubricants and greases. *Biobased Lubricants and Greases*, 1–11. https://doi.org/10.1002/9780470971956

Horner, D. (2002). Recent trends lubricants in environmentally friendly. *Journal Synthetic Lubrication*, *18*(4), 327–347.

Issariyakul, T., & Dalai, A. K. (2014). Biodiesel from vegetable oils. *Renewable and Sustainable Energy Reviews*, *31*, 446–471. https://doi.org/10.1016/j.rser.2013.11.001

Joshi, P., Sharma, N., & Manab Sarma, P. (2015). Assessment of biomass potential and current status of bio-fuels and bioenergy production in India. *Current Biochemical Engineering*, *3*(1), 4–15. https://doi.org/10.2174/2212711902999150923155545

Karmakar, G., Ghosh, P., & Sharma, B. K. (2017). Chemically modifying vegetable oils to prepare green lubricants. *Lubricants*, *5*(4), 1–17. https://doi.org/10.3390/lubricants5040044

Karmee, S. K., & Chadha, A. (2005). *Preparation of biodiesel from crude oil of Pongamia pinnata. 96*, 1425–1429. https://doi.org/10.1016/j.biortech.2004.12.011

Legrand J., D. K. (1998). High-performing lubricants based on renewable resources. *Journal of Agro Food Industries Hi- Technology*, *9*, 16–18.

Liu, C., Liu, J., Ma, L., & Rong, L. (2014). Preparation of novel high-temperature polyol esters from vegetable oils. *Journal of Chemistry*, *2014*(3). https://doi.org/10.1155/2014/802732

Liu, Z., Biresaw, G., Biswas, A., & Cheng, H. N. (2018). Effect of polysoap on the physical and tribological properties of soybean oil-based grease. *JAOCS, Journal of the American Oil Chemists' Society*, *95*(5), 629–634. https://doi.org/10.1002/aocs.12069

Mobarak, H. M. (2013). Non-edible oil as a source of bio-lubricant for industrial applications: A review. *International Journal of Engineering Science and Innovative Technology*, *2*(1), 299–305.

Naik, S., Goud, V. V., Rout, P. K., Jacobson, K., & Dalai, A. K. (2010). Characterization of Canadian biomass for alternative renewable biofuel. *Renewable Energy*, *35*(8), 1624–1631. https://doi.org/10.1016/j.renene.2009.08.033

Rabelo, S. N., Ferraz, V. P., Oliveira, L. S., & Franca, A. S. (2015). FTIR analysis for quantification of fatty acid methyl esters in biodiesel produced by microwave-assisted transesterification. *International Journal of Environmental Science and Development*, *6*(12), 964–969. https://doi.org/10.7763/ijesd.2015.v6.730

Salih, N., Salimon, J., & Jantan, F. N. (2013). Synthesis and characterization of palm kernel oil based trimethylolpropane ester. *Asian Journal of Chemistry*, *25*(17), 9751–9754. https://doi.org/10.14233/ajchem.2013.15299

Salih, N., Salimon, J., Yousif, E., & Abdullah, B. M. (2013). Biolubricant basestocks from chemically modified plant oils: Ricinoleic acid based-tetraesters. *Chemistry Central Journal*, *7*(1). https://doi.org/10.1186/1752-153X-7-128

Sanni, S. E., Emetere, M. E., Efeovbokhan, V. E., & Udonne, J. D. (2017). Process optimization of the transesterification processes of palm kernel and soybean oils for lube oil synthesis. *International Journal of Applied Engineering Research*, *12*(14), 4113–4129.

Schneider, M. P. (2006). Plant-oil-based lubricants and hydraulic fluids. In *Journal of the science of food and agriculture*. https://doi.org/10.1002/jsfa.2559

Shaba, M. I., Mohammed, A., Musa, D. S., Audu, B. A., & Katamba, Y. B. (2018). Synthesis and optimization process of ethylene glycol-based bio-lubricant from palm kernel oil (PKO). *Journal of Renewable Energy and Environment*, *5*(2), 1–9 Journal.

Sharma, B. K., Adhvaryu, A., Liu, Z., & Erhan, S. Z. (2006). Chemical modification of vegetable oils for lubricant applications. *JAOCS, Journal of the American Oil Chemists' Society*, *83*(2), 129–136. https://doi.org/10.1007/s11746-006-1185-z

Soni, S., & Agarwal, M. (2014). Lubricants from renewable energy sources – a review. *Green Chemistry Letters and Reviews*, *7*(4), 359–382. https://doi.org/10.1080/17518253.2014.959565

Syaima, M. T. S., Zamratul, M. I. M., Noor, I. M., & Rifdi, W. M. W. (2014). Development of bio-lubricant from Jatropha curcas oils. *Int'l Journal of Research in Chemical, Metallurgical and Civil Engineering*, *1*(1), 3.

Wagner, H., Luther, R., & Mang, T. (2001). Lubricant base fluids based on renewable raw materials: Their catalytic manufacture and modification. *Applied Catalysis A: General*, *221*(1–2), 429–442. https://doi.org/10.1016/S0926-860X(01)00891-2

Chapter 6

Lubricants derived from 2D materials

Kavitha Sharanappa Gudadur, Avinash Kumar, Abhishek Kumar, Ashwani Kumar, Amit Choudhari, Ashish Kumar Gupta, and Nusrat Chowdhury

6.1 INTRODUCTION

6.1.1 2D materials

The emergence of 2D materials reshapes materials science to their atomic structure and remarkable properties, inspiring innovation globally across multiple disciplines. This surge represents a transformative phase in material exploration, promising groundbreaking advancements in technology [1]. The atomic-scale thickness of 2D materials grants exceptional mechanical flexibility and high surface-to-volume ratios, fueling innovative applications in flexible electronics, optoelectronics, and biomedical devices. Additionally, their tunable properties enable customization for various applications, paving the way for transformative advancements in technology [2–4]. Beyond graphene, other notable 2D materials include transition metal dichalcogenides (TMDs) such as molybdenum disulfide (MoS_2) and tungsten diselenide (WSe_2), which exhibit unique electronic and optical properties, as well as layered materials like boron nitride (BN) and black phosphorus (BP) as shown in Figure 6.1. Each of these materials possesses distinct characteristics that make them suitable for diverse applications ranging from nanoelectronics, lubrication, and photonics to energy storage and catalysis.

6.1.2 2D material as lubricant

Energy can be wasted greatly during the friction process due to heat buildup and dissipation. Serious issues brought on by friction-induced wear include surface degradation, decreased device performance, and decreased dependability. For instance, wear and friction in passenger cars lose around one-third of the energy used [5, 6]. A vast range of applications, including microscale devices, bio-related devices like teeth and implants, traditional bearings, pistons, and gears are severely impacted by friction and wear [7]. The most effective technique to reduce wear and friction is lubrication. Ancient Egyptian frescoes [8] describe the usage of water or vegetable oil

DOI: 10.1201/9781003535447-6

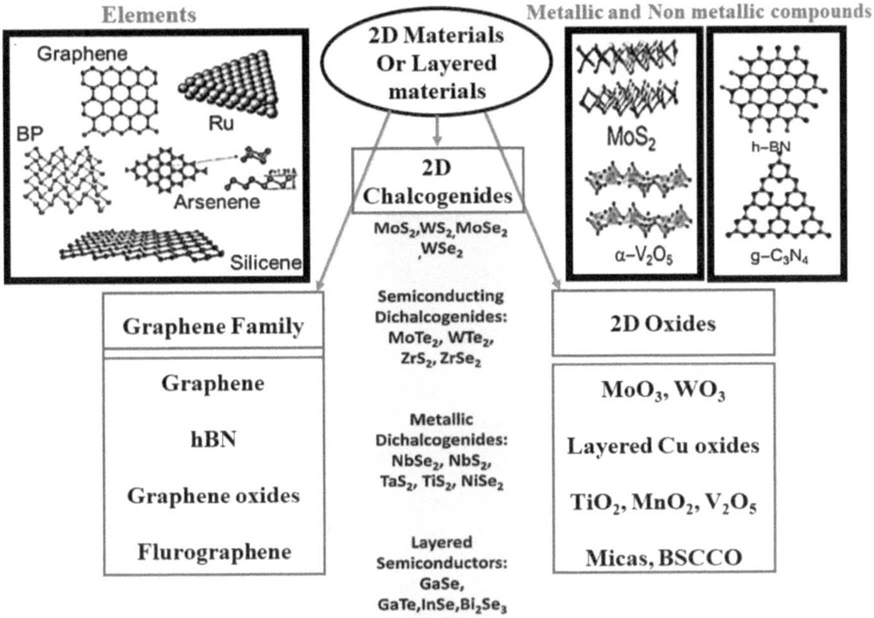

Figure 6.1 Structure and categorization of various 2D materials.

as lubricants as early as 4,000 years ago. Naturally, the field of lubrication research has advanced significantly in terms of our capacity to regulate wear and friction. Two-dimensional (2D)-derived lubricants, such as graphene and molybdenum disulfide (MoS_2), can be used in biomedical applications to reduce friction and wear in biotribological systems [9, 10].

Ongoing research aims to enhance lubricant performance due to existing options falling short. Advanced lubricants typically combine additives with base oil, where the base oil primarily determines key features. Additives, though a smaller part, play a crucial role in imparting new qualities or compensating for deficiencies. These additives include antioxidants, extreme pressure agents, viscosity modifiers, anti-foam agents, and corrosion inhibitors, tailored for specific application needs [11–13]. The ongoing energy crisis has led to various challenges, including a significant surge in the need for advanced lubricant additives characterized by superior friction reduction capabilities. Moreover, Figure 6.2 shows the key features of 2D material as lubricant.

Graphite and MoS_2 are renowned for their layered structures, making them ideal candidates as solid lubricants [14, 15]. 2D nanostructures, bonded by weak van der Waals forces, facilitate easy sliding between layers, reducing friction. While monolayer structures possess strong covalent bonds, imparting high modulus strength, they also boast larger specific

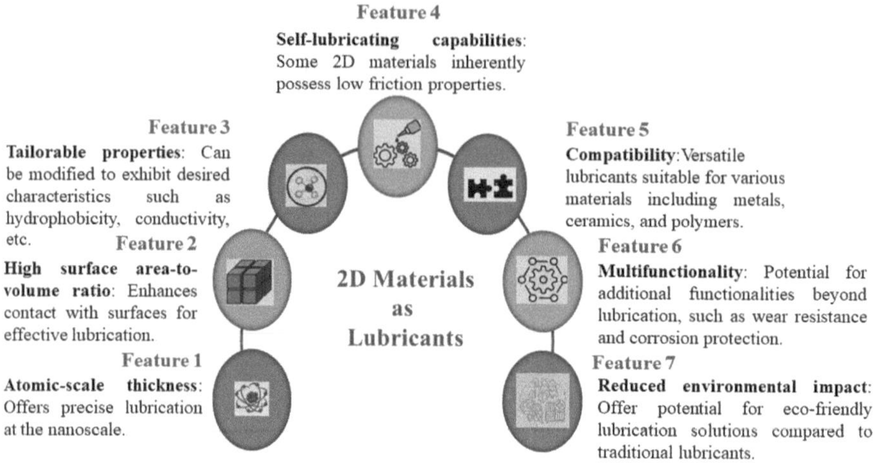

Figure 6.2 Key features of 2D material as lubricant.

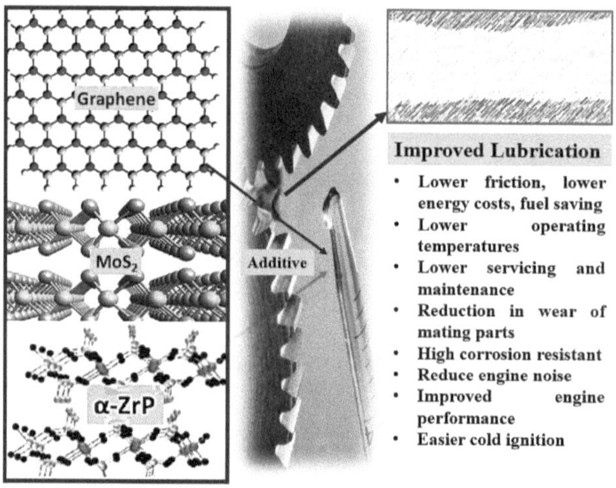

Figure 6.3 2D nanosheets improve lubrication by acting as a lubricant additive.

surface areas, minimizing direct contact and friction between surfaces when absorbed on substrates [16–18] (Figure 6.3).

Three broad categories can be used to classify the 2D materials under investigation as lubricants: metal dichalcogenides, the family of graphene, and “others”. Because of graphene's exceptional mechanical, chemical, and electrical properties, it has been the subject of much investigation, both theoretical and experimental [19, 20]. In a series of papers published in 2011, the tribological characteristics of derivatives of graphene and the family of graphene as lubricant additives were first described [21, 22].

Graphene-based nanoparticles serve as an effective oil lubricant additive and can also be added to grease [23, 24] and water-based media [21, 25–28]. Because graphene family nanoparticles can separate the contact and form a protective coating on the surface, they can effectively improve lubricating performance in a variety of base media [29]: graphene-based engine oil nanofluids for tribological applications [22, 25]. Rhombohedral structures and the MX_2 stoichiometry are characteristics of metal dichalcogenides. It has long been known that the two most prevalent metal dichalcogenides, MoS_2 and WS_2, are superior solid lubricants and lubricant additives [30–33]. More recently, it has been shown that other metal dichalcogenides, like $TiSe_2$ and $NbSe_2$, operate satisfactorily as additives following [34, 35].

Lately, additional layered materials have also been investigated for use as lubricant additives. Silicates, phosphates, and oxides with structure (layered) are examples of recently studied 2D nanomaterials [16, 36, 37]. The tribological activity of Y_2O_3 as a lubricant additive in a variety of lubricating media, such as water, oil, and grease, has been thoroughly investigated by authors and collaborators [16, 36–38]. These materials' activity has been shown in experiments to decrease viscosity and enhance the functionality of base lubricants when added. These nanoparticles show comparatively enhanced van der Waals force between any two layers, which are adjacent to graphene-like nanomaterials and MoS_2, and exfoliation is less likely [39]. 2D nanoparticles have discovered a novel technique for reducing friction: they may alter fluid flow.

This chapter discusses the performance of these three classes of 2D materials as lubricants, focusing on how well they each reduce friction and mechanisms that lead to an improvement in lubricating performance. The 2D nanomaterials added to liquid or grease lubricant are the ones covered in this manuscript. The majority have a single layer or several layers with thicknesses varying from 1–2 nm to tens of nanometers [38, 40]. These layered materials' tribological performance can be managed using surface modification approaches [41]. It is true that several adjustments have improved the anti-friction and wear-reduction performance [41, 42].

6.2 LUBRICANTS DERIVED FROM 2D MATERIALS: GRAPHENE FAMILY

Graphene, the thinnest manmade substance, boasts a hexagonal structure of tightly packed carbon atoms, offering outstanding mechanical, electrical, thermal, and tribological properties. Its layered structure infiltrates contact interfaces, reducing wear and friction. Even small graphene quantities between surfaces decrease energy loss, improving mechanical system durability [43–48]. Table 6.1 shows the graphene tribological performance as lubricant additives. Lubricants derived from 2D materials like graphene are increasingly used in drug-eluting devices to ensure smooth operation

Table 6.1 Graphene tribological performance as lubricant additives

2D nanosheets	Initial stock	Method of modification	Tribological efficacy		References
			COF	Wear	
Graphene (many micrometers in size, few layered structure) 0.6–0.8	PAO_4	Span-80, a dispersing agent	0.18–0.3 (0.01 wt%)	0.8–0.9 (wear rate, µm³/Nm) 0.05–0.1 (0.01wt%)	[40]
Graphene oxide (GO) (many layered structure; size in hundreds of nanometers)	Paraffin that is neutral	Treatment with ammonia solution and boric acid; reinforcement with TiO_2 nanoparticles	Around 0.076–0.056 (0.15% w/v)	~0.73 (diameter of wear scar, mm) ~0.35 w/v (0.15%)	[51]
GO (structure with numerous layers and different µm sizes)	Paraffin wax	Ni nanoparticle deposition aided by supercritical CO_2	0.095 0.064 (or 0.08 wt%)	~0.5 (diameter of worn scar, mm) ~0.28 (0.08 wt%)	[52]
GO (many layered structure; size in hundreds of nanometers)	Paraffin oil	ZrO_2 nanoparticles were placed	0.26–0.12 wt%	Wear rate, µm³/Nm) = 33.7 × 2.15 (0.06%)	[53]
Fluorinated graphene (several layers, several micrometers in size)	Water	Altered with urea	About 0.41–0.39 mg/ml	33 (rate of wear, µm³/Nm) 12 (1 mg/ml)	[54]

and controlled drug release, benefiting from their biocompatibility and unique mechanical properties [49, 50].

6.2.1 Oil-based lubricant

Graphene-related compounds, characterized by high π–π interaction, pose challenges in solubility, hindering their use as oil additives due to precipitation and aggregation in lubricants. Despite efforts to enhance dispersion, the focus shifts toward modifying these compounds to improve lubricating qualities rather than solely emphasizing dispersity [43].

6.2.1.1 Techniques

In addition to chemical modification, graphite oxide exfoliation by solar-concentrated electromagnetic radiation can also significantly increase graphene dispersion [29, 55, 56]. By removing the oxygen element from the initial graphite oxide, this novel process gives the product the hydrophobic characteristics that facilitate considerably easier dispersion in oil-based lubricants. After a month of engine oil treatment, graphene generated by solar exfoliation was successfully diffused and stabilized by Eswaraiah et al. [29]. Figure 6.4 illustrates the engine oil tribological characteristics and improvement in that was noted following modification. The modification fills the nano- and microgaps in the surfaces of rubbing to generate a continuous coating of lubricates. Fan et al. generated GO nanosheets modified with alkyl imidazolium ionic liquid using a straightforward mixing procedure [57]. At temperatures as high as 200°C, the modified GO

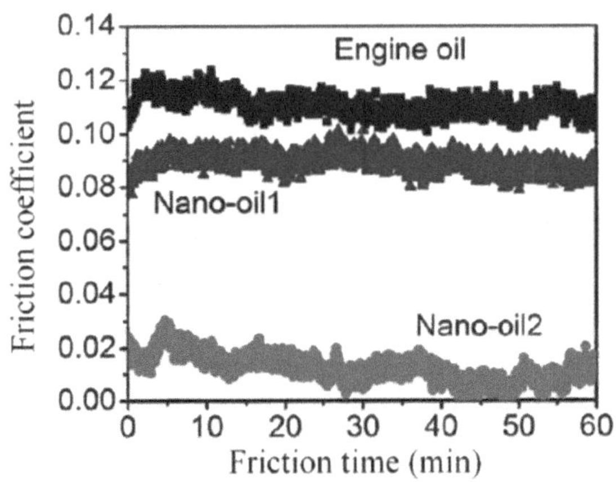

Figure 6.4 Sun-exfoliated graphene reduces friction in comparison to engine oil. Engine oil (nano-oil 2, 0.025 mg/ml) and solar-exfoliated graphene. Reprint (adapted) with permission from (*ACS Appl. Mater. Interfaces*, vol. 3, no. 11, pp. 4221–4227). Copyright (2011) *American Chemical Society* [29].

demonstrated noticeably better lubricating ability in multialkylated cyclopentane base oils. Friction and wear are reduced through the creation of a tribofilm made of fluorinated oil components and graphene.

6.2.2 Water or hydraulic-based lubricant

Water serves as an inexpensive and environmentally friendly solvent for lubricant additives, including graphene family compounds, due to their easy shearing between layers. However, successful utilization depends on the additives' hydrophilicity and ability to disperse in water. Granular organic matter (GO), rich in functional groups like hydroxyl, carboxylic, and epoxide, offers promise as it readily dissolves in water.

When Song et al. used nanosheet GO as a lubricant in pure water, they discovered that the addition decreased wear and friction [21]. After creating GO monolayer nanosheets and dispersing them in water, Kinoshita et al. conducted a tribological experiment, and the findings showed that GO-containing water outperforms PAO oil in terms of lubrication at contact pressures of up to 2.2 GPa [25]. Additional research has proven that when added to water-based liquids, GO nanosheets can significantly minimize wear and friction by forming a protective coating on the rubbing surface [58, 59].

To make use of the lubricity of other graphene compounds, the graphene must be uniformly dispersed in water or a solvent. Applied directly in watery conditions, their extreme hydrophobicity makes them unsuitable. Chemical functionalization [60–64], which involves adding polar groups to the chemical structure [65–67], can be used to tackle this problem. This section outlines various methods for creating graphene aqueous dispersions as additives. Figure 6.5 illustrates the model of the water's friction mechanism with reduced graphene oxide (rGO) and GO additions.

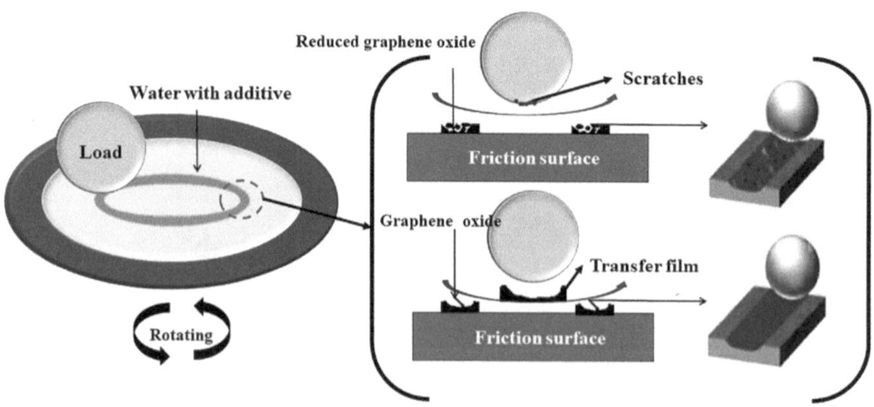

Figure 6.5 Model of the water's friction mechanism with reduced graphene oxide (rGO) and GO additions.

6.2.3 Transition metal dichalcogenides (TMDCs)

With the chemical formula of MX_2, where X is a chalcogen (Se, S, or Te) and M is the transition metal element, TMDCs are a well-defined family both structurally and chemically. MX_2 materials have a 2D shape that reduces friction and wear at the interface. One TMDC that has attracted a lot of attention from researchers is MoS_2.

6.2.3.1 Morphological effect

MoS_2 nanoflowers and nanosheets were produced by Vattikuti et al., who then compared the lubricating properties of each [68]. In comparison to the nanoflowers, the nanosheets demonstrated better friction-reducing and anti-wear qualities when used as lubrication in engine oil [69].

The production of a counter portion tribofilm that improves the lubricant properties was verified by a three-dimensional topography study [68]. The efficiency of WS_2 and MoS_2 as lubricant additives has been demonstrated by several investigations; however, the variables affecting their tribological behavior are still unclear. There are multiple contradicting reports in the literature. Kogovšek et al. found that particle material has a greater influence on tribological performance than size or morphology when evaluating seven different TMDCs as additives in PAO oil, i.e., WS_2 and MoS_2 with varying particle sizes [69, 70].

The lubricating effectiveness of PAO was investigated by Dassenoy and colleagues in relation to the morphology or size of MoS_2 (inorganic fullerene-like structures) nanoparticles [71, 72]. They discovered that while MoS_2 that are spherical perform better in friction/wear reduction than MoS_2 that are hexagonal, MoS_2's size has no effect on friction [71]. Because they help MoS_2 nanoparticles exfoliate, structural flaws have been shown to have a greater impact on tribofilm formation and friction reduction than size and shape [71, 72]. Additional comprehensive research is necessary to investigate the parameters that influence TMDCs' lubricating performance.

6.2.3.2 Task-specific applications

Extreme working conditions, characterized by high loads and temperatures, pose challenges for lubricant design, necessitating exceptional thermal stability. Materials like WS_2 and MoS_2 exhibit unique thermal stability, enabling low coefficients of friction (COF) at high pressures and temperatures. Research by An et al. revealed WS_2's superior stability over 400°C compared to MoS_2, with WS_2 maintaining stable COF in hard alloy-on-carbon steel lubrication. Shi et al. demonstrated that MoS_2 and Ti_3SiC_2 form self-lubricating composites with NiAl, providing synergistic lubrication up to 800°C [73]. At temperatures below 400°C, the addition of MoS_2 had a more noticeable decreasing effect on COF, while Ti_3SiC_2 was more important at higher temperatures. Due to thermal stress, the MoS_2 was gradually pushed out and dispersed uniformly throughout the worn surfaces as the temperature increased.

As oxidation occurs during sliding, an oxide layer that is rich in Mo–Ti–Si oxides emerges at the interfaces, impeding direct frictional contact. The breakdown of this Mo–Ti–Si–oxides layer (loosely structured) led to a decline in Mo content with increasing temperature [73]. Incorporating transition metal dichalcogenides (TMDCs) like MX_2 enhances resilience and performance in harsh environments due to their unique lubricating properties. Rubbing action between MX_2 and metal substrates forms a chemical tribofilm, reducing friction. Experimental analyses using X-ray Photoelectron Spectroscopy (XPS) and Focused Ion Beam Secondary Ion Mass Spectrometry (FIB/SIMS) confirmed significant W presence with WS_2 introduction, indicating its efficacy in reducing friction [74]. The average Hertz contact pressure in these experiments was 0.94 GPa. A comprehensive model (see Figure 6.6(a, b)) was proposed to elucidate the friction reduction at high contact pressures and to characterize the composition of the tribofilm. WS_2 nanosheets effectively filled concave regions, resulting in a smoother surface, thereby mitigating boundary friction and providing a protective tribofilm [74]. Furthermore, it was observed that MX_2 nanomaterials filled surface irregularities, contributing to surface smoothing and friction reduction [74]. Moreover, the addition of MX_2 as a lubricant additive significantly improves tribological performance and is provided in Table 6.2.

6.3 FRICTION AND WEAR REDUCTION MECHANISM

To encourage a wider use of 2D materials as lubricant additives, a great deal of research has been done to understand their underlying mechanics. In general, there are four components to the development of lubricant protection: (1) penetrating the sliding surface contact area; (2) creating a tribofilm; (3) filling in the contact area's gaps and pits; and (4) influencing the fluid's viscosity and drag. Figure 6.7 [29] provides a graphic expression of the four suggested methods. These methods heavily rely on the 2D materials that were previously addressed [27, 32, 77].

6.3.1 Reaching the sliding surface contact region and effectively lubricating

The flat structure of 2D materials allows easy interaction and nanoparticle incorporation, enabling them to permeate surface gaps and reduce friction. Under pressure and shear stress, the layers of multilayer 2D materials slide

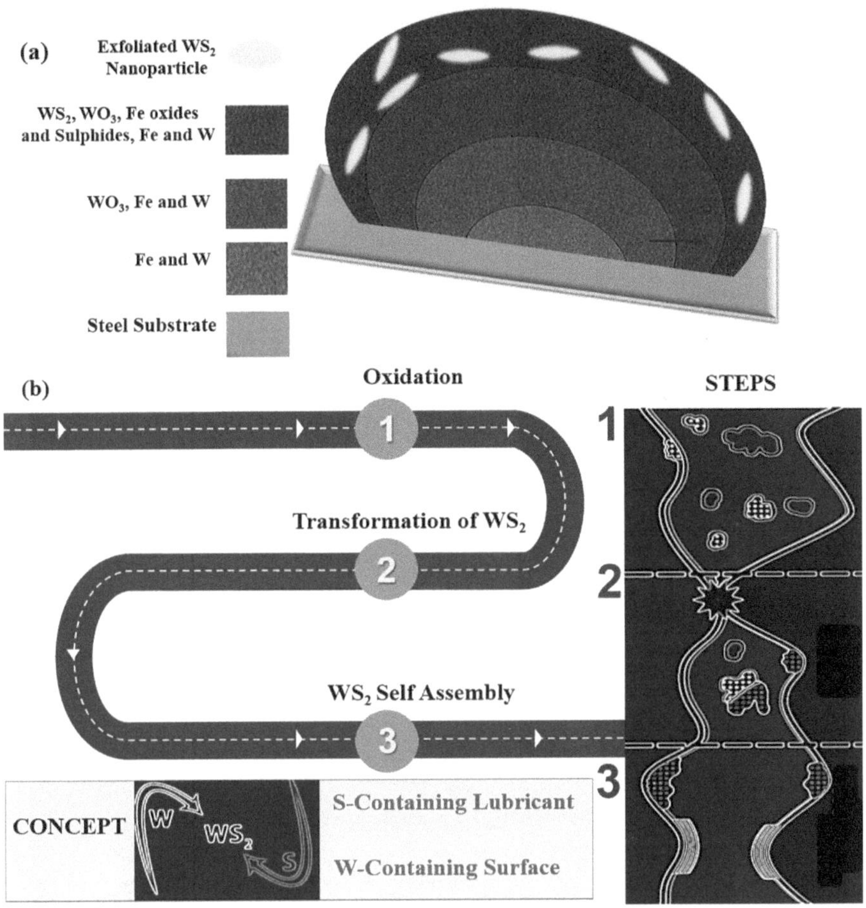

Figure 6.6 (a) Potential chemically generated tribofilm composition and structure with WS_2 addition. (b) Demonstrating theory of targeted tribofilm generation using the best solid lubricants, such as WS_2.

over each other, creating effective lubrication by minimizing solid contact and reducing friction [18, 78].

6.3.2 Formation of tribofilm

During lubrication, a tribofilm forms on sliding surfaces, undergoing two phases. Initially, the high surface energy of the film allows it to adsorb onto the substrate, forming a protective physical film and reducing friction and wear. As sliding continues, the physical film breaks, triggering a tribochemical reaction between the substrate and lubricant, replacing the physical film with a new, more durable tribofilm, and enhancing tribological performance significantly [27, 32, 33, 77, 79–82].

Table 6.2 The addition of MX_2 as a lubricant additive significantly improves tribological performance

2D nanosheets	Initial stock	Method of modification	Tribological efficacy COF	Tribological efficacy Wear	References
MoS_2 (structure with many layers, size of several μm)	Oil for hydraulics		0.23 (100°C) 0.05 (1 wt%, 100°C)	75°C (wear volume, $mm^3 \times 10^{-5}$) 12 (1 wt%, 100°C)	[39]
$NbTe_2$ (structure with many layers, size of 5–10 μm)	Paraffin oil	Agent dispersant (Span-80)	0.15 0.08 (2 wt%)	Nine (wear rate, $mm^3/Nm \times 10^{-6}$) 3.5 (2 wt%: 100 rpm, 10 N)	[75]
MoS_2 (many layered structure; size range: 100 nm to several micrometers)	M8V base oil	Dispersant	0.113 0.115 (5 wt%)	74.8 (volume of wear, $\mu m^3 \times 10^{-3}$) 15.1 wt%	[76]
MoS_2, a multiple layered structure with a dimension of hundreds of nanometers	PAO		0.07 0.032 (0.1 wt%)		[68]

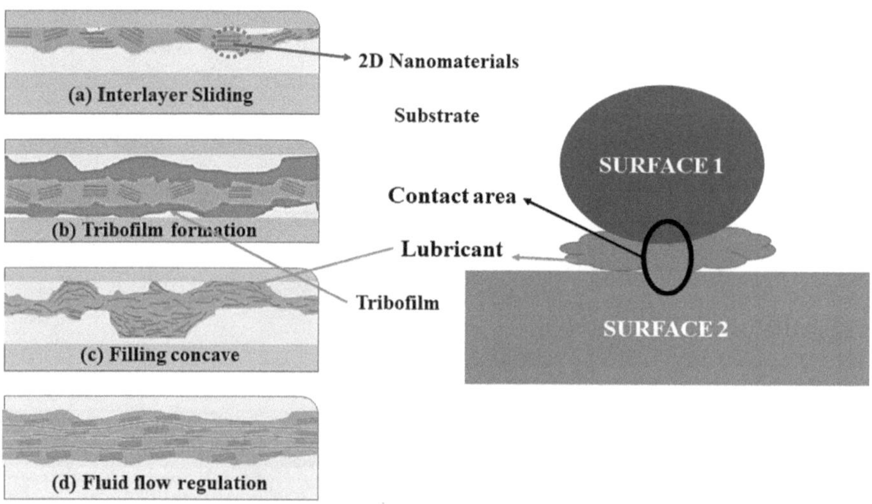

Figure 6.7 Strategies for reducing wear and friction with (a) 2D nanosheets involve their infiltration into sliding surfaces, (b) formation of a tribofilm, (c) filling of pits and gaps in the contact area, and (d) modulation of fluid viscosity and drag.

6.3.3 Filling the gaps and concave areas

As a result of intrinsically restricted manufacturing methods, no tribopair is flawless. At the microscale, the interface is not smooth. When rubbing, macro-asperities can result in significant contact pressure. Concave regions on the rubbing surfaces can be filled in with 2D nanomaterials, particularly those anchored with nanoparticles, to smooth them out [29, 34, 83]. There are fewer asperities and less localized contact pressure on a smoother surface. Wear is caused by plastic deformation, which is less likely to occur under lower pressure. Furthermore, the concave regions' 2D nanomaterials aid in the formation of a continuous oil layer, which improves lubrication even [34].

6.3.4 Influencing the viscosity and drag of the fluid

Hydrodynamic lubrication relies on viscosity to determine load-bearing capacity and friction force between rubbing surfaces. Additives that modify fluid flow and drag, affecting viscosity, are effective in this scenario. 2D nanomaterials align along the fluid flow, thus reducing viscosity and momentum transfer between fluid layers and enhancing lubrication efficiency [84–88].

6.4 CURRENT CHALLENGES OF 2D MATERIAL AS LUBRICANT ADDITIVES

Using 2D materials as lubricants holds promise for advancing technology but faces challenges. Some exhibit high friction coefficients due to interlayer interactions or surface roughness, leading to increased energy consumption and wear. Additionally, limited durability under mechanical stress or harsh conditions poses obstacles, necessitating solutions to fully exploit their potential [85–95]. The need for frequent replacement and maintenance arises from the deterioration of lubricating properties over time with certain 2D materials. Moreover, the low load-bearing capacity of some 2D lubricants due to their atomic-scale thickness limits their suitability for heavy-duty industrial applications. Additionally, chemical instability in certain 2D materials poses a significant concern for their long-term performance [96–104]. Also, shear-thinning behavior in some 2D materials poses challenges by reducing viscosity under shear stress, affecting load-bearing capacity and lubricant film thickness [105–110]. Interdisciplinary efforts in materials science, tribology, and surface engineering are essential to address challenges associated with 2D materials as lubricants. Research endeavors focusing on fundamental mechanisms and strategies to enhance durability, load-bearing capacity, and thermal stability are crucial. Innovative approaches like surface modification, nanocomposite formulations, and additive engineering can mitigate limitations and unlock the full potential of 2D materials in industrial applications [28, 33, 105, 111–114]. Moreover, Table 6.3 shows the other challenge of using 2D materials as lubricants.

6.5 FUTURE PROSPECTIVES

Recognizing the need for improved lubricant performance, researchers are exploring modifications and functionalization treatments to meet demands for efficient operation under extreme conditions while minimizing friction, wear, and maintenance. This shift toward highly functional nanomaterials like graphene, hexagonal boron nitride (h-BN) monolayers, and transition metal dichalcogenides (TMDs) offers promising alternatives to conventional lubricants [122, 123].

The evolution of traditional dry lubricants into functional nanomaterials brings advantages like simple production and environmentally friendly large-scale synthesis. However, to ensure sustained performance under demanding conditions, these lubricants must exhibit thermal and chemical stability, along with anticorrosion properties over time. Achieving these often involves adding additives, offering a simple, effective, and economical solution to enhance lubricant performance under extreme conditions [28, 33, 111–114].

An area of particular interest is the study of the formation mechanisms of tribofilms in 2D materials, which plays a pivotal role in establishing and understanding this type of lubrication. In many cases, the effectiveness of these lubricants is attributed to the synergistic effects of multiple mechanisms, highlighting the importance of a comprehensive understanding of their tribological behavior [79, 84, 124–128].

Researchers have also begun exploring combinations of two or more 2D materials as lubricant additives, recognizing that the structure, morphology, concentration, and individual properties of each additive can significantly influence the tribological behavior of the resulting lubricant. For instance, studies have demonstrated the efficacy of combining graphene and MoS_2 or molybdenum disulfide film and graphene to form solid–liquid or solid lubricants with excellent anti-friction and anti-wear properties [129]. Similarly, the use of Ti_3C_2/graphene hybrids has shown promising results in improving the tribological properties of marine nanolubricants [130].

In addition to experimental investigations, molecular dynamics (MD) simulations have emerged as powerful tools for understanding lubrication mechanisms at the nanoscale [131, 132]. These simulations provide valuable insights into the frictional behavior and lubrication mechanisms of atomically thin nanofilms, such as graphene, h-BN, and MoS_2, under

Table 6.3 Challenge of using 2D materials as lubricants

Sl. No	*Property*	*Challenge*	*Impact on performance*	*Potential solutions*	*References*
1	Friction coefficient	High sensitivity to environment	Decreases efficiency	Surface modification	[2]
2	Wear rate	Limited durability	Increases maintenance	Nanocomposite lubricants	[115]
3	Load-bearing capacity	Low load-bearing capability	Decreases operational range	Reinforcement techniques	[116]
4	Chemical stability	Susceptible to degradation	Decreases longevity	Protective coatings	[117]
5	Temperature sensitivity	Performance degradation at high temperatures	Limits applications	Thermal management systems	[2]
6	Shear strength	Weak interlayer bonding	Increases risk of failure	Interlayer crosslinking	[118]
7	Lubrication efficiency	Poor spreading and adherence	Decreases lubrication effectiveness	Functionalized additives	[119]
8	Tribocorrosion resistance	Prone to corrosion in harsh environments	Increases material degradation	Corrosion inhibitors	[120]
9	Contaminant interaction	Affinity to particulate contaminants	Increases friction and wear	Self-cleaning coatings	[121]
10	Surface roughness	Sensitivity to surface roughness	Impairs performance on rough surfaces	Surface polishing techniques	[115]
11	Oxidation susceptibility	Prone to oxidation in air	Decreases stability	Oxygen-free environments	[119]
12	Hydrophobicity	Limited water repellency	Decreases performance in wet conditions	Hydrophobic coatings	[116]
13	Nanoscale wear mechanisms	Enhanced wear at nanoscale	Increases wear rate	Nanostructured lubricants	[120]
14	Surface energy	High surface energy	Increases adhesion	Surface functionalization	[119]
15	Mechanical compatibility	Mismatch with substrate materials	Causes delamination	Interfacial adhesion promoters	[118]

various conditions [133–140]. Studying factors like surface roughness and interaction strength alongside water molecules helps understand friction reduction and wear protection mechanisms. Future research could prioritize understanding lubrication mechanisms and developing self-lubricating composites with optimized properties, thus addressing challenges in processing and stability. Leveraging materials science and tribology advancements can unlock 2D materials' full potential as lubricants, thus driving innovation in lubrication technology [130–135].

This combined approach of optimizing lubricant performance and gearbox design ensures that gearboxes operate efficiently and reliably over a long service life [136–141]. By understanding the complex interplay between lubricants, contact mechanics, and thermal management, gearbox engineers can create high-performance systems that meet the demands of modern applications. Optimizing thermal contact between gear teeth and the lubricant film minimizes frictional heating [142–156, 157].

6.6 SUMMARY

Human exploration into challenging environments demands machinery lubricants capable of withstanding extreme conditions like high vacuum, temperature, pressure, and motion speed. Developing next-generation lubricants involves incorporating additives, especially focusing on 2D nanomaterials like those from the graphene family. Challenges persist due to poor dispersion and stability of 2D nanomaterials within lubricants, necessitating modifications to enhance dispersity and impart new properties such as improved lubricating ability and thermal stability. Success in developing these lubricants requires a systematic approach, an understanding of how alterations impact tribological behaviors. Tribofilm formation during rubbing significantly contributes to enhanced lubricity observed with 2D nanomaterials, requiring further exploration. Various theories explain the reduction of wear and friction by 2D nanomaterials, emphasizing the need for extensive investigation into physicochemical reactions and interactions between additives and lubricants. Researchers explore combinations of more than two 2D materials as lubricants, considering factors like type, concentration, and unique characteristics of each constituent. Both synergistic and antagonistic effects may arise, underscoring the need for experimental evidence. Challenges in producing 2D nanomaterials with precise layer thicknesses require advancements in manufacturing processes for accurate comparisons of lubricating efficacy. Assessing lubricant performance should extend beyond friction and wear to include qualities like load capacity, heat resistance, and vacuum stability. This comprehensive evaluation is crucial as demand for task-specific lubricants grows with human exploration into challenging frontiers.

REFERENCES

1. H. Zhang, "Introduction: 2D materials chemistry," *Chemical Reviews*, vol. 118, no. 13, pp. 6089–6090, 2018.
2. R. Amendola and A. Acharjee, "Introduction to 2D materials," in *Machine Learning in 2D Materials Science*, CRC Press, pp. 7–32, 2023.
3. R. Sameer, "Length dependence thermal conductivity of zinc-selenide (ZnSe) and zinc telluride (ZnTe)—a combined first principles and frequency domain thermoreflectance (FDTR) study," 2022.
4. A. J. Austin *et al.*, "High-temperature atomic layer deposition of GaN on 1D nanostructures," pp. 1–16, 2020.
5. K. Holmberg, P. Andersson, and A. Erdemir, "Global energy consumption due to friction in passenger cars," *Tribology International*, vol. 47, pp. 221–234, 2012.
6. O. Carvalho, M. Buciumeanu, S. Madeira, D. Soares, F. S. Silva, and G. Miranda, "Optimization of AlSi–CNTs functionally graded material composites for engine piston rings," *Materials & Design*, vol. 80, pp. 163–173, 2015.
7. K. Holmberg, P. Andersson, N.-O. Nylund, K. Mäkelä, and A. Erdemir, "Global energy consumption due to friction in trucks and buses," *Tribology International*, vol. 78, pp. 94–114, 2014.
8. M. Nosonovsky, "Oil as a lubricant in the ancient Middle East," *Tribology Online*, vol. 2, no. 2, pp. 44–49, 2007.
9. A. Kumar, A. Kumar, A. Choudhari, A. K. Gupta, and N. Chowdhury, "Biotribology," Springer Cham, 2024.
10. A. Kumar, A. Kumar, A. Kumar, A. Choudhari, A. K. Gupta, and N. Chowdhury, "Introduction to biotribology: a science of surface interaction," in *Applications of Biotribology in Biomedical Systems*, Springer, 2024.
11. L. R. Rudnick, *Lubricant Additives: Chemistry and Applications*, CRC Press, 2009.
12. P. C. Hamblin, U. Kristen, and D. Chasan, "Ashless antioxidants, copper deactivators and corrosion inhibitors: their use in lubricating oils," *Lubrication Science*, vol. 2, no. 4, pp. 287–318, 1990.
13. H. A. Spikes, "Additive-additive and additive-surface interactions in lubrication," *Lubrication Science*, vol. 2, no. 1, pp. 3–23, 1989.
14. S. Watanabe, J. Noshiro, and S. Miyake, "Tribological characteristics of WS2/MoS2 solid lubricating multilayer films," *Surface and Coatings Technology*, vol. 183, nos. 2–3, pp. 347–351, 2004.
15. A. Savan, E. Pflüger, P. Voumard, A. Schröer, and M. Simmonds, "Modern solid lubrication: recent developments and applications of MoS2," *Lubrication Science*, vol. 12, no. 2, pp. 185–203, 2000.
16. X. He *et al.*, "α-Zirconium phosphate nanoplatelets as lubricant additives," *Colloids and Surfaces A: Physicochemical and Engineering Aspects*, vol. 452, pp. 32–38, 2014.
17. K.-S. Kim *et al.*, "Chemical vapor deposition-grown graphene: the thinnest solid lubricant," *ACS Nano*, vol. 5, no. 6, pp. 5107–5114, 2011.
18. D. Berman, A. Erdemir, and A. V. Sumant, "Graphene: a new emerging lubricant," *Materials Today*, vol. 17, no. 1, pp. 31–42, 2014.

19. R. Raccichini, A. Varzi, S. Passerini, and B. Scrosati, "The role of graphene for electrochemical energy storage," *Nature Materials*, vol. 14, no. 3, pp. 271–279, 2015.
20. F. Bonaccorso *et al.*, "Graphene, related two-dimensional crystals, and hybrid systems for energy conversion and storage," *Science*, vol. 347, no. 6217, p. 1246501, 2015.
21. H.-J. Song and N. Li, "Frictional behavior of oxide graphene nanosheets as water-base lubricant additive," *Applied Physics A*, vol. 105, pp. 827–832, 2011.
22. J. Lin, L. Wang, and G. Chen, "Modification of graphene platelets and their tribological properties as a lubricant additive," *Tribology Letters*, vol. 41, pp. 209–215, 2011.
23. X. Fan, Y. Xia, L. Wang, and W. Li, "Multilayer graphene as a lubricating additive in bentone grease," *Tribology Letters*, vol. 55, pp. 455–464, 2014.
24. T. Missala *et al.*, "Study on tribological properties of lubricating grease with additive of graphene," in *Progress in Automation, Robotics and Measuring Techniques: Volume 3 Measuring Techniques and Systems*, 2015, pp. 181–187.
25. H. Kinoshita, Y. Nishina, A. A. Alias, and M. Fujii, "Tribological properties of monolayer graphene oxide sheets as water-based lubricant additives," *Carbon*, vol. 66, pp. 720–723, 2014.
26. D. V. Kosynkin *et al.*, "Graphene oxide as a high-performance fluid-loss-control additive in water-based drilling fluids," *ACS Applied Materials & Interfaces*, vol. 4, no. 1, pp. 222–227, 2012.
27. A. Kumar, S. Keerti, J. Jain, S. Sinha, S. Tekumalla, and M. Gupta, "Investigations of wear response of pure Mg and Mg-0.4 Ce-Y2O3/ZnO nanocomposites using a single and repeated scratch tests," *Tribology Transactions*, vol. 61, no. 5, pp. 951–959, 2018.
28. A. Kumar, A. Datta, A. Kumar, and A. Kumar, "Recent advancements and future trends in next-generation materials for biomedical applications," in *Advanced Materials for Biomedical Applications*, CRC Press, 2022, pp. 1–19.
29. V. Eswaraiah, V. Sankaranarayanan, and S. Ramaprabhu, "Graphene-based engine oil nanofluids for tribological applications," *ACS Applied Materials & Interfaces*, vol. 3, no. 11, pp. 4221–4227, 2011.
30. N. M. Renevier, V. C. Fox, D. G. Teer, and J. Hampshire, "Performance of low friction MoS_2/titanium composite coatings used in forming applications," *Materials & Design*, vol. 21, no. 4, pp. 337–343, 2000.
31. C. Donnet, J. M. Martin, T. Le Mogne, and M. Belin, "Super-low friction of MoS2 coatings in various environments," *Tribology International*, vol. 29, no. 2, pp. 123–128, 1996.
32. S. D. Leventini, B. S. Martin-Gutierrez, A. Kumar, A. S. Mittman, S. M. Kim, and A. Martini, "Tactile perception of vellum quantified by friction and surface roughness," *Tribology Letters*, vol. 70, no. 4, p. 127, 2022.
33. A. Kumar, *et al.*, "Physics of laser—matter interaction in laser-based manufacturing," in *Laser-based Technologies for Sustainable Manufacturing*, CRC Press, 2023, pp. 45–54.
34. Q. Shi *et al.*, "Synthesis and tribological properties of Ti-doped $NbSe_2$ nanoparticles," *Chalcogenide Letters*, vol. 11, no. 5, 2014.

35. R. G. Soto-Castillo, G. S. Chaubey, and G. Diloyan, "Industrial lubricant including metal chalcogenide particles and phosphorus-based additive." Google Patents, Jul-2019.
36. W. Dai, B. Kheireddin, H. Gao, Y. Kan, A. Clearfield, and H. Liang, "Formation of anti-wear tribofilms via α-ZrP nanoplatelet as lubricant additives," *Lubricants*, vol. 4, no. 3, p. 28, 2016.
37. H. Xiao, W. Dai, Y. Kan, A. Clearfield, and H. Liang, "Amine-intercalated α-zirconium phosphates as lubricant additives," *Applied Surface Science*, vol. 329, pp. 384–389, 2015.
38. X. He, H. Xiao, J. P. Kyle, E. J. Terrell, and H. Liang, "Two-dimensional nanostructured Y_2O_3 particles for viscosity modification," *Applied Physics Letters*, vol. 104, no. 16, 2014.
39. J. Zhao, Y. He, Y. Wang, W. Wang, L. Yan, and J. Luo, "An investigation on the tribological properties of multilayer graphene and MoS2 nanosheets as additives used in hydraulic applications," *Tribology International*, vol. 97, pp. 14–20, 2016.
40. D. Zheng, Z. Cai, M. Shen, Z. Li, and M. Zhu, "Investigation of the tribology behaviour of the graphene nanosheets as oil additives on textured alloy cast iron surface," *Applied Surface Science*, vol. 387, pp. 66–75, 2016.
41. H.-J. Song, X.-H. Jia, N. Li, X.-F. Yang, and H. Tang, "Synthesis of α-Fe_2O_3 nanorod/graphene oxide composites and their tribological properties," *Journal of Materials Chemistry*, vol. 22, no. 3, pp. 895–902, 2012.
42. X. Yang, T. Mei, J. Yang, C. Zhang, M. Lv, and X. Wang, "Synthesis and characterization of alkylamine-functionalized graphene for polyolefin-based nanocomposites," *Applied Surface Science*, vol. 305, pp. 725–731, 2014.
43. D. Berman, A. Erdemir, and A. V. Sumant, "Reduced wear and friction enabled by graphene layers on sliding steel surfaces in dry nitrogen," *Carbon*, vol. 59, pp. 167–175, 2013.
44. W. Zhai *et al.*, "Grain refinement: a mechanism for graphene nanoplatelets to reduce friction and wear of Ni3Al matrix self-lubricating composites," *Wear*, vol. 310, nos. 1–2, pp. 33–40, 2014.
45. D. Berman, A. Erdemir, and A. V. Sumant, "Few layer graphene to reduce wear and friction on sliding steel surfaces," *Carbon*, vol. 54, pp. 454–459, 2013.
46. S. A. Amit Choudhari, Pavan Rayar, Sunil Shimpi, and Neel Pawar, "Design and development of vacuum frying machine for the production of high-qua lityFriedProducts.pdf."
47. A. Klemenz, L. Pastewka, S. G. Balakrishna, A. Caron, R. Bennewitz, and M. Moseler, "Atomic scale mechanisms of friction reduction and wear protection by graphene," *Nano Letters*, vol. 14, no. 12, pp. 7145–7152, 2014.
48. D. Berman, S. A. Deshmukh, S. K. R. S. Sankaranarayanan, A. Erdemir, and A. V Sumant, "Macroscale superlubricity enabled by graphene nanoscroll formation," *Science*, vol. 348, no. 6239, pp. 1118–1122, 2015.
49. A. K. Gupta, A. Kumar, A. Choudhari, A. Kumar, and N. Chowdhury, "Composites for drug eluting devices: emerging biomedical applications, in *Applications of Biotribology in Biomedical Systems*, Springer, 2024.
50. N. Chowdhury, S. Faisal, A. Kumar, A. Kumar, A. Choudhari, and A. K. Gupta, "Characteristics of hydrogel properties in the advancement of biotribology," in *Applications of Biotribology in Biomedical Systems*, Springer, 2024.

51. V. Jaiswal, Kalyani, S. Umrao, R. B. Rastogi, R. Kumar, and A. Srivastava, "Synthesis, characterization, and tribological evaluation of TiO_2-reinforced boron and nitrogen co-doped reduced graphene oxide based hybrid nanomaterials as efficient antiwear lubricant additives," *ACS Applied Materials & Interfaces*, vol. 8, no. 18, pp. 11698–11710, 2016.
52. Y. Meng, F. Su, and Y. Chen, "A novel nanomaterial of graphene oxide dotted with Ni nanoparticles produced by supercritical CO_2-assisted deposition for reducing friction and wear," *ACS Applied Materials & Interfaces*, vol. 7, no. 21, pp. 11604–11612, 2015.
53. Q. Zhou *et al.*, "Preparation of a reduced graphene oxide/zirconia nanocomposite and its application as a novel lubricant oil additive," *RSC Advances*, vol. 5, no. 111, pp. 91802–91812, 2015.
54. H. P. Mungse, N. Kumar, and O. P. Khatri, "Synthesis, dispersion and lubrication potential of basal plane functionalized alkylated graphene nanosheets," *RSC Advances*, vol. 5, no. 32, pp. 25565–25571, 2015.
55. V. Eswaraiah, S. S. J. Aravind, and S. Ramaprabhu, "Top down method for synthesis of highly conducting graphene by exfoliation of graphite oxide using focused solar radiation," *Journal of Materials Chemistry*, vol. 21, no. 19, pp. 6800–6803, 2011.
56. S. S. J. Aravind, R. I. Jafri, N. Rajalakshmi, and S. Ramaprabhu, "Solar exfoliated graphene–carbon nanotube hybrid nano composites as efficient catalyst supports for proton exchange membrane fuel cells," *Journal of Materials Chemistry*, vol. 21, no. 45, pp. 18199–18204, 2011.
57. X. Fan and L. Wang, "High-performance lubricant additives based on modified graphene oxide by ionic liquids," *Journal of Colloid and Interface Science*, vol. 452, pp. 98–108, 2015.
58. O. Elomaa, V. K. Singh, A. Iyer, T. J. Hakala, and J. Koskinen, "Graphene oxide in water lubrication on diamond-like carbon vs. stainless steel high-load contacts," *Diamond and Related Materials*, vol. 52, pp. 43–48, 2015.
59. Y. Liu, X. Wang, G. Pan, and J. Luo, "A comparative study between graphene oxide and diamond nanoparticles as water-based lubricating additives," *Science China Technological Sciences*, vol. 56, pp. 152–157, 2013.
60. S.-Z. Zu and B.-H. Han, "Aqueous dispersion of graphene sheets stabilized by pluronic copolymers: formation of supramolecular hydrogel," *Journal of Physical Chemistry C*, vol. 113, no. 31, pp. 13651–13657, 2009.
61. S. Stankovich, R. D. Piner, X. Chen, N. Wu, S. T. Nguyen, and R. S. Ruoff, "Stable aqueous dispersions of graphitic nanoplatelets via the reduction of exfoliated graphite oxide in the presence of poly (sodium 4-styrenesulfonate)," *Journal of Materials Chemistry*, vol. 16, no. 2, pp. 155–158, 2006.
62. K. Jo *et al.*, "Stable aqueous dispersion of reduced graphene nanosheets via non-covalent functionalization with conducting polymers and application in transparent electrodes," *Langmuir*, vol. 27, no. 5, pp. 2014–2018, 2011.
63. S. Liu, J. Tian, L. Wang, H. Li, Y. Zhang, and X. Sun, "Stable aqueous dispersion of graphene nanosheets: noncovalent functionalization by a polymeric reducing agent and their subsequent decoration with Ag nanoparticles for enzymeless hydrogen peroxide detection," *Macromolecules*, vol. 43, no. 23, pp. 10078–10083, 2010.
64. D. Li, M. B. Müller, S. Gilje, R. B. Kaner, and G. G. Wallace, "Processable aqueous dispersions of graphene nanosheets," *Nature Nanotechnology*, vol. 3, no. 2, pp. 101–105, 2008.

65. M. Lotya, P. J. King, U. Khan, S. De, and J. N. Coleman, "High-concentration, surfactant-stabilized graphene dispersions," *ACS Nano*, vol. 4, no. 6, pp. 3155–3162, 2010.
66. L. Guardia *et al.*, "High-throughput production of pristine graphene in an aqueous dispersion assisted by non-ionic surfactants," *Carbon*, vol. 49, no. 5, pp. 1653–1662, 2011.
67. M. Lotya *et al.*, "Liquid phase production of graphene by exfoliation of graphite in surfactant/water solutions," *Journal of the American Chemical Society*, vol. 131, no. 10, pp. 3611–3620, 2009.
68. S. V. P. Vattikuti and C. Byon, "Synthesis and characterization of molybdenum disulfide nanoflowers and nanosheets: nanotribology," *Journal of Nanomaterials*, vol. 2015, p. 9, 2015.
69. M. Kalin, J. Kogovšek, and M. Remškar, "Mechanisms and improvements in the friction and wear behavior using MoS2 nanotubes as potential oil additives," *Wear*, vol. 280, pp. 36–45, 2012.
70. J. Kogovšek and M. Kalin, "Various MoS2-, WS2-and C-based micro-and nanoparticles in boundary lubrication," *Tribology Letters*, vol. 53, pp. 585–597, 2014.
71. P. Rabaso *et al.*, "Boundary lubrication: influence of the size and structure of inorganic fullerene-like MoS2 nanoparticles on friction and wear reduction," *Wear*, vol. 320, pp. 161–178, 2014.
72. I. Lahouij, B. Vacher, J.-M. Martin, and F. Dassenoy, "IF-MoS2 based lubricants: influence of size, shape and crystal structure," *Wear*, vol. 296, nos. 1–2, pp. 558–567, 2012.
73. X. Shi *et al.*, "Synergetic lubricating effect of MoS2 and Ti3SiC2 on tribological properties of NiAl matrix self-lubricating composites over a wide temperature range," *Materials & Design*, vol. 55, pp. 93–103, 2014.
74. M. Ratoi, V. B. Niste, J. Walker, and J. Zekonyte, "Mechanism of action of WS 2 lubricant nanoadditives in high-pressure contacts," *Tribology Letters*, vol. 52, pp. 81–91, 2013.
75. J. Dong *et al.*, "Facile synthesis of ultrathin NbTe2 nanosheets for enhanced tribological properties as a lubricant additive," *Crystal Research and Technology*, vol. 51, no. 11, pp. 671–680, 2016.
76. V. An, Y. Irtegov, and C. de Izarra, "Study of tribological properties of nanolamellar WS 2 and MoS 2 as additives to lubricants," *Journal of Nanomaterials*, vol. 2014, p. 188, 2014.
77. A. Kumar, "Advancements in emerging superlubricity: a review of the atomistic models, simulation techniques and their applications to explore the state of ultra-low friction," *Materials Today: Proceedings*, vol. 42, no. 4, 2021.
78. M. R. Hilton, R. Bauer, S. V. Didziulis, M. T. Dugger, J. M. Keem, and J. Scholhamer, "Structural and tribological studies of MoS2 solid lubricant films having tailored metal-multilayer nanostructures," *Surface and Coatings Technology*, vol. 53, no. 1, pp. 13–23, 1992.
79. A. Kumar, M. Byadwal, A. Kumar, and A. Kumar, "Laser micromachining in biomedical industry," in *Laser-based Technologies for Sustainable Manufacturing*, CRC Press, 2023, pp. 169–206.
80. S. R. Shrivastava Prateek *et al.*, "Investigation of torsional rod to minimize vibration in automobile using ANSYS," in *2020 1st International Conference on Innovative Research in Applied Science, Engineering and Technology (IRASET)*, 2020, pp. 1–6.

81. A. K. P. S. Saurabh Gupta Ruchika Saini, "Performance analysis of gudgeon pin of various cross sections by FEM," *International Journal of Recent Technology and Engineering*, vol. 8, no. 5, pp. 4569–4573, 2020.
82. A. K. Mehra, R. Saini, and A. Kumar, "The effect of fibre contents on mechanical and moisture absorption properties of gourd sponge/coir fibre reinforced epoxy hybrid composites," *Composites Communications*, vol. 25, p. 100732, 2021.
83. K. Zhang *et al.*, "Synthesis and tribological properties of Ti-doped WSe2 nanoflakes," *Chalcogenide Letter*, vol. 12, p. 51, 2015.
84. M. Singh, A. Kumar, and A. R. Khan, "Capillary as a liquid diode," *Physical Review Fluids*, vol. 5, no. 10, p. 102101, 2020.
85. A. Kumar, S. Datta, and D. Kalyanasundaram, "Liquid slippage in con fined flows: effect of periodic micropatterns of arbitrary pitch and amplitude," in *ASME 2016 5th Micro/Nanoscale Heat and Mass Transfer International Conference (MNHMT2016),* Biopolis, Singapore, 2016, vol. 1, no. 1.
86. A. Kumar, S. Datta, and D. Kalyanasundaram, "Reduction of hydraulic friction in confined flows by laser texturing: experiments and theoretical validation," in *ASME 2018 16th International Conference on Nanochannels, Microchannels, and Minichannels, ICNMM 2018*, 2018.
87. A. Kumar, S. Datta, and D. Kalyanasundaram, "Confinement effects on effective slip of patterned surfaces," in *Fluids Engineering Division Summer Meeting*, 2017, vol. 58066, p. V01CT23A004.
88. A. Kumar, A. Parihar, U. Panda, and D. S. Parihar, "Microfluidics-based point-of-care testing (POCT) devices in dealing with waves of COVID-19 pandemic: the emerging solution," *ACS Applied Bio Materials*, vol. 5, no. 5, pp. 2046–2068, 2022.
89. E. L. Castellanos-Leal, A. Osuna-Zatarain, and A. Garcia-Garcia, "Frictional properties of two-dimensional nanomaterials as an additive in liquid lubricants: current challenges and potential research topics," *Lubricants*, vol. 11, no. 3, p. 137, 2023.
90. A. Kumar, S. Datta, and D. Kalyanasundaram, "Liquid slippage in confined flows: effect of periodic micropatterns of arbitrary pitch and amplitude," in *ASME 2016 5th International Conference on Micro/Nanoscale Heat and Mass Transfer, MNHMT 2016*, vol. 1, 2016.
91. Y. Subramanian *et al.*, "Structural, photoabsorption and photocatalytic characteristics of BiFeO3-WO3 nanocomposites: an attempt to validate the experimental data through SVM-based Artificial Intelligence (AI)," *Journal of Electronic Materials*, pp. 1–11, 2023.
92. Y. Subramanian *et al.*, "Artificial intelligence technique based performance estimation of solid oxide fuel cells," *Materials Today: Proceedings*, 2021.
93. S. Chakraborty, A. K. Gupta, D. Roy, and A. Basumallick, "Studies on nano-metal dispersed Cu-Cr matrix composite," *Materials Letters*, vol. 257, no. September, p. 126739, 2019.
94. D. Roy, S. Chakraborty, A. K. Gupta, A. Basu Mallick, and C. C. Koch, "Synergistic effect of Nb and Zr addition in thermal stabilization of nanocrystalline Cu synthesized by ball milling," *Materials Letters*, vol. 271, p. 127780, 2020.
95. S. Chakraborty, A. K. Gupta, D. Roy, and A. Basu Mallick, "Nanomechanical properties of mechanically alloyed and spark plasma sintered W-nanoparticulate dispersed Cu-Nb alloys," *Materials Letters*, vol. 274, p. 128004, 2020.

96. V. Parmar *et al.*, "Oxidation facilitated antimicrobial ability of laser micro-textured titanium alloy against gram-positive Staphylococcus aureus for biomedical applications," *Journal of Laser Applications*, vol. 30, no. 3, 2018.
97. A. Dhanasekaran, V. R. Y. Subramanian, and A. Kumar, "Deformation analysis for BFO thin films with different thickness by finite element method," in *ASEAN International Conference on Energy and Environment,* Universiti Brunei Darussalam, Brunei, 2021.
98. A. Dhanasekaran *et al.*, "Residual stress measurement in thin films using ansys approach," in *AIP Conference Proceedings*, vol. 2968, no. 1, 2023.
99. A. Parihar, A. Kumar, U. Panda, R. Khan, D. S. Parihar, and R. Khan, "Cryopreservation: a comprehensive overview, challenges, and future perspectives," *Advanced Biology*, p. 2200285, 2023.
100. A. Kumar, S. Datta, and D. Kalyanasundaram, "Liquid slippage in confined flows: effect of periodic micropatterns of arbitrary pitch and amplitude," *Journal of Heat Transfer*, vol. 140, no. 1, 2018.
101. Y. Subramanian *et al.*, "Artificial intelligence technique based performance estimation of solid oxide fuel cells," in *International Conference on Nanoelectronics, Nanophotonics, Nanomaterials, Nanobioscience & Nanotechnology,* Mangalam College of Engineering, Kottayam, Kerala, 2021.
102. V. Parmar, A. Kumar, G. V. Prakash, S. Datta, and D. Kalyanasundaram, "Investigation, modelling and validation of material separation mechanism during fiber laser machining of medical grade titanium alloy Ti6Al4V and stainless steel SS316L," *Mechanics of Materials*, vol. 137, 2019.
103. A. Kumar, U. Panda, V. K. Patel, and R. Kant, "Laser-assisted fabrication of polymers by pushing down the limit of resolution," *Trends in Fabrication of Polymers and Polymer Composites*, pp. 1–3, 2022.
104. A. Kumar *et al.*, "Errata: optimization of laser machining process for the preparation of photomasks, and its application to microsystems fabrication," *Journal of Micro/Nanolithography, MEMS, and MOEMS*, vol. 13, no. 1, 2014.
105. H. A. Zaharin, M. J. Ghazali, N. Thachnatharen, F. Ezzah, R. Walvekar, and M. Khalid, "Progress in 2D materials based nanolubricants: a review," *FlatChem*, vol. 38, p. 100485, 2023.
106. A. Parihar *et al.*, "3D Printing: advancement in biogenerative engineering to combat shortage of organs and bioapplicable materials," *Regenerative Engineering and Translational Medicine*, pp. 1–27, 2021.
107. S. Kumar, R. K. Verma, A. Kumar, and V. K. Patel, "Importance of chemically treated natural fibers in the fabrication of natural fiber reinforced polymer composites," *Trends in Fabrication of Polymers and Polymer Composites*, pp. 10–11, 2022.
108. A. Kumar, A. Pathak, A. Kumar, and A. Kumar, "Physics of laser–matter interaction in laser-based manufacturing," in *Laser-based Technologies for Sustainable Manufacturing*, CRC Press, 2023, pp. 45–54.
109. A. Kumar, A. K. Sharma, T. V. K. Gupta, and J. K. Katiyar, "Influence of hexagonal boron nitride additive nanocutting fluid on the machining of AA6061-T6 alloy using minimum quality lubrication," *Proceedings of the Institution of Mechanical Engineers, Part E: Journal of Process Mechanical Engineering*, p. 09544089221110980, 2022.

110. S. Chakravarthy *et al.*, "Design of an endoscopic haptic display system using an integrated ring-actuator," In 1st International and 16th National Conference on Machines and Mechanisms (iNaCoMM2013), 2013.
111. A. Pathak, A. Kumar, A. Kumar, and A. Kumar, "Application of laser technology in the mechanical and machine manufacturing industry," in *Laser-based Technologies for Sustainable Manufacturing*, CRC Press, 2023, pp. 107–155.
112. A. Datta, A. Kumar, A. Kumar, A. Kumar, and V. P. Singh, "Advanced materials in biological implants and surgical tools," in *Advanced Materials for Biomedical Applications*, CRC Press, 2022, pp. 21–43.
113. A. Kumar, A. Kumar, and A. Kumar, "Introduction to optics and laser-based manufacturing technologies," in *Laser-based Technologies for Sustainable Manufacturing*, CRC Press, 2023, pp. 1–43.
114. A. Kumar, A. Kumar, and A. Kumar, *Laser-based Technologies for Sustainable Manufacturing*. CRC Press, 2023.
115. W. Zhang *et al.*, "Soluble, exfoliated two-dimensional nanosheets as excellent aqueous lubricants," *ACS Applied Materials & Interfaces*, vol. 8, no. 47, pp. 32440–32449, 2016.
116. J. C. Spear, B. W. Ewers, and J. D. Batteas, "2D-nanomaterials for controlling friction and wear at interfaces," *Nano Today*, vol. 10, no. 3, pp. 301–314, 2015.
117. L. Liu *et al.*, "Research progress in application of 2D materials in liquid-phase lubrication system," *Materials*, vol. 11, no. 8, p. 1314, 2018.
118. Z. Chen, Y. Liu, S. Gunsel, and J. Luo, "Mechanism of antiwear property under high pressure of synthetic oil-soluble ultrathin MoS2 sheets as lubricant additives," *Langmuir*, vol. 34, no. 4, pp. 1635–1644, 2018.
119. A. Rosenkranz, Y. Liu, L. Yang, and L. Chen, "2D nano-materials beyond graphene: from synthesis to tribological studies," *Applied Nanoscience*, vol. 10, no. 9, pp. 3353–3388, 2020.
120. C. Kim, M.-A. Yoon, B. Jang, J.-H. Kim, and K.-S. Kim, "A review on transfer process of two-dimensional materials," *Tribology and Lubricants*, vol. 36, no. 1, pp. 1–10, 2020.
121. J. Taha-Tijerina *et al.*, "Thermal conductivity performance of 2D h-BN/MoS2/-hybrid nanostructures used on natural and synthetic esters," *Nanomaterials*, vol. 10, no. 6, p. 1160, 2020.
122. A. R. Kamali, *Green Production of Carbon Nanomaterials in Molten Salts and Applications*. Springer, 2020.
123. Z. Liang, S. Wang, K. Zhu, Y. Chen, F. Wei, and D. Chen, "Enhancing the tribological properties and corrosion resistance of graphene-based lubricating grease via ultrasonic-assisted ball milling," *Colloids and Surfaces A: Physicochemical and Engineering Aspects*, vol. 633, p. 127889, 2022.
124. A. Kumar, A. K. Sharma, and J. K. Katiyar, "State-of-the-art in sustainable machining of different materials using Nano Minimum Quality Lubrication (NMQL)," *Lubricants*, vol. 11, no. 2, p. 64, 2023.
125. R. K. Singh, A. Kumar, R. Kant, A. Gupta, E. Suresh, and S. Bhattacharya, "Design and fabrication of 3-dimensional helical structures in polydimethylsiloxane for flow control applications," *Microsystem Technologies*, vol. 20, no. 1, 2014.
126. R.K.Singh,A.Kumar,R.Kant,A.Gupta,E.Suresh,andS.Bhattacharya,"Design and fabrication of 3-dimensional helical structures in polydimethylsiloxane

for flow control applications," *Microsystem Technologies*, vol. 20, no. 1, pp. 101–111, 2014.

127. A. Kumar and U. Panda, "Microfluidics-based devices and their role on point-of-care testing," in *Biosensor Based Advanced Cancer Diagnostics*, Elsevier, 2022, pp. 197–224.
128. A. Kumar, A. Parihar, S. N. Basha, and U. Panda, "Clinically available/under trial drugs and vaccines for treatment of SARS-COV-2," in *Computational Approaches for Novel Therapeutic and Diagnostic Designing to Mitigate SARS-CoV2 Infection*, Elsevier, 2022, pp. 451–488.
129. K. C. Mutyala, Y. A. Wu, A. Erdemir, and A. V. Sumant, "Graphene-MoS2 ensembles to reduce friction and wear in DLC-Steel contacts," *Carbon*, vol. 146, pp. 524–527, 2019.
130. H. A. Zaharin, M. J. Ghazali, A. K. Rasheed, M. Khalid, and Y. Otsuka, "Tribological performance of hybrid Ti3C2/Graphene additive on out-board engine oil," in *Proceedings of the Malaysian International Tribology Conference*, pp. 146–153, 2020.
131. D. Roy, S. Chakraborty, A. K. Gupta, A. BasuMallick, R. O. Scattergood, and C. C. Koch, "Synergistic effect of Nb and Zr additions on the structure-property relationships of nanocrystalline Cu processed by mechanical alloying and hot pressing," *Journal of Alloys and Compounds*, vol. 854, p. 157174, 2021.
132. D. Roy, S. Pal, C. S. Tiwary, A. K. Gupta, P. N. Babu, and R. Mitra, "Stable nano-crystalline structure attainment and strength enhancement of Cu base alloy using bi- modal distributed tungsten dispersoids," *Philosophical Magazine*, pp. 1–21, 2021.
133. C. Xu, Z. Ye, and P. Egberts, "Friction hysteretic behavior of supported atomically thin nanofilms," *NPJ 2D Materials and Applications*, vol. 7, no. 1, p. 1, 2023.
134. S. Chakravarthy, K. Avinash, G. Ramu, G. K. Ananthasuresh, "Design of an endoscopic haptic display system using an integrated ring-actuator," in *1st International and 16th National Conference on Machines and Mechanisms, iNaCoMM 2013*, 2013.
135. P. Vats, K. K. Gajrani, and A. Kumar, "Laser-based additive manufacturing," in *Laser-based Technologies for Sustainable Manufacturing*, CRC Press, 2023, pp. 67–83.
136. D. Dias, J. Bons, A. Kumar, M. H. Kabir, and H. Liang, "Forever chemicals, per- and polyfluoroalkyl substances (PFAS), in Lubrication." *Lubricants*, vol. 12, no. 4, p. 114, 2024.
137. A. Kumar, S. Datta, and D. Kalyanasundaram, "Permeability and effective slip in confined flows transverse to wall slippage patterns," *Physics of Fluids*, vol. 28, no. 8, 2016.
138. A. Kumar *et al.*, "Biomedical study of femur bone fracture and healing," in *Advanced Materials for Biomedical Applications*, CRC Press, 2022, pp. 235–250.
139. A. Kumar, A. Kumar, A. Kumar, A. Choudhari, A. K. Gupta, and S. Faisal, "Introduction to biotribology: a science of surface interaction," in *Applications of Biotribology in Biomedical Systems*, Ed. A. Kumar, A. Kumar, and A. Kumar, Springer Press, Springer Nature Group, 2024.
140. A. Kumar *et al.*, "Optimization of laser machining process for the preparation of photomasks, and its application to microsystems fabrication," *Journal of Micro/Nanolithography, MEMS, and MOEMS*, vol. 12, no. 4, 2013.

141. A. Kumar, S. Rana, Y. Gori, and N. K. Sharma, "Thermal contact conductance prediction using FEM based computational techniques," in *Advanced Computational Methods in Mechanical and Materials Engineering*, CRC Press, 2021, pp. 183–220. DOI: 10.1201/9781003202233-13.
142. A. Kumar and P. P. Patil, "Free vibration and connecting bolt constraint based FEA analysis of heavy vehicle medium duty transmission gearbox housing made from AISI 4130 Alloy," in *Mathematics Applied to Engineering*, Academic Press Elsevier, 2017, pp. 133–146. DOI: 10.1016/B978-0-12-810998-4.00007-7.
143. A. Kumar, A. Dwivedi, H. Jaiswal, and P. P. Patil, 2015. "Material-based vibration characteristic analysis of heavy vehicle transmission gearbox casing using finite element analysis," in *Springer Book Series: Advances in Intelligent Systems and Computing*, vol. 308, pp. 527–533. DOI: 10.1007/978-81-322-2012-1_56.
144. F. Ahmad, V. Tomer, A. Kumar, and P. P. Patil. "FEA simulation based thermo-mechanical analysis of tractor exhaust manifold," in *Springer Book Series: Lecture Notes in Mechanical Engineering: CAD/CAM, Robotics and Factories of the Future*, pp. 173–181, 2016. DOI: 10.1007/978-81-322-2740-3_18.
145. A. Kumar, Y. Gori, N. Dutt, Y. K. Singla, and A. Maurya, Editors, *Advanced Computational Methods in Mechanical and Materials Engineering*, CRC Press, 2021. DOI: 10.1201/9781003202233.
146. K. K. Gajrani, A. Prasad, and A. Kumar, Editors, *Advances in Sustainable Machining and Manufacturing Processes*, CRC Press, 2022. DOI: 10.1201/9781003284574.
147. S. Singh, S. K. Behura, A. Kumar, and K. Verma, Editors, *Nanomanufacturing and Nanomaterials Design: Principles and Applications*, CRC Press, 2022. DOI: 10.1201/9781003220602.
148. V. K. Sharma, A. Kumar, M. Gupta, V. Kumar, D. K. Sharma, and S. Sharma, Editors, *Additive Manufacturing in Industry 4.0: Methods, Techniques, Modeling, and Nano Aspects*, CRC Press, 2022. DOI: 10.1201/9781003360001.
149. P. Patil, Y. Gori, A. Kumar, and M. R. Tyagi, "Experimental analysis of tribological properties of polyisobutylene thickened oil in lubricated contacts," *Tribology International*, vol. 159, 106983, pp. 1–7, 2021. DOI: 10.1016/j.triboint.2021.106983.
150. P. P Patil and A. Kumar, "Dynamic structural and thermal characteristics analysis of oil lubricated multi speed transmission gearbox: variation of load, rotational speed and convection heat transfer," *Iranian Journal of Science and Technology Transactions of Mechanical Engineering (ISTM)*, vol. 41, pp. 281–291, 2017. DOI: 10.1007/s40997-016-0063-z.
151. N. Chowdhury, S. Faisalb, A. Kumar, A. Choudhari, A. K. Gupta, A. Kumar, and A. Kumar, "Characterization of hydrogel properties in the advancement of biotribology," in *Applications of Biotribology in Biomedical Systems*, Ed. A. Kumar, A. Kumar, and A. Kumar, Springer Press, Springer Nature Group, 2024.
152. A. Pal, R. Gour, A. Kumar, A. Kumar, A. Kumar, and N. Chowdhury, "Tribological measurements of human skin," in *Applications of Biotribology in Biomedical Systems*, Ed. A. Kumar, A. Kumar, and A. Kumar, Springer Press, Springer Nature Group, 2024.

153. A. Kumar, P. Gajjar, K. S. Gudadur, A. Kumar, A. K. Gupta, A. Choudhari, N. Chowdhury, and A. Kumar, "Tribological hurdles in biomedical manufacturing: a comprehensive examination," in *Applications of Biotribology in Biomedical Systems*, Ed. A. Kumar, A. Kumar, and A. Kumar, Springer Press, Springer Nature Group, 2024.
154. M. Heidari, S. Thangavel, P. Rahmanivahid, M. Khashehchi, A. Kumar, A. Kumar, and A. Kumar, "Navigating the landscape: cutting-edge biomedical manufacturing techniques," in *Applications of Biotribology in Biomedical Systems*, Ed. A. Kumar, A. Kumar, and A. Kumar, Springer Press, Springer Nature Group, 2024.
155. R. Gour, N. Baliyan, A. Pal, A. Kumar, A. Kumar, and A. Kumar, "Animal tribology," in *Applications of Biotribology in Biomedical Systems*, Ed. A. Kumar, A. Kumar, and A. Kumar, Springer Press, Springer Nature Group, 2024.
156. A. Choudhari, A. K. Gupta, A. Kumar, A. Kumar, A. Gupta, N. Chowdhury, and A. Kumar, "Wear and friction mechanism study in knee and hip rehabilitation: a comprehensive review," in *Applications of Biotribology in Biomedical Systems*, Ed. A. Kumar, A. Kumar, and A. Kumar, Springer Press, Springer Nature Group, 2024.
157. D. Dias, J. Bons, A. Kumar, M. Humaun Kabir, and Hong Liang, "Forever chemicals, per- and polyfluoroalkyl substances (PFAS), in lubrication," *Lubricants*, vol. 12, no. 4, p. 114, 2024.

Chapter 7

Two-dimensional nanomaterial as lubricants

Pramod Belkhode and Pratibha Agrawal

7.1 INTRODUCTION

Two-dimensional (2D) materials used as lubricants have had a great advantage due to their properties and applications in different fields such as electronics and optoelectronics (1–3), catalysis (4, 5), batteries (6–8), solar cells (9–11), and other medical applications (12, 13). The use of those materials as a lubricant or additive has been growing in the last few years. Related data were collected through scholarly articles using 2D lubricants, and the number of related publications obtained is shown in Figure 7.1.

The engineering applications of nanoscience and nanotechnology have direct implications for industries and everyday life. Consequently, an abundance of nanomaterials, nanodevices, and nanosystems have been created and implemented for various engineering objectives, all to improve the human condition. Similarly, nanofluids are utilized in a diverse array of thermal engineering applications. Nanofluid is a suspension of nanoparticles adulterated and disseminated within a liquid medium. The introduction of nanoparticles into base fluids, including distilled water, induces alterations in physical characteristics, including thermal conductivity, viscosity, and density. Among all physical properties, the dispersion of nanoparticles has the most rapid effect on the thermal conductivity of the basal fluid. As of yet, the augmentation of thermal conductivity remains an unexplored domain of study among academics, scientists, and researchers.

7.1.1 Nanotechnology

Nanotechnology is the technological application of nanoscience. In addition to the technological significance of nanoscience, nanotechnology is surrounded by tremendous enthusiasm. There have been extravagant assertions regarding faster processors, inexpensive product manufacturing, and medical advancements. Products including thermal undergarments, tennis racquets, self-cleaning automobiles, paint, food, and cosmetics are anticipated to incorporate nanotechnology. Billions of dollars are being invested by governments in nanoscience research. It was estimated that \$8.6 billion was the

DOI: 10.1201/9781003535447-7

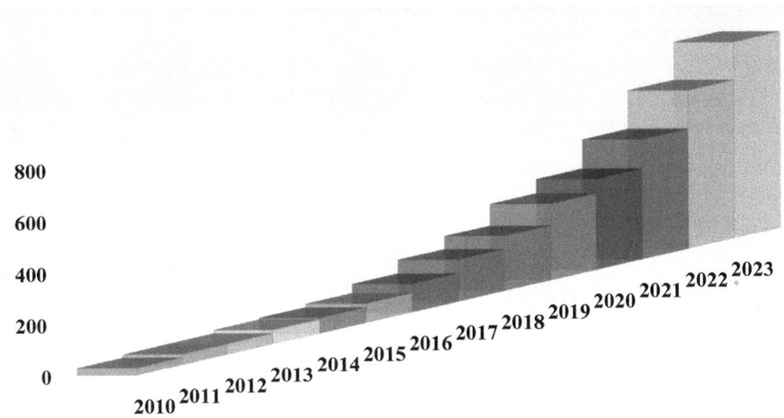

Figure 7.1 Documents published between 2010 and 2023 related to 2D lubricants.

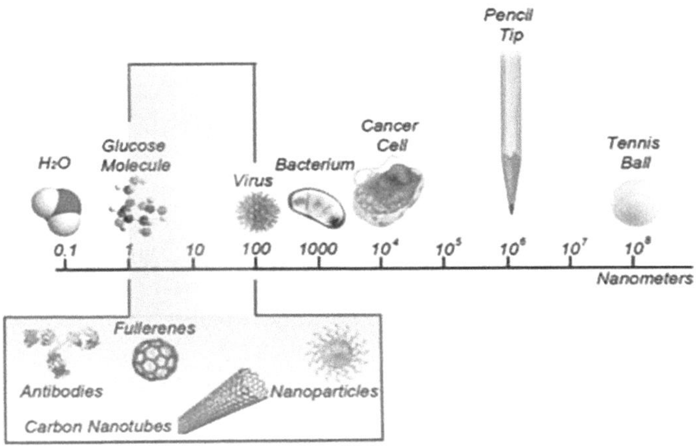

Figure 7.2 Particle size of different objects on the nanoscale.

global investment in nanotechnology research and development in 2004. The National Nanotechnology Initiative refers to the American endeavor in nanotechnology (14). During the period 2002–2006, the European Union allocated €1300 million to nanotechnology research (15) in recognition of its significance as a research domain. A distinct comparison of the nanoscale particle sizes of various objects is illustrated in Figure 7.2.

7.1.2 Nanomaterials

Physical properties of a system in a specific crystallographic direction are typically altered when the spatial dimension is decreased or particles or quasi-particles are confined in that direction. Therefore, an additional categorization of nanostructured systems and materials is fundamentally predicated on the number of dimensions that fall within the nanometer scale: (a) 3D systems, such as structures commonly composed of consolidated equiaxed crystallites, operate in a three-dimensional space; (b) 2D systems include filamentary structures in which the length significantly exceeds the cross-sectional dimensions; (c) one-dimensional (1D) systems, such as layered or laminate structures, operate in a one-dimensional space; and (d) zero-dimensional (0D) structures include nanopores and nanoparticles.

7.1.3 Applications of nanomaterials

Presently, nanoscale materials find application in a multitude of sectors. For instance, carbon black particles impart wear resistance to rubber tires, while nanofibers serve as reinforcements and insulators in composites. Iron oxide, meanwhile, is utilized to produce the magnetic material found in audiovisual tapes and disk drives. Nano-zinc oxides and titania, on the other hand, function as sunblocks against ultraviolet rays. The utilization of nanoscale particles and nano-thin material layers is being employed to enhance the conductivity, strength, and weight of products, among other applications (16–18). Magnetic recording tapes, computer hard disks, bumper stickers, solid-state compasses, glare-reducing and protective coatings for windows and eyeglasses, metal-cutting tools, automobile catalytic converters, durable tennis balls, dental bonding agents, ink, wound, and burn dressings, and so forth, are some examples of commercially available products that utilize nanotechnology. Nanoscale diagnostic and therapeutic devices are a major nanomedicine application. Nanoscale gadgets or nanorobots can carry therapeutic substances, identify early-stage illnesses, and rectify metabolic or genetic problems. The greatest obstacle for medical applications is "miniaturization": the development of new instruments capable of molecular-level tissue analysis, sensors smaller than a cell that enable the observation of ongoing functions, and miniature machines that traverse the human body in pursuit of pathogens and neutralize chemical toxins (19).

7.1.4 Nanofluids

In contrast, nanofluids present numerous benefits in comparison to single-phase pure fluids and microparticle suspensions. Nanofluids significantly mitigate the challenges associated with particle sedimentation caused by gravity, obstruction of microchannel passages, and erosion of tube material in heat exchangers. Moreover, nanofluids generate stable suspensions in which nanoparticles are uniformly dispersed throughout the base fluid. The thermophysical properties of fluids used in single-phase heat transfer, including water, lubricants, and glycols, are widely recognized and documented in relevant literature and handbooks. There is limited research on the thermophysical properties of two-phase nanofluids. As an alternative and next-generation liquid for the transfer of thermal energy, nanofluids can be utilized instead of pure single-phase fluids as heat transfer fluids in heat exchangers. In the realm of heat transfer, nanofluid finds utility in various sectors, including process and chemical engineering, automotive radiators, solar water heating, refrigeration, and electronic device chilling. Nanofluids promote heat transfer by allowing large heat fluxes to flow through, reducing heat exchanger cost and size. This conserves energy and materials.

7.1.5 Nano lubricants

The term "nano lubricant" refers to solid-like particles measuring in the nanometer range in diameter from a few nanometers to 0.1 μm. These particles possess noteworthy tribological characteristics. They demonstrate a reduction in friction and wear when incorporated into a coating or utilized as an additive to lubricating base oil, as demonstrated in specialized laboratory experiments. Spheroidal nano lubricants, such as carbon fullerenes (C60), carbon onions, and inorganic fullerenes, derived from metal dichalcogenides or over-based detergents, are preferred. In addition, they can take the form of carbon nanotubes, which are cylindrical in shape or in cones, disks, or filaments. The engineering applications of nanoscience and nanotechnology have direct implications for industries and everyday life. Subsequently, an abundance of nanomaterials, nanodevices, and nanosystems have been created and implemented for various engineering objectives, all to improve the human condition. Similarly, nanofluids are utilized in a diverse array of thermal engineering applications. Nanofluid is a suspension of nanoparticles adulterated and disseminated within a liquid medium. The introduction of nanoparticles into base fluids, including distilled water, induces alterations in physical characteristics, including thermal conductivity, viscosity, and density. Among all physical properties, the dispersion of nanoparticles has the most rapid effect on the thermal conductivity of the basal fluid. The augmentation of thermal conductivity remains an unexplored domain of investigation among academics, scientists, and researchers (20, 21). At present, a substantial quantity of scholarly articles are being published about this topic, indicative of the discernible expansion and significance of heat transfer enhancement technology. This is because numerous modern applications, including automobile/industrial cooling, plate heat exchangers, shell and tube heat exchangers, compact heat exchangers, and double pipe heat exchangers, rely on thermal conductivity enhancement. At this time, the significance of energy conservation motivates businesses to implement a variety of techniques in their facilities. This enhanced functionality and energy savings are accomplished primarily through the modification of conventional heat exchangers. The utilization of nanofluids enables the fabrication of efficient heat exchangers in a compact configuration. Materials science is being influenced by nanotechnology, which produces materials with novel properties such as ease of cleaning, among others. According to research, the dispersion of nanoparticles in base fluids substantially improves the thermal conductivity and other critical properties of the fluids. Numerous researchers from around the world have investigated these effects using a variety of nanostructures and allotropes of carbon, including fullerenes, single-walled and multi-walled carbon nanotubes, metal oxides, and diamond. Nanotechnology is a field that incorporates materials science with the chemistry and physics of nanostructures to define the complexity of fundamental and engineering sciences. Nanofluids are a fundamental subfield of nanotechnology that finds extensive utility in various domains. Numerous contemporary applications—including automobile and industrial cooling, plate heat exchangers, shell and tube heat exchangers, compact heat exchangers, and double pipe heat exchangers—rely on thermal conductivity improvement.

7.2 NANOMATERIALS

Physical properties of a system in a specific crystallographic direction are typically altered when the spatial dimension is decreased or particles or quasi-particles are confined in that direction. Therefore, an additional categorization of nanostructured systems and materials is fundamentally predicated on the number of dimensions that fall within the nanometer scale: (a) 3D systems, such as structures commonly composed of consolidated equiaxed crystallites, operate in three dimensions; (b) 2D systems, such as filamentary structures where the length significantly exceeds the cross-sectional dimensions, operate in two dimensions; (c) 1D systems, such as layered or laminate structures, operate in one dimension; (d) 0D structures consist of nanopores and nanoparticles as shown in Figure 7.3 (22, 23).

Presently, nano-sized materials find application in a multitude of sectors. For instance, carbon black particles impart wear resistance to rubber tires, nanofibers serve as reinforcements and insulators in composites, iron oxide is the constituent of the magnetic material found in audiovisual tapes and disk drives, and titania and nano-zinc oxides function as sunblocks against ultraviolet rays. The utilization of nanoscale particles and nano-thin material layers is being employed to enhance the conductivity, strength, and weight of products, among other applications. Commercial nanotechnology products include magnetic recording tapes, computer hard disks, bumper stickers, solid-state compasses, protective and glare-reducing coatings for eyeglasses and windows, automobile catalytic converters, metal-cutting tools, dental bonding agents, durable tennis balls, burn and wound dressings, ink, and more. The area has grown rapidly due to promising nanotechnology applications in health and biology. Nanoscale diagnostic and

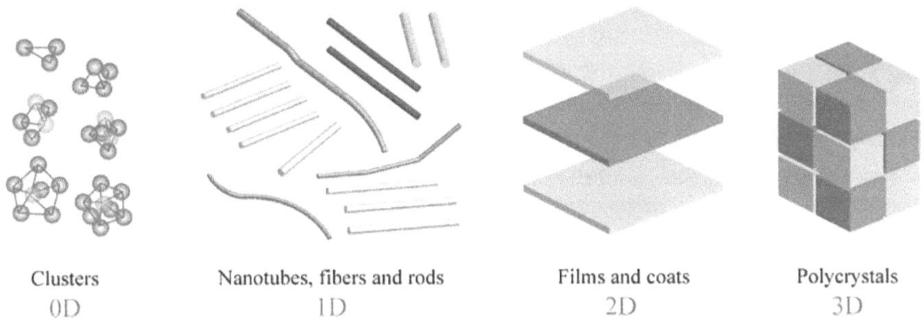

Figure 7.3 Types of nanostructured materials based on the dimensions of their structural elements are zero-dimensional (0D), one-dimensional (1D), two-dimensional (2D), and three-dimensional (3D) nanomaterials.

therapeutic devices are a major nanomedicine application. Nanoscale gadgets or nanorobots may carry therapeutic substances, identify early-stage illnesses, and rectify metabolic or genetic problems. The greatest obstacle for medical applications is "miniaturization": the development of new instruments capable of molecular-level tissue analysis, sensors smaller than a cell that enable the observation of ongoing functions, and miniature machines that traverse the human body in pursuit of pathogens and neutralize chemical toxins (24). In the coming years, scientists anticipate the development of novel commercial applications for nanotechnology. Among these are the following: advanced pharmaceuticals medical diagnostic instruments like cancer taggers and "lab-on-a-chip" diagnostics for doctors; medication delivery systems such as implanted devices that automatically dispense and detect drug levels; cooling chips or wafers to replace compressors in cars, refrigerators, and air conditioners, and other equipment without chemicals or moving components; airborne chemical and toxin sensors; sun fuel cells and portable electricity for cheap, green energy.

7.3 CHARACTERISTICS OF NANOMATERIALS

When the size of particles is decreased from the macroscopic to the nanoscale, an enormous transformation in every property is observed. The term for this phenomenon is the "size effect". Polycrystalline materials are commonly conceptualized as composites comprising two distinct phases: bulk grains and grain boundary phases. An increase in the relative portion of the grain boundary phase in polycrystalline results from a reduction in grain size, thereby inducing alterations in all polycrystalline properties. Size effects of this nature are referred to as intrinsic (I) or inherent size effects. They manifest as a result of particle size reduction. Extrinsic (E) or induced size effects are determined by the phenomena that occur when reduced particles are subjected to external fields of electricity, magnetism, electromagnetics, acoustics, radiation, temperature, or chemistry. A summary of the effect of material size on various properties is provided in Table 7.1.

The study of heat transmission is a critical component of thermal engineering. When developing the thermal design of heat exchangers, the choice of an appropriate heat transfer fluid for heat dissipation is a critical factor. One of the critical parameters that influence the cost and dimensions of heat exchanger systems is the heat transfer fluid. Traditional fluids such as water and oil possess restricted capacity for heat transfer. Research groups worldwide are increasingly acknowledging the critical nature of developing novel fluids that possess enhanced heat transfer properties. The progress achieved in the field of nanotechnology has enabled the production of metal and metal oxide particles at the nanoscale. Nanoparticles are regarded as materials of the next generation with potential thermal transfer applications.

Nanofluid refers to any base liquid or fluid that contains nanoparticles in a suspended state. Nanofluids, which consist of a solid–liquid composition in two phases, are regarded as heat transfer fluids of the next generation. Recent years have seen the emergence of nanofluids as potentially effective thermofluids for heat transfer applications. It is anticipated that the thermal conductivity of nanofluids will surpass that of the base liquids. The

Table 7.1 Summary table exhibiting effects of size

Property	*Influence of particle size reduction on properties of nanoparticles*	*Research reference*
Structural	Lower or higher lattice parameter Transformations of structure	(26–28)
Thermal	Lower melting point Lower phase transition temperatures Lower melting entropy Phonon spectrum softening	(29)
Mechanical	Increased hardness, strength, and fracture ductility Superplasticity starts Increased wear resistance	(30)
Thermodynamical	Increased heat capacity Increased thermal expansion Lower Debye temperature Phase stabilization at high temperature	(31, 32)
Magnetic	Increase or reduce coercion Lower Curie temperature Rising paramagnetism in ferromagnetics Rising enormous magnetoresistance Maximum magnetoresistance temperature rise Ferromagnetics' magnetic permeability increases	(33)
Electrical	Conductivity increase for nanometals Nanodielectric conductivity rises Ferroelectric dielectric inductivity increases	(34)

introduction of particulate on the micron scale into conventional heat transfer fluids has been a practice documented as early as 1962 by Hamilton and Crosser as stated by Murshed et al. (25). However, the sedimentation produced by a two-phase suspension of microparticles impedes the free passage of fluid by clogging the channel and causing erosion of the tube material.

In contrast, nanofluids present numerous benefits in comparison to single-phase pure fluids and microparticle suspensions. Nanofluids significantly mitigate the challenges associated with particle sedimentation caused by gravity, obstruction of microchannel passages, and erosion of tube material in heat exchangers. Moreover, nanofluids generate stable suspensions in which nanoparticles are uniformly dispersed throughout the base fluid. The thermophysical properties of fluids used in single-phase heat transfer, including water, lubricants, and glycols, are widely recognized and documented in relevant literature and handbooks. There is limited research on the thermophysical properties of two-phase nanofluids. The precise characterization of nanofluid properties is an essential requirement to ascertain their heat transfer coefficient. The heat transfer performance of a specific nanofluid is subsequently utilized to determine its suitability for heat transfer applications. As an alternative and next-generation liquid for the transfer of thermal energy, nanofluids can be utilized instead of pure single-phase fluids as heat transfer fluids in heat exchangers. In the realm of heat transfer, nanofluid finds utility in various sectors, including process and chemical engineering, automotive radiators, solar water heating, refrigeration, and electronic device chilling. The primary aim of achieving heat transfer enhancement through the utilization of nanofluids is to facilitate the passage of high heat fluxes, thereby minimizing the expenditure and dimensions of the heat exchangers. This, in turn, promotes energy and material conservation.

Das et al. (35) identified in their research that the thermal conductivity of heat transfer fluids can be substantially enhanced by the addition of highly conductive particles. Furthermore, the viscosity of nanofluids is volumetrically sensitive; it exhibits a direct correlation with the nanoparticle concentration in the base fluid. Duan et al. (36) established that an increase in volume concentration results in a concomitant escalation in relative viscosity. A 60% increase was detected in the re-ultrasonicated nanofluids at a volume concentration of 5% when compared to the basal fluid. Additionally, this report deduces that the nanofluids exhibited greater nanoparticle aggregation before re-ultrasonication. Sridhara et al. provided an exhaustive synopsis of the most recent advancements in nanofluids based on Al_2O_3. The stability of Al_2O_3-based nanofluids, as well as their enhanced thermal conductivity, viscosity, and heat transfer properties, was the primary concern of this review. This review proposes that the utilization of Al_2O_3 nanoparticles, spanning in size from 13 nm to 302 nm, can result in a 2% to 36% increase in thermal conductivity.

Wang et al. (37, 38) conducted an exhaustive investigation into the heat transfer and fluid flow properties of nanofluids in forced and convection flows and identified their potential research applications shortly. The kinetic theory posits that the kinetics of aggregates is determined by the induced micro-convection and stochastic motion of nanoparticles. The stability of nanofluids is a critical factor in the improvement of thermal conductivity. The effective thermal conductivity increases as the nanofluid is stabilized according to the findings of this study. Kleinstreuer et al. comment on the inconsistency between thermal conductivity measurement methodologies and experimental data. Furthermore, these studies give rise to inquiries regarding whether the atypical increase in thermal conductivity of nanofluid occurs solely as a result of variations in temperature, nanoparticle size and shape, and aggregation state. In conclusion, the aforementioned study asserts that precise benchmark experiments are necessary to comprehend the correct mechanism underlying the improvement of thermal conductivity. Earlier researchers also (39, 40) shed light on several additional variables that influence the thermal conductivity of nanofluids, including size, shape, nanoparticle concentration, temperature, and pH. Nsofor (41) established that the thermal conductivities of nanoparticles in liquid suspensions are greater than those of the basal fluids. Liu et al. demonstrate that thermal conductivity is significantly influenced by the stability of nanofluids, in addition to concentration, size, and temperature. Excellent stability of nanofluid can be attained through the appropriate application of surfactants or the addition of the correct amount of ammonia. According to Hong et al., the thermal properties of nanofluids are influenced by a variety of factors, including particle size, surfactant effect, and particle dispersion. The thermal properties of Fe and TiO_2 nanofluids were investigated in this report about their volume fractions (42).

Timofeeva et al. devised an innovative method for decreasing the viscosity of alumina nanofluid while preserving its thermal conductivity. This report emphasizes the significance of the surface charge of nanoparticles about viscosity, which can be manipulated by modifying the pH of the nanofluid (43). Agglomeration and clustering were identified by Wong et al. as significant issues negatively impact the performance of nanofluids. As a result, concentrated research endeavors are required to reduce this undesired aggregation. The stability characteristics of the ZnO/glycol water nanofluids were optimized by Cabaleiro et al. The research was conducted utilizing dynamic light scattering. An excellent correlation is observed among experimental data represented by the dimensionless Nusselt number (44).

Yarmand et al. synthesized graphene with silver functional groups. We investigated the thermophysical properties, heat transfer efficacy, and friction factor of nanofluids based on nanoplatelets. The findings of the inquiry indicate that the effective thermal conductivity and efficiency of heat transfer are enhanced in comparison to the basal fluid. In contrast, the Nusselt number is enhanced by 32.7% when the friction factor is penalized 1.08 times (45). The authors Garoosi et al. (46) conducted a simulation to determine the optimal volume fraction of nanoparticles at each Richardson number to achieve the highest heat transfer rate. Ali et al. developed a nanofluid composed of ZnO to improve the heat transfer capabilities of an automobile radiator. The maximum effective heat transfer enhancement of 46% was observed at a volumetric concentration of 0.2% in the base fluid. Solangi et al. described the various techniques used to assess the stability of nanofluids as well as the parameters that influence that stability. Additionally, this study highlights the influence of Reynolds number, mass concentration, and particle size on the heat transfer characteristics of fluids (47).

7.4 APPLICATION OF NANOFLUIDS IN COOLING SYSTEMS

This segment provides a synopsis of recent reports concerning the utilization of nanofluids in cooling systems. The majority of researchers concur that a low volume fraction of nanoparticles is sufficient to attain optimal cooling system performance. However, there are several issues and obstacles associated with the mechanisms of nanofluid-based cooling systems. This field of study is currently in its nascent phase and requires additional progress.

The thermal energy storage properties of Cu–H_2O nanofluids as a phase change material for cooling systems were investigated by Wang et al. The assessment of a thermal energy storage mechanism was conducted through the examination of the contact angle and thermal conductivity of nanofluids. The addition of 0.1 wt% Cu nanoparticles resulted in a 19.2% reduction in the total freezing time, indicating that Cu–H_2O nanofluids may have utility in thermal energy storage (48). The thermoelectric cooling efficacy of electronic devices was investigated by Ahammed et al. using nanofluid in a mini-channel heat exchanger with multiple ports. The findings indicate that the local Nusselt number experiences an increase of 23.92% and 19.22% for concentrations of 0.1% and 0.2%, respectively, in comparison to water at a Reynolds number of 1000 and an input power of 400 W.

The efficacy of single-phase heat transfer and pressure drop of Al_2O_3–water nanofluids in micro-pin-fin heat exchangers was numerically analyzed by Mohammadi et al. The impact of nanofluid volume fraction and nanoparticle diameter on total entropy generation and coefficient of performance (COP) was investigated in this study (49). The efficacy of a counter-flow microchannel heat exchanger utilizing nanofluid as a cooling medium was investigated by Hasan et al. For this investigation, Cu–water and Al_2O_3–water nanofluids are utilized. The utilization of nanofluids has been found to enhance the efficiency and cooling capabilities of a coolant by their extremely fine particles and small volume fraction (50). The lattice Boltzmann model for single-phase fluids was devised by Mehta et al. through the coupling of the functions representing density and temperature distribution. This study concludes that the utilization of nanoparticles dispersed in deionized water increases the rate of heat transfer (51).

7.5 TWO-DIMENSIONAL LUBRICANTS

Energy consumption, industrial expansion, and greenhouse gas emissions increase in tandem with the rapid growth of the global population. Few methods are addressing the thermomechanical perpetrators, which contribute to the global impact through friction and wear: for instance, by optimizing the performance of lubricants and implementing the most recent surface design through the integration of state-of-the-art materials and additives. It could potentially save millions of euros and reduce energy consumption loss by as much as 45%. Therefore, it is imperative to prioritize advanced technologies that reduce friction and enhance wear resistance on a global scale. To achieve this, there is a significant demand for comprehensive scientific research.

By differentiating two contact surfaces in the machine, high-quality lubricants can reduce friction and wear, thereby addressing the machine's energy waste and rising carbon dioxide re-emission problem. Lubrication is the optimal solution to this issue. Typically, solid lubricants, liquid lubricants, and semi-solid lubricants are utilized to enhance tribological performance, facilitate smooth operation, protect critical components, and prolong the life of appliances. The specific working environment and application of the appliance determine which lubricants are utilized. Lubricants reduce friction and wear. Lubricants that are less efficient or have function limitations are generally hazardous to human health and the environment. For instance, lubricants containing zinc dialkyl dithiophosphates (ZDDP) derived from phosphorus oxide, sulfur oxide, and ash are less effective and pose risks to human health (52).

Innovating lubricating fluids, two-dimensional material-based nano lubricants, contain nanoscale material with a two-dimensional structure. These two-dimensional material-based nano lubricants are composed of graphene and other identical substances that are also composed of one or two atoms. Their greatest advantage over conventional lubricants is that they provide superior lubrication performance.

7.5.1 Properties of graphene

Honeycomb-structured graphene is used as an additive to improve the performance of lubricants. Usually, graphene oxide having a similar type of structure offers good additive pathways, but at the same time, the presence of abundant oxygen on the surface disperses it into the water layer uniformly.

But, it's challenging to maintain the same dispersed state in an oil medium making it less suitable to act as a lubricant. To improve its performance and dispersibility, modified graphene which was specially functionalized can be used. Those graphene-based nanomaterials can be usually synthesized by both one-step and multistep strategies. In a single-step reaction, crystal nuclei of the desired material can be grown using suitable precursors on graphene nanosheets. The material can also be grown of desired dimensions and nanotubes, nanoparticles, nanorods, nanoplates, or nanoflowers can be obtained as per the end-user requirements. The oxygen group at the surface attracts many organic functional groups also by physical or chemical reactions. Graphene oxide/polyethylene glycol (GO/PEG) composite is one such example of this process (53). But many times, it becomes difficult to control the structures at the micro-level by this method, so a multistep strategy is preferred. In this ideal required composition of lubricant can be gained. In this method, decorated functionalized precursor is synthesized separately and then grafted on the surface by a suitable method.

The nanocomposites can be supported wrapped or even grafted on the graphene sheets. Zubir et al. (54) combined carbon nanotube (CNTs) and carbon nanofiber (CNFs) with graphene nanosheets via π–π interactions. This results in improved lubricant properties of graphene oxide sheets for future applications. Jaiswal et al. studied improved lubricant properties for heteroatom (N, P, S, B)-induced graphene oxide in paraffin base oil.

Fluorinated graphene is a 2D layered structured material that can reduce friction and wear as a lubricating material, but its use is restricted in water-based lubricants due to its hydrophobic nature. Min et al. (55) recently obtained a fluorinated graphene oxide water-based lubricant additive, in which numerous oxygen-containing functional groups on the surface greatly promoted its dispersibility in water, hence enhancing its efficiency. Some of the functionalized graphene oxides are represented in Figure 7.4.

7.5.2 Boron nitride

Boron nitride (BN) has cubic and hexagonal structures. The cubic structure has a rigid diamond-like structure and hence cannot be used as a lubricant, but hexagonal BN (h-BN) has a lamellar structure. And thus, it offers suitable candidature as a 2D substrate for lubrication. It can be used as an additive to liquid lubricant or as a solid lubricant itself. This also offers a similar structure to graphene with poor dispersibility. Water-dispersible functionalized boron nitride stands to resolve the dispersibility issue. It possesses higher oxidation resistance, good thermal resistivity with no electrical conductivity, and better chemical stability making its effective use under severe environmental conditions also. Hexagonal boron nitride (h-BN) retains its lubrication properties up to 5000°F (2760°C) in an inert or reducing environment and up to 1600°F (870°C) in an oxidizing atmosphere (56). Most importantly, h-BN nanomaterials have excellent biocompatibility, so their

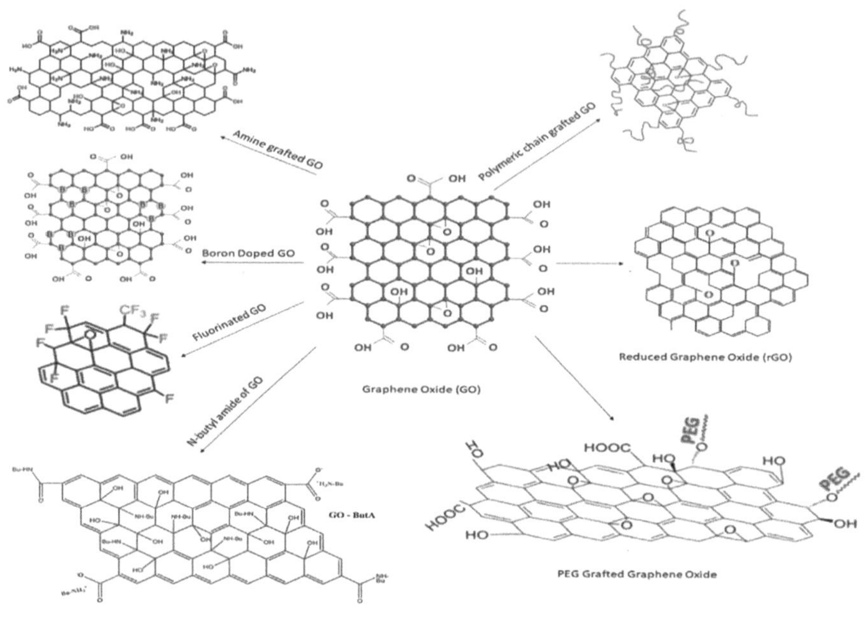

Figure 7.4 Functionalized graphene oxides are used as 2D lubricants.

use as a green lubricant will not damage the environment and hence widely supported in practical applications.

Hydroxylated boron nitride-grafted carboxylic silicone oil (h-BN-OH@CS) and amino boron nitride-grafted carboxylic silicone oil (h-BN-NH_2@CS) were reported by Yumeng et al. (57). Addition in different wt% of BN affects the properties of silicone grease as shown in Figure 7.5.

The offered lubricant-friendly properties are as follows:

- Outstanding mechanical strength
- Minimal friction
- An excessive thermal conductivity

When graphene material diffuses at the nanoscale into conventional lubricating fluid, these combined materials can form a protective film-like surface between moving surfaces, resulting in reduced friction and enhanced lubrication efficiency through the use of two-dimensional material-based nano lubricants. By capitalizing on the identity of two-dimensional materials, the focus is on exceeding universal performance and durability.

Profitability-enhancing measures include minimizing energy consumption, extending the life of machinery components, and increasing universal efficiency. The primary application of two-dimensional material-based nano lubricants is to develop dependable avenues for producing high-performance

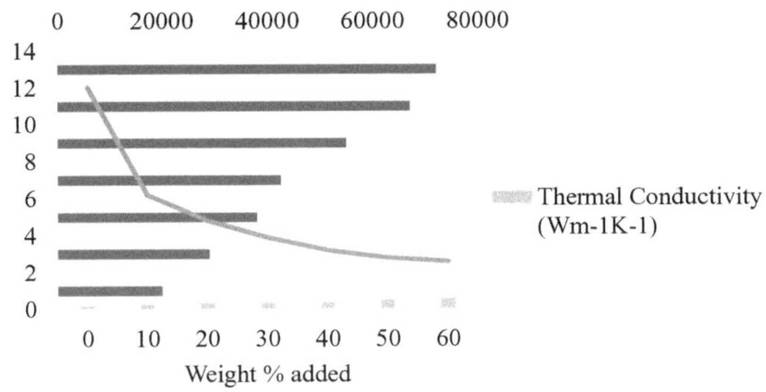

Figure 7.5 Properties of silicone grease on addition in different wt% of BN.

lubricants in various industrial sectors. In addition to its prospective lubricant additives, two-dimensional material-based lubricants have the following special applications:

- **Automotive industry:** Engine lubrication: internal combustion engines can benefit from the use of two-dimensional nano lubricants to reduce friction between moving elements, thereby increasing fuel economy and extending engine life. The implementation of these nano lubricants in transmission systems improves the efficiency of the gears and diminishes their wear, thereby facilitating a more seamless operation and extending their lifespan.
- **Machining and manufacturing:** The integration of two-dimensional substances into cutting fluids and lubricants employed in machining operations is feasible. This contributes to the reduction of tool wear, the enhancement of surface polish, and the optimization of metalworking operations as a whole.
- **Bearings and gears:** The utilization of nano lubricants in gearboxes and bearings, among other manufacturing equipment, can reduce friction and improve dependability.
- **Wind turbines:** The utilization of nano lubricants in the gear systems of wind turbines has the potential to enhance operational efficiency and reduce maintenance demands, especially in remote or arduous surroundings. Improving the lubrication of rotating components within power facilities has the potential to result in enhanced operational efficiency and diminished energy dissipation.
- **Micro- and nano-electromechanical systems (MEMS/NEMS):** Two-dimensional nano lubricants have the potential to significantly reduce friction and attrition in microelectromechanical systems, thereby contributing to the durability and dependability of such components.
- **Aircraft engines:** By minimizing the total weight of lubrication systems and increasing lubrication efficiency, nano lubricants can be

utilized in aircraft engines to decrease friction and enhance lubrication efficiency, thereby contributing to fuel economy.

- **Biomedical implants:** To decrease wear and friction in biomedical devices and implants, two-dimensional nano lubricants may be used. This could extend the life and enhance the functionality of these devices inside the human body.

7.6 CONCLUSIONS

Tribology is closely related to physics in which the size of a particle influences the lubrication property. Nanoparticles in various shapes and forms when successfully used with lubricating oils improved tribological performance. Two-dimensional material-based nano lubricants have a wide range of applications, and current research is investigating new avenues in several industries to improve productivity, cut down on energy usage, and prolong the life of mechanical components. It will be of environmental benefit also to use nano lubricants without S and P. Yet the mechanism is complex to understand. The use of nano lubricants is an innovative strategy in the tribological field and engine efficiency, which is of great importance for saving energy, diminishing harmful emissions, and increasing the life of engine components.

REFERENCES

1. Addou R, Wallace RM. Using photoelectron spectroscopy in the integration of 2D materials for advanced devices. *Journal of Electron Spectroscopy and Related Phenomena*. 2019.
2. Singh NB, Shukla SK. Properties of two-dimensional nanomaterials. *Two-Dimensional Nanostructures for Biomedical Technology: A Bridge between Material Science and Bioengineering*. 2019.
3. Zibouche N, Philipsen P, Kuc A. Strong variation of electronic properties of MoS 2 and WS 2 nanotubes in the presence of external electric fields. *Journal of Physical Chemistry C*. 2019.
4. Miao X, Li Z, Liu S, Wang J, Yang S. MXenes in tribology: Current status and perspectives. *Advanced Powder Materials*. 2023.
5. Jiao L, Tong J, Wu Y, Hu Y, Wu H, Li D, et al. Self-assembly of supraparticles on a lubricated-superamphiphobic patterned surface. *Applied Surface Science*. 2022.
6. Shah R, Gashi B, González-Poggini S, Colet-Lagrille M, Rosenkranz A. Recent trends in batteries and lubricants for electric vehicles. *Advances in Mechanical Engineering*. 2021.
7. Mateti S, Sultana I, Chen Y, Kota M, Rahman MM. Boron nitride-based nanomaterials: Synthesis and application in rechargeable batteries. *Batteries*. 2023.

8. Fan Z, Wang Y, Xie Z, Wang Z. Investigation on performance of battery thermal management system using spray cooling. *International Journal of Energy Research*. 2022.
9. Ghotekar S, Pansambal S, Lin KYA, Pore D, Oza R. Recent advances in synthesis of CeVO4 nanoparticles and their potential scaffold for photocatalytic applications. *Topics in Catalysis*. 2023.
10. Li D, Zhu J, Yu J, Liu J, Chen R, Liu Q, et al. Enhanced anticorrosion and antifouling properties of lubricant-infused pyramidal polydimethylsiloxane coating. *Macromolecular Materials and Engineering*. 2023.
11. Premalatha K, Keerthivasan T, Gayathiri A, Rubini V. Design of EV charging station using renewable energy sources. In *2nd International Conference on Advancements in Electrical, Electronics, Communication, Computing and Automation*. 2023.
12. Jason YJJ, How HG, Teoh YH, Chuah HG. A study on the tribological performance of nanolubricants. *Processes*. 2020.
13. Aljeaidi M, Keen C, Bell JS, Cooper T, Robson L, Tan ECK. Dry eyes, ocular lubricants, and use of systemic medications known or suspected to cause dry eyes in residents of aged care services. *International Journal of Environmental Research and Public Health*. 2020.
14. National nanotechnology Initiative. *Journal of the National Cancer Institute*. 2001.
15. Hasmin NA, Matmin J, Basir SA. National nanotechnology policies in Malaysia, European Union and United States: A comparative analysis. *Quantum Journal of Social Sciences and Humanities*. 2022.
16. Deshpande BD, Agrawal PS, Yenkie MKN. Nanoparticles aided AOP for degradation of p-nitro benzoic acid. In: *Materials Today: Proceedings*. 2019.
17. Deshpande BD, Agrawal PS, Yenkie MKN, Dhoble SJ. Prospective of nanotechnology in degradation of waste water: A new challenges. *Nano-Structures and Nano-Objects*. 2020.
18. Deshpande BD, Agrawal PS, Yenkie MKN, Dhoble SJ. Nanoparticle-aided AOP for treatment of benzoic acid. In: *Water Pollution Sources and Purification: Challenges and Scope*. 2022.
19. Harrison P, Valavanis A. Quantum wells, wires and dots. *Quantum Wells, Wires and Dots*. 2016.
20. Gong X, Zhang Y, Zhao W, Duan Y, Wu H, Zhang Z, et al. Remarkably enhanced thermal properties of tobacco granules with high-thermal-conductivity nanoparticles. *Heliyon*. 2023.
21. Coetzee D, Venkataraman M, Militky J, Petru M. Influence of nanoparticles on thermal and electrical conductivity of composites. *Polymers*. 2020.
22. Zhang X, Kang L, Zhu M. Density functional theory study of low-dimensional (2D, 1D, 0D) boron nitride nanomaterials catalyzing acetylene acetate reaction. *International Journal of Molecular Sciences*. 2022.
23. Lv Y, Min L, Niu F, Qin Z, Zhang M, Zhao B, et al. Conductive nanomaterials with different dimensions for flexible piezoresistive sensors: From selectivity to applications. *Advanced Materials Technologies*. 2023.
24. Galler KM, Weber M, Korkmaz Y, Widbiller M, Feuerer M. Inflammatory response mechanisms of the dentine–pulp complex and the periapical tissues. *International Journal of Molecular Sciences*. 2021.

25. Murshed SMS, Leong KC, Yang C. Enhanced thermal conductivity of TiO_2—water based nanofluids. *International Journal of Thermal*. 2005.
26. Gleiter H. Nanostructured materials: Basic concepts and microstructure. *Acta Materialia*. 2000.
27. Gleiter H. Nanostructured materials: Basic concepts, microstructure and properties. *ChemInform*. 2004.
28. Gleiter H. Nanostructured materials: Basic concepts, microstructure and properties. In: *Nano Science and Technology: Novel Structures and Phenomena*. 2003.
29. de Mello JDB, Binder C, Probst SMH, Klein AN. In situ generated turbostratic 2D graphite: A new way to obtain high-performance self-lubricating iron-based composites. In: *Self-Lubricating Composites: Second Edition*. 2022.
30. de Mello JDB, Binder C, Probst SMH, Klein AN. In situ generated turbostratic 2D graphite: A new way to obtain high-performance self-lubricating iron-based composites. In: *Self-Lubricating Composites*. 2018.
31. Arora S, Srivastava S. Nanofluids: Future industrial coolants. In: *AIP Conference Proceedings*. 2011.
32. Al-Sharafi A, Sahin AZ, Yilbas BS, Das SK, Choi SUSUSUS, Yu W, et al. Chapter 3. Preparation and estimation of nanofluid properties. *Journal of Heat and Mass Transfer*. 2016.
33. Reddy MK, Muralidharan K, Chandra RS, Varma PLK, Venugopal K, Vignesh RV, et al. Influence of nanophase particles on the physical, chemical and tribological characteristics of SAE15W40. *Journal of Bio- and Tribo-Corrosion*. 2023.
34. Tae-Keun H, Ho-Soon Y. Nanoparticle-dispersion-dependent thermal conductivity in nanofluids. *Journal of the Korean Physical Society*. 2005.
35. Das S, Ghosh MM. A review of the preparation, stability and thermal conductivity of nanofluids. In: *An Essential Guide to Thermal Conductivity*. 2021.
36. Yuan C, Duan B, Li L, Xie B, Huang M, Luo X. Thermal conductivity of polymer-based composites with magnetic aligned hexagonal boron Nitride Platelets. *ACS Applied Materials & Interfaces*. 2015.
37. Wei X, Yin Y, Qin B, Wang W, Ding J, Lu J. Preparation and enhanced thermal conductivity of molten salt nanofluids with nearly unaltered viscosity. *Renewable Energy*. 2020.
38. Mao M, Lou D, Wang D, Younes H, Hong H, Chen H, et al. Ti3C2Tx MXene nanofluids with enhanced thermal conductivity. *Chemical Thermodynamics and Thermal Analysis*. 2022.
39. Jeong J, Li C, Kwon Y, Lee J, Kim SH, Yun R. Particle shape effect on the viscosity and thermal conductivity of ZnO nanofluids. *International Journal of Refrigeration*. 2013.
40. Ahmadi MH, Mirlohi A, Alhuyi Nazari M, Ghasempour R. A review of thermal conductivity of various nanofluids. *Journal of Molecular Liquids*. 2018.
41. Nsofor E. Recent patents on nanofluids (Nanoparticles in Liquids) heat transfer. *Recent Patents on Mechanical Engineering*. 2012.
42. Hong TK, Yang HS. Nanoparticle-dispersion-dependent thermal conductivity in nanofluids. *Journal of the Korean Physical Society*. 2005.

43. Timofeeva EV, Routbort JL, Singh D. Particle shape effects on thermophysical properties of alumina nanofluids. *Journal of Applied Physics.* 2009.
44. Cabaleiro D, Colla L, Agresti F, Lugo L, Fedele L. Transport properties and heat transfer coefficients of ZnO/(ethylene glycol + water) nanofluids. *International Journal of Heat and Mass Transfer.* 2015.
45. Yarmand H, Gharehkhani S, Ahmadi G, Shirazi SFS, Baradaran S, Montazer E, et al. Graphene nanoplatelets-silver hybrid nanofluids for enhanced heat transfer. *Energy Conversion and Management.* 2015.
46. Garoosi F, Jahanshaloo L, Garoosi S. Numerical simulation of mixed convection of the nanofluid in heat exchangers using a Buongiorno model. *Powder Technology.* 2015.
47. Solangi KH, Kazi SN, Luhur MR, Badarudin A, Amiri A, Sadri R, et al. A comprehensive review of thermo-physical properties and convective heat transfer to nanofluids. *Energy.* 2015.
48. Wang XJ, Li XF, Xu YH, Zhu DS. Thermal energy storage characteristics of Cu-H_2O nanofluids. *Energy.* 2014.
49. Mohammadian SK, Zhang Y. Analysis of nanofluid effects on thermoelectric cooling by micro-pin-fin heat exchangers. *Applied Thermal Engineering.* 2014.
50. Hasan MI, Rageb AMAR, Yaghoubi M. Investigation of a counter flow microchannel heat exchanger performance with using nanofluid as a coolant. *Journal of Electronics Cooling and Thermal Control.* 2012.
51. Mehta A, Patel N, Tantia D, Jha N. Mini heat exchanger using Al_2O_3-water based nano fluid. *International Journal of Mechanical Engineering & Technology.* 2013.
52. Fang H, Li Y, Li H, Zhang S, Ding Q, Yu W, et al. Tribological behavior at electronic scales for an oil-soluble ionic liquid with extremely low effective addition and highly load bearing capacity. *Chemical Engineering Journal.* 2023.
53. Gupta B, Kumar N, Panda K, Kanan V, Joshi S, Visoly-Fisher I. Role of oxygen functional groups in reduced graphene oxide for lubrication. *Scientific Report.* 2017.
54. Zubir NA, Yacou C, Motuzas J, Zhang X, Diniz Da Costa JC. Structural and functional investigation of graphene oxide-Fe_3O_4 nanocomposites for the heterogeneous Fenton-like reaction. *Scientific Report.* 2014.
55. Min C, He Z, Song H, Liang H, Liu D, Dong C, et al. Fluorinated graphene oxide nanosheet: A highly efficient water-based lubricated additive. *Tribology International.* 2019.
56. Abdollah MF Bin, Amiruddin H, Jamallulil AD. Experimental analysis of tribological performance of palm oil blended with hexagonal boron nitride nanoparticles as an environment-friendly lubricant. *International Journal of Advanced Manufacturing Technology.* 2020.
57. Shi Y, Hamsen C, Jia X, Kim KK, Reina A, Hofmann M, et al. Synthesis of few-layer hexagonal boron nitride thin film by chemical vapor deposition. *Nano Letters.* 2010.

Chapter 8

Role of lubrication in electrical systems

Challenges and emerging solutions

Bipasa Bimalendu Patra, Shivaji Suryabhan Aher, Ashwani Kumar, Avinash Kumar, Abhishek Kumar, Sivasakthivel Thangavel, and Anamika Gupta

8.1 INTRODUCTION

Lubricants are vital to relatively major aspects of electrical engineering, the reason being that they guarantee the reliability, robustness, and efficiency in the operation of types of machinery that are mechanical such as generators, transformers, motors, and switches. In the domain of electrical engineering, where mechanical constituents work in unison in the generation, transmission, and distribution of power in electrical form, lubricants serve as vital components. Ranging from optimal and sustaining performance, reducing friction and wear, and extending the lifespan of equipment entirely are subjected to proper lubrication, whether it is a part of the mechanisms that are complex of electrical switches or the gear being revolving of power plants. The essential role that lubricants convey in electrical engineering is meticulously examined in this chapter, including its usage, challenges, and prospective future trends [1–3]. This chapter provides a detailed analysis of lubricants relative to the applications in electrical engineering, challenges faced, and future possibilities of innovations. Methodologies are explained in relation to exploring the part that sufficient lubrication plays in maximizing the performance of the equipment, preventing corrosion and wear, reducing losses due to friction, and enhancing energy efficiency. It also scrutinizes the various necessities and variables that cause an impact on the lubricant choice in several electrical applications, highlighting noteworthy complications in developments in emerging lubricant technology. Finally, it considers future guidelines and roadmap for research and development with the target of developing lubricating systems that are tailored to the varying needs of electrical engineering [4–5].

Lubricants have a complex but essential part to play in the effective operation mechanism of mechanical constituents within the complex machinery of electrical domain of engineering. They are varied and range in complexity from switches that are complex and transformer complexities to robust

 DOI: 10.1201/9781003535447-8

generators and motors that develop the foundation of the generation of power, transmission, and distribution systems applied in an extensive variety of industrial and many more applications. The serving base for these essential parts' supreme performance, sturdiness, and trustworthiness is the management of the correct lubricants and the vigilant selection. Inspect the communications that are complex among electric motor parts, a vital constituent in numerous industrial and commercial backgrounds. In the case of bearings, gears, and shafts to guarantee smooth and robust performance, important components of the mechanical structure of the motor must be adequately lubricated. The lubricants assist a number of purposes when used accurately on these fragments. By serving as an obstruction and minimizing the friction between dynamic parts, they assist in extending component lifespan, lowering wear and tear, and sustaining operational efficiency over time. Additionally, while operational the heat produced is dissipated by lubricants, resisting high temperatures that may lead to degradation thermally and premature failure [6–8]. Lubricants are vital constituents when it comes to the dependability of motors and safeguard their integrity and even in terrible climatic conditions by preventing rust and corrosion, thus ensuring operation for a long term. Insulating oils serve as coolants on the inner side of the core of the transformer, where sturdy fields that are magnetic lead the currents to flow and hence to cope with the rise in temperatures. This permits dispersal of the heat and maintenance of ideal working temperatures. These oils also serve the purpose of electrical insulation, which prevents the electrical currents from breaking down and ensures that the effective transfer of power is safe. Lubricants are necessarily silent and also vital allies in the complicated electrical domain apparatus that permit operations that are smooth with the long-term durability of vital components. In the ever-changing world of electrical power systems, their selection is to be careful, including installation, and upkeep are keystones in the mission for improved apparatus performance, dependability, and sustainability [9, 10].

8.2 ROLE OF LUBRICATION IN ELECTRICAL ENGINEERING

Lubricants play a vital role in the field of applications in electrical engineering. These include the following:

- **Reduction in friction:** Through reduction in frictional losses in rotating machinery, including motors and generators, lubricants thereby increase the energy efficiency and also minimize the operating costs. One utmost vital tool for minimizing friction in systems that are mechanical is none other than lubrication. A thin film is formed by the lubricants which isolate surfaces that come into contact with each other when being in use, including the gears, shafts, and also the bearings. By acting as a cushion, this layer minimizes friction by preventing direct contact between the parts that are moving. This results in the mechanical function to be much more efficient and causes minimal resistance and hence uses a relatively small amount of energy to overcome forces due to friction. For the components if the wear and tear on them is reduced, this friction minimization enhances the efficacy and also an extended lifespan. Keeping all these considerations, lubrication is crucial for efficiency enhancement and improvement, improvement in energy economy, and thus mechanical systems are maintained operational through frictional reduction [11].

- **Wear prevention:** Lubricants are used to create a protective layer on surfaces like bearings, gears, and shafts that come into touch with one another during operation. By acting as a barrier, this layer lessens abrasion and deterioration and the direct contact between the surfaces. As a result, over time, the components endure reduced wear and tear, extending the equipment's lifespan and dependability. Lubricants can also help spread impurities and lessen corrosion's effects, which further protect against premature wear. Essentially, by preventing excessive wear and ensuring optimal performance, lubrication plays a critical role in maintaining the integrity and functionality of mechanical systems [12].
- **Enhancement of reliability:** Lubricants create a protective film on parts like shafts, gears, and bearings that helps to lessen wear and friction during operation. By acting as a barrier and avoiding direct metal-to-metal contact, this protective layer reduces the possibility of failure or damage. Lubricants can also aid in the prevention of contamination and corrosion, guaranteeing that the components stay in ideal condition throughout time. The equipment functions more dependably as a result, requiring less maintenance and downtime. In summary, by reducing wear, preventing corrosion, and preserving optimal performance, lubrication significantly contributes to the increased reliability of mechanical systems [13].
- **Heat dissipation facilitation:** Lubricants aid in distributing and dissipating heat produced during operation by forming a thin protective layer when applied to parts like shafts, gears, and bearings. Heat can be effectively transferred away from important components, thanks to the thermal barrier that this layer provides. Lubricants aid in preventing excessive heat buildup, which can cause thermal degradation and early equipment failure by lowering friction and metal-to-metal contact. Furthermore, by facilitating the transfer of heat away from the componentry, lubricants can aid in raising the effectiveness of cooling systems. In conclusion, lubrication helps mechanical systems handle

heat effectively, resulting in maximum performance and dependability [14].

8.3 LUBRICANTS IN DIFFERENT ELECTRICAL SYSTEMS

The particular requirements and operational environments of electrical equipment influence the choice of lubricants. It is essential to use high-performance lubricants with strong load-bearing capability, outstanding oxidation resistance, and thermal stability for electric motors and generators [9]. These characteristics are necessary for the proper operation of the machinery, which involves high temperatures and mechanical strains. Insulating oils serve as coolants and electrical insulators in transformers. These lubricants have high conductivity and low friction coefficients, which are essential for reducing contact resistance, averting arcing, and extending the component's operating life. Different kinds of electric motors and generators in industrial environments highlight the importance of lubrication to guarantee their efficient operation [15–16]. Transformers next to insulating oils used in electrical transformers highlight the function of lubricants in heat dissipation and electrical insulation (Figures 8.1–8.3).

Gears, bearings, and other mechanical parts of electrical equipment need the right lubricant to reduce wear and friction. Various types of lubricants, such as mineral oils, synthetic lubricants, and bio-based mixtures [17], showcase their various chemical makeup and characteristics. Table 8.1 summarizes the different types of lubricants in electrical engineering

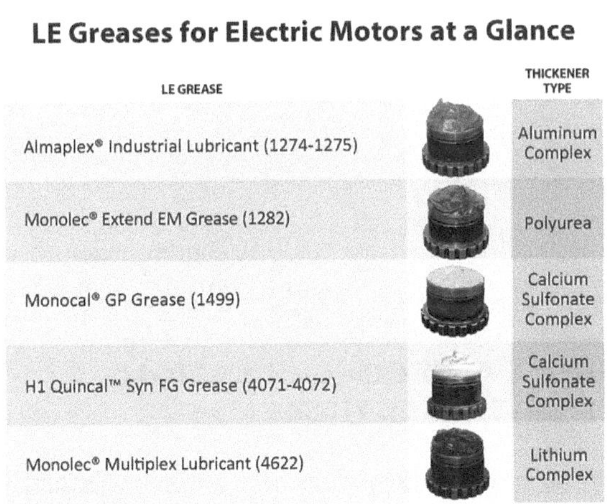

Figure 8.1 Lubrication greases.

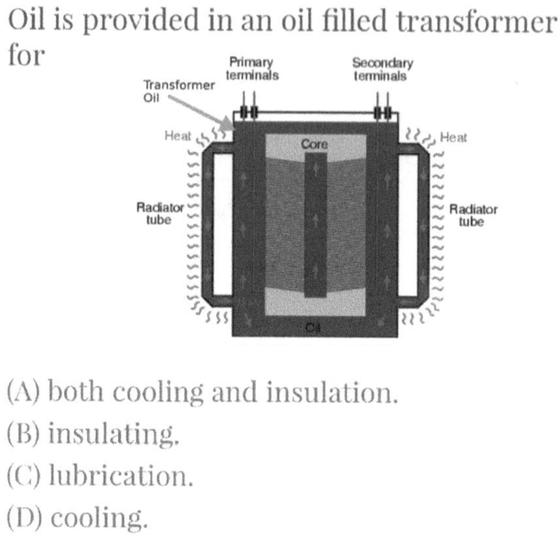

Figure 8.2 Lubricating oil in transformers.

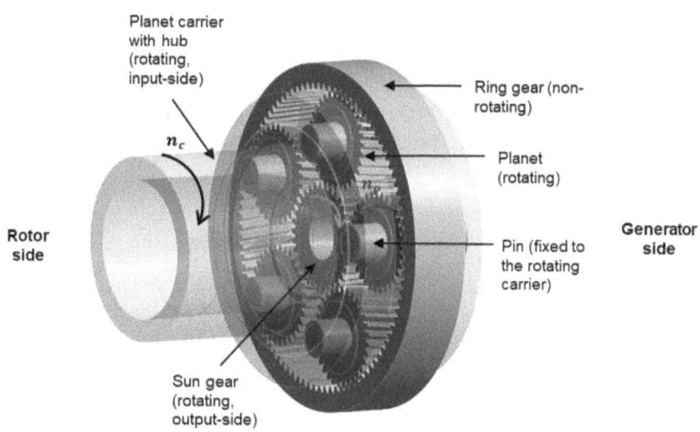

Figure 8.3 Lubrication in gears.

ranging from mineral oils, synthetic lubricants, silicone lubricants, bio-based lubricants, greases, nanolubricants, and dry film lubricants.

8.4 CHALLENGES AND CONSIDERATIONS IN LUBRICANT SELECTION

Although lubrication has many benefits, choosing and maintaining the right lubricants for electrical engineering applications present a number of difficulties.

Table 8.1 The different types of lubricants in electrical engineering [32, 35, 39]

Type of lubricant	*Key points*
Mineral oils	Traditional lubricants derived from crude oil, offering good lubricating properties and compatibility with various materials.
Synthetic lubricants	Engineered lubricants with tailored chemical compositions and performance characteristics, offering enhanced thermal stability, oxidation resistance, and longevity.
Bio-based lubricants	Environmentally friendly lubricants derived from renewable resources such as plant oils or animal fats, offering biodegradability and reduced environmental impact.
Silicone lubricants	Lubricants based on silicone oil, known for their high thermal stability, wide temperature range, and compatibility with rubber and plastics.
Greases	Semi-solid lubricants consisting of a base oil thickened with a soap or other agent, suitable for applications requiring adhesion and resistance to water washout [2, 3].
Nanolubricants	Lubricants containing nanoparticles or other nanostructures, offering improved lubricating properties such as reduced friction and wear.
Dry film lubricants	Lubricants in the form of solid films or coatings applied to surfaces, providing lubrication without the need for liquid lubricants.

- **Compatibility:** To prevent deterioration and adverse reactions, lubricants must be compatible with the materials used in electrical equipment, including metals, polymers, and elastomers. There are a number of difficulties and factors to take into account while choosing lubricants, especially compatibility. A careful selection of lubricants is necessary to guarantee their compatibility with the metals, polymers, and elastomers utilized in the equipment. Neglecting to choose an appropriate lubricant may cause the materials to deteriorate, which could have negative effects and jeopardize the equipment's longevity and functionality. Thus, before choosing a lubricant, it is essential to carefully evaluate how well it works with the particular components that are in the system. Engineers may choose the right lubricant by taking these issues and factors into account, which will ultimately improve the equipment's dependability and efficiency [18].
- **Contamination management:** Dust, particulates, and moisture are examples of contaminants that can reduce the effectiveness of lubricants and jeopardize the dependability of equipment. Therefore, it becomes essential to implement efficient contamination control procedures. Among the many difficulties and factors to be taken into account when choosing a lubricant is controlling contamination. Lubricants must always be kept free from impurities such as moisture, dust, and particles because these things might reduce their efficiency and endanger the dependability of equipment. Implementing

precautions to keep pollutants out of the lubricating system and eliminating those that do is necessary for effective contamination management. This calls for using appropriate filtration systems, putting in place safe handling and storage procedures, and carrying out routine maintenance and inspections. Engineers can reduce the risks of contamination and preserve the integrity and functionality of the lubrication system and the equipment it supports by taking these issues and concerns into account [19].

- **Environmental compliance:** Environmental impact must be reduced by the use of eco-friendly lubricants with non-toxic and biodegradable compositions, since concerns about sustainability and the growth of environmental regulations have made this necessary. Environmental compliance presents several difficulties and factors to be taken into account while choosing lubricants. Selecting lubricants that adhere to strict environmental standards is essential in light of the increased emphasis on sustainability and the proliferation of environmental legislation. To reduce their environmental impact, this entails choosing lubricants with non-toxic compositions and biodegradable formulations. Through the resolution of these issues and concerns, engineers can choose lubricants in a way that minimize their influence on the environment and promote sustainable practices [20–27].

8.5 CURRENT RESEARCH AND DEVELOPMENT IN LUBRICANTS FOR ELECTRICAL SYSTEMS

The topic of lubricants in electrical engineering has seen recent developments and continuous research that have proved crucial in pushing the frontiers of innovation, solving problems, and improving the longevity and performance of mechanical systems. Engineers and researchers are always looking for new ways to improve lubrication techniques, as the need for electrical equipment that is more dependable and efficient grows.

Graphene, carbon nanotubes, and metal oxides are examples of nanoparticles that can be added to lubricant formulations to lower wear, friction, and energy usage in electrical gear. Predictive maintenance approaches and condition monitoring are the key fields of research. Real-time monitoring of lubricant condition and equipment health is made possible by advancements in sensor technology, data analytics, and machine learning algorithms. This enables proactive maintenance techniques to avoid equipment failures and maximize lubrication practices. Industries can decrease downtime, lower maintenance costs, and increase the longevity of electrical equipment by putting predictive maintenance systems into place [28–36].

In addition, scientists are exploring innovative lubrication approaches and procedures to enhance lubricant dispersal and application in electrical systems. This involves creating precise lubrication systems that provide essential components with the proper amount of lubricant, guaranteeing dependability and peak performance. In order to reduce environmental effects and adhere to regulatory standards, research is also being done on the use of ecologically friendly lubricants with low toxicity and biodegradability. In conclusion, the future of mechanical systems is being influenced by the ongoing study and present advancements in the lubricating sector of electrical engineering. Researchers and engineers are developing advanced lubrication formulae, implementing predictive maintenance schedules, and exploring novel lubricating techniques in an effort to improve the efficiency, sustainability, and dependability of electrical equipment. As technology advances, lubricants will become more crucial than ever to extending the life and efficiency of electrical equipment [37–40].

8.6 EMERGING TRENDS AND FUTURE DIRECTIONS

The lubricant industry in electrical engineering is always changing to meet the needs of newer technology and environmental objectives. Notable new trends and directions for the future are as follows:

- **Advancements in the field of synthetic and bio-based lubricants:** Research efforts are focused on creating synthetic and bio-based lubricants with enhanced performance characteristics, such as increased thermal stability and reduced environmental impact. There are some noteworthy new trends and directions for developments in the field of synthetic and bio-based lubricants. Scientists and business experts are putting more and more effort into creating lubricants made of synthetic chemicals and renewable resources that have improved performance characteristics. It is anticipated that research in this area will carry on in the future, with an emphasis on improving the manufacturing and formulation procedures for synthetic and bio-based lubricants. This entails maximizing the material's chemical makeup, boosting compatibility with current materials, and strengthening overall performance attributes [41–45]. Furthermore, it is projected that the demand for more environmentally friendly and sustainable solutions will lead to an expansion in the integration of synthetic and bio-based lubricants into a wider range of industrial applications. In general, the continuous progress in bio-based and synthetic lubricants shows potential to tackle issues with environmental compliance, sustainability, and performance in lubrication technology. The lubricants industry may continue to develop and change, satisfying the changing needs of many sectors and fostering environmental stewardship, by embracing these new trends and future directions [46–49].
- **Integration with nanotechnology:** In electrical engineering applications, the incorporation of nanomaterials, such as nanoparticles,

nanolubricants, and nanocomposites, has the potential to improve wear resistance, reduce friction, and strengthen lubricant efficacy. Future directions and promising new developments in several sectors, including lubrication technology, are presented by the integration of nanotechnology. The application of nanomaterials, such as nanoparticles, nanolubricants, and nanocomposites, is being investigated by academics and business leaders as a way to improve lubricant performance and solve problems in mechanical systems [20]. Moreover, lubricant additives or base oils can have their qualities customized for certain uses, thanks to nanocomposites, which include nanoparticles in their structure. Researchers can create lubricants with greater resistance to oxidation and corrosion, improved load-carrying capacity, and enhanced thermal stability by carefully regulating the size, shape, and composition of nanoparticles [9, 10]. Future developments in lubricating technology are anticipated to result from advances in nanotechnology. Scientists are investigating new ways to create lubricant compositions with enhanced performance characteristics and functionalize nanoparticles [11]. Furthermore, real-time monitoring of lubricant state and equipment health may be made possible by the development of nanoscale sensors and monitoring systems, which might result in more preventive maintenance practices and increased dependability [50–58].

- **Progress in condition monitoring and predictive maintenance:** Developments in sensor technology and predictive analytics make it possible to monitor lubricant state in real time, which opens the door to proactive maintenance plans that maximize lubrication routines and prevent equipment breakdowns. Additionally, the analysis of sensor data to find patterns and trends suggestive of possible problems or failure modes is made possible by developments in machine learning algorithms and data analytics. In addition, the goal of integrating predictive maintenance solutions is to reduce downtime and improve maintenance schedules. Predictive maintenance algorithms may more accurately predict equipment failures by utilizing data from condition monitoring systems. This enables preemptive interventions to save expensive breakdowns and unscheduled downtime. Future developments in sensor technology, data analytics, and artificial intelligence are anticipated to propel further improvement in condition monitoring and predictive maintenance. Scholars are now investigating unique methodologies to incorporate condition monitoring systems into pre-existing equipment and infrastructure, as well as new ways to improve the precision and dependability of predictive maintenance models. In general, advancements in predictive maintenance and condition monitoring have enormous potential to maximize asset performance, save maintenance costs, and increase equipment reliability. Industries may increase operational efficiency and production while lowering the risk of equipment breakdowns and downtime by adopting these new trends and future directions [59–60].

8.7 DISCUSSION

In electrical engineering settings, lubricants are essential for ensuring the reliable and seamless operation of mechanical components. Understanding the many requirements and challenges associated with lubrication enables engineers and researchers to develop creative solutions that maximize equipment performance, reduce energy consumption, and mitigate environmental impact. Continued R&D efforts are essential for propelling lubricant technology forward and guaranteeing that it will always adapt to the changing needs of electrical engineering. The field's unwavering dedication to innovation and continuous progress is highlighted by its unrelenting pursuit of sustainability and excellence. In the field of electrical engineering, lubricants are essential components that perform vital tasks that guarantee the dependable and seamless functioning of mechanical parts. During this investigation, we have examined the many functions of lubricants in several areas of electrical engineering, including wear prevention, heat dissipation facilitation, and reliability improvement [61–64]. When selecting lubricants for electrical engineering applications, researchers and engineers need to take into account things like environmental requirements, contaminant management, and material compatibility. To tackle these obstacles, meticulous attention to detail, a comprehensive evaluation of operational circumstances, and adherence to sustainable principles are necessary. Future developments and innovations in the field of lubricants in biomedical engineering appear promising. The field of lubrication technology is being shaped by new developments including the creation of synthetic and bio-based lubricants, the incorporation of nanotechnology, and advancements in condition monitoring and predictive maintenance. These developments address environmental impact and sustainability issues in addition to enhancing equipment performance and dependability in biomedical tribology. Lubricants will continue to evolve as long as research and development activities are maintained, ensuring that they fulfill the changing demands of electrical engineering's goal of excellence and sustainability.

REFERENCES

1. Yang, H., Liu, K., Wang, Y., Liu, H., Man, J., and Sun, B. (2017). Electronic controlling on nanotribological properties of a textured surface by laser processing. *J. Spectrosc.* 7920652. 1–7 . doi: 10.1155/2017/7920652.

2. Kumar, A., Kumar, A., and Kumar, A., eds. (2023). *Laser based technologies for sustainable manufacturing.* Boca Raton, FL: CRC Press. doi: 10.1201/9781003402398.
3. Kumar, A., Pathak, A., Kumar, A., and Kumar, A. (2023). Physics of laser-matter interaction in laser-based manufacturing. In *Laser-based technologies for sustainable manufacturing,* A. Kumar, A. Kumar, and A. Kumar (eds.), pp. 45–54. Boca Raton, FL: CRC Press. doi: 10.1201/9781003402398-2.
4. Pathak, A., Kumar, A., Kumar, A., Kumar, A., and King M., F. L. (2023). Application of laser technology in the mechanical and machine manufacturing industry. In *Laser-based technologies for sustainable manufacturing,* A. Kumar, A. Kumar, and A. Kumar (eds.), pp. 107–155. Boca Raton, FL: CRC Press. doi: 10.1201/9781003402398-6.
5. Kumar, A., Byadwal, M., Kumar, A., Kumar, A., and King M., F. L. (2023). Laser micromachining in biomedical industry. In *Laser-based technologies for sustainable manufacturing,* A. Kumar, A. Kumar, and A. Kumar (eds.), pp. 169–206. Boca Raton, FL: CRC Press. doi: 10.1201/9781003402398-8.
6. Callen, H. B. (1998). *Thermodynamics and an introduction to thermostatistics.* Hoboken, NJ: American Association of Physics Teachers.
7. Cann, P. (2007). Grease lubrication of rolling element bearings—role of the grease thickener. *Lubr. Sci.* 19, 183–196. doi: 10.1002/ls.39.
8. Casserly, E., Langlais, T., Springer, S. P., Kumar, A., and Mallory, B. J. L. M. (2018). The effect of base oils on thickening and physical properties of lubricating greases. *Euro. Lubric. Ind. Magazine* 144, 32–37.
9. Chen, Y., and Liang, H. (2019). Tribological evaluation of electrical resistance of lubricated contacts. *J. Tribol.* 142, 114502. doi: 10.1115/1.4045578.
10. Chen, Y., Renner, P., and Liang, H. (2019a). Dispersion of nanoparticles in lubricating oil: A critical review. *Lubricants* 7, 7. doi: 10.3390/lubricants7010007.
11. Dai, W., Kheireddin, B., Gao, H., and Liang, H. (2016). Roles of nanoparticles in oil lubrication. *Tribol. Int.* 102, 88–98. doi: 10.1016/j.triboint.2016.05.020.
12. Dassenoy, F. (2019). Nanoparticles as additives for the development of high performance and environmentally friendly engine lubricants. *Tribol. Online* 14, 237–253. doi: 10.2474/trol.14.237.
13. Fan, X., and Wang, L. (2014). Tribological properties of conductive lubricating greases. *Friction* 2, 343–353. doi: 10.1007/s40544-014-0062-2.
14. King M., F. L., Kumar, A., Kumar, A., and Kumar, A. (2023). Introduction to optics and laser-based manufacturing technologies. In *Laser-based technologies for sustainable manufacturing,* pp. 1–43. Boca Raton, FL: CRC Press. doi: 10.1201/9781003402398-1.
15. Hunt, G. (2017). New perspectives on the temperature dependence of lubricant additives on copper corrosion. *SAE Int. J. Fueis Lubr.* 10, 521–527. doi: 10.4271/2017-01-0891.
16. Hunt, G. J., Gahagan, M. P., and Peplow, M. A. (2017). Wire resistance method for measuring the corrosion of copper by lubricating fluids. *Lubric. Sci.* 29, 279–290. doi: 10.1002/ls.1368.
17. Jiang, Z., Fang, J., Chen, Y., Wu, J., Liu, P., Gu, K., et al. (2018). A review of lubricating mechanisms from the perspective of tribo-electrophysics and tribo-electrochemistry. *Sci. China Technol. Sci.* 61, 1–11. doi: 10.1007/s11431-017-9291-3.

18. Jin, H., Andritsch, T., Tsekmes, I. A., Kochetov, R., Morshuis, P. H., and Smit, J. J. (2014). Properties of mineral oil based silica nanofluids. *IEEE Trans. Dielectr. Electr. Insul.* 21, 1100–1108.
19. Kolodziejczyk, L., Rojas, T., Martinez-Martinez, D., Fernandez, A., and Sanchez Lopez, J. (2007). Surface-modified Pd nanoparticles as a superior additive for lubrication. *J. Nanoparticle Res.* 9, 639–645.
20. Korcek, S., Sorab, J., and Johnson, J. R. K. (2000). Automotive lubricants for the next millennium. *Ind. Lubr. Tribol.* 52, 209–220.
21. Kumar, A., Datta, S., and Kalyanasundaram, D. (2018). Reduction of hydraulic friction in confined flows by laser texturing: Experiments and theoretical validation. In *ASME 2018 16th International Conference on Nanochannels, Microchannels, and Minichannels*, ICNMM 2018.
22. Kumar, A., Keerti, S., Jain, J., Sinha, S., Tekumalla, S., and Gupta, M. (2018). Investigations of wear response of pure Mg and Mg-0.4 Ce-Y2O3/ZnO nanocomposites using a single and repeated scratch tests. *Tribol. Trans.* 61, 951–959.
23. Kwak, Y., Cleveland, C., Adhvaryu, A., Fang, X., Hurley, S., and Adachi, T. (2019). Understanding base oils and lubricants for electric drive train applications. Richmond, VA: SAE Technical Paper. Report No.: 0148-7191. doi: 10.4271/2019-01-2337.
24. Naddaf, A., and Heris, S. Z. (2018). Experimental study on thermal conductivity and electrical conductivity of diesel oil-based nanofluids of graphene nanoplatelets and carbon nanotubes. *Int. Commun. Heat Mass Transfer* 95, 116–122. doi: 10.1016/j.icheatmasstransfer.2018.05.004.
25. Nakanishi, Y., Honda, T., Kasamura, K., Nakashima, Y., Nakano, K., Kondo, K., et al. (2016). Bio-inspired shaft seal in coolant pump for electric vehicles. *2016 IEEE International Conference on Renewable Energy Research and Applications (ICRERA)*, Birmingham, UK, 2016, 112–117. doi: 10.1109/ICRERA.2016.7884449.
26. Narita, K., and Takekawa, D. (2019). Lubricants technology applied to transmissions in hybrid electric vehicles and electric vehicles. Chiba: SAE Technical Paper Report No.: 0148-7191.
27. Passut, C. A. (2013). Engine lubricants. In *Encyclopedia of tribology*, Q. J. Wang and Y.-W. Chung (eds.), pp. 977–981. Boston, MA: Springer.
28. Kumar, A., and Panda, U. (2022) Microfluidics-based devices and their role on point-of-care testing. In *Biosensor Based Advanced Cancer Diagnostics*, Raju Khan, Arpana Parihar, and Sunil K. Sanghi (eds.), pp. 197–224. Elsevier. https://doi.org/10.1016/B978-0-12-823424-2.00011-9.
29. Rawat, S. S., and Harsha, A. P. (2019). Current and future trends in Grease Lubrication. In *Automotive Tribology. Energy, Environment, and Sustainability*, Katiyar, J., Bhattacharya, S., Patel, V., Kumar, V. (eds.). Springer, Singapore. https://doi.org/10.1007/978-981-15-0434-1_9.
30. Rawat, S. S., and Harsha, A. (2019). Current and future trends in grease lubrication. In *Automotive tribology*, J. Katiyar, S. Bhattacharya, V. Patel, and V. Kumar (eds.), pp. 147–182. Singapore: Springer.
31. Reed, J. K., Ludois, D. C., Mason, W. P., and Wolf, D. (2017). Rotary capacitor for shunting high frequency bearing currents and reducing EMI in electric machinery. U.S. Patent No. 9,653,970B2. Madison, WI: C-Motive Technologies Inc.

32. Rensselar, J. (2010). Biobased lubricants: Gearing up for a green world. *Tribol. Lubr. Technol.* 66, 32–34.
33. Romanenko, A., Muetze, A., and Ahola, J. (2016). Effects of electrostatic discharges on bearing grease dielectric strength and composition. *IEEE Trans. Ind. Appl.* 52, 4835–4842. doi: 10.1109/TIA.2016.2596239.
34. Parihar, A., Kumar, A., Panda, U., Khan, R., Parihar, D. S., and Khan, R. (2023). Cryopreservation: A comprehensive overview, challenges, and future perspectives. *Adv. Biol.* 7, 2200285.
35. Stockton, T. R. (1983). *Transmission lubrication and motor cooling system.* Washington, DC: U.S. Patent and Trademark Office.
36. Syahir, A. Z., Zulkifli, N. W. M., Masjuki, H. H., Kalam, M. A., Alabdulkarem, A., Gulzar, M., et al. (2017). A review on bio-based lubricants and their applications. *J. Clean. Prod.* 168, 997–1016. doi: 10.1016/j.jclepro.2017.09.106.
37. Vladescu, S.-C., Marx, N., Fernández, L., Barceló, F., and Spikes, H. (2018). Hydrodynamic friction of viscosity-modified oils in a journal bearing machine. *Tribol. Lett.* 66, 127. doi: 10.1007/s11249-018-1080-4.
38. Walther, H. C., and Holub, R. A. (2014). Lubrication of electric motors as defined by IEEE standard 841-2009, shortcomings and potential improvement opportunities. In *2014 IEEE Petroleum and Chemical Industry Technical Conference (PCIC)*, pp. 91–98. San Francisco, CA. doi: 10.1109/PCICon.2014.6961872.
39. Wang, Y., Jones, S., Dai, A., and Liu, G. (2014). Reliability enhancement by integrated liquid cooling in power IGBT modules for hybrid and electric vehicles. *Microelectron. Reliabil.* 54, 1911–1915. doi: 10.1016/j.microrel.2014.07.037.
40. Parihar, A., Pandita, V., Kumar, A, Parihar, D. S., Puranik, N., Bajpai, T., and Khan, R. (2021). 3D printing: Advancement in biogenerative engineering to combat shortage of organs and bioapplicable materials. *Regen. Eng. Transl. Med.* 8, 1–27.
41. Parmar, V., Kumar, A., Mani Sankar, M., Datta, S., Vijaya Prakash, G., Mohanty, S., and Kalyanasundaram, D. (2018). Oxidation facilitated antimicrobial ability of laser micro-textured titanium alloy against gram-positive Staphylococcus aureus for biomedical applications. *J. Laser. Appl.* 30, 032001.
42. Parmar, V., Kumar, A., Prakash, G. V., Datta, S., and Kalyanasundaram, D. (2019). Investigation, modelling and validation of material separation mechanism during fiber laser machining of medical grade titanium alloy Ti6Al4V and stainless steel SS316L. *Mech. Mater.* 137, 103125.
43. Shrivastava Prateek, S. R., Kumar, A., others. (2020). Investigation of torsional rod to minimize vibration in automobile using ANSYS. In *2020 1st International Conference on Innovative Research in Applied Science, Engineering and Technology (IRASET)*, pp. 1–6.
44. Whitby, R. D. (2018). Oils for hybrid electric vehicles. *Tribol. Lubr. Technol.* 74, 88.
45. Willwerth, A., and Roman, M. (2013). Electrical bearing damage—a lurking problem in inverter-driven traction motors. In *2013 IEEE Transportation Electrification Conference and Expo (ITEC)*, Michigan, IN. doi: 10.1109/ITEC.2013.6573484.

46. Xie, G., Cui, Z., Si, L., and Guo, D. (2017). Destabilization of lubrication oil micropool under charged conditions. *Ind. Lubr. Tribol.* 69, 59–64. doi: 10.1108/ILT-11-2015-0184.
47. Singh, M., Kumar, A., and Khan, A. R. (2020) Capillary as a liquid diode. *Phys. Rev. Fluids.* 5, 102101.
48. Singh, R. K., Kumar, A., Kant, R., Gupta, A., Suresh, E., and Bhattacharya, S. (2014). Design and fabrication of 3-dimensional helical structures in polydimethylsiloxane for flow control applications. *Microsyst. Technol.* 20, 101–111.
49. Subramanian, Y., Gajendiran, J., Veena, R., Azad, A. K., Sabarish, V. C. B., Muhammed Ali, S. A., Kumar, A., and Gubendiran, R. K. (2023). Structural, photoabsorption and photocatalytic characteristics of BiFeO3-WO3 nanocomposites: An attempt to validate the experimental data through SVM-based artificial intelligence (AI). *J. Electron. Mater.* 52, 1–11.
50. Dias, D., Bons, J., Kumar, A., Kabir, M. H., and Liang, H. (2024). Forever chemicals, per- and polyfluoroalkyl substances (PFAS), in lubrication. *Lubricants* 12 (4), 114.
51. Yang, Y., Ali, K. A., Roeleveld, J., and Emadi, A. (2016). State-of-the-art electrified powertrains—hybrid, plug-in, and electric vehicles. *Int. J. Powertrains* 5, 1–29. doi: 10.1504/IJPT.2016.075181.
52. Yang, Y., Bilgin, B., Kasprzak, M., Nalakath, S., Sadek, H., Preindl, M., et al. (2016). Thermal management of electric machines. *IET Electr. Syst. Transp.* 7, 104–116. doi: 10.1049/iet-est.2015.0050.
53. Yu, Z.-Q., and Yang, Z.-G. (2011). Fatigue failure analysis of a grease-lubricated roller bearing from an electric motor. *J. Fail. Anal. Prev.* 11, 158–166. doi: 10.1007/s11668-010-9422-z.
54. Zin, V., Barison, S., Agresti, F., Colla, L., Pagura, C., and Fabrizio, M. (2016). Improved tribological and thermal properties of lubricants by graphene based nano-additives. *RSC Adv.* 6, 59477–59486. doi: 10.1039/C6RA12029F.
55. Kumar, A., Kumar, A., and Kumar, A., eds. (2024). *Applications of biotribology in biomedical systems.* Cham: Springer. doi: 10.1007/978-3-031-58327-8.
56. Kumar, A., Kumar, A., Kumar, A., Choudhari, A., Gupta, A. K., and Faisal, S. (2024). Introduction to biotribology: A science of surface interaction. In *Applications of biotribology in biomedical systems*, A. Kumar, A. Kumar, and A. Kumar (eds.), pp. 1–26. Cham: Springer. doi: 10.1007/978-3-031-58327-8_1.
57. Chowdhury, N., Faisalb, S., Kumar, A., Choudhari, A., Gupta, A. K., Kumar, A., and Kumar, A. (2024). Characterization of hydrogel properties in the advancement of biotribology. In *Applications of biotribology in biomedical systems*, A. Kumar, A. Kumar, and A. Kumar (eds.), pp. 27–78. Cham: Springer. doi: 10.1007/978-3-031-58327-8_2.
58. Pal, A., Gour, R., Kumar, A., Kumar, A., Kumar, A., and Chowdhury, N. (2024). Tribological measurements of human skin. In *Applications of biotribology in biomedical systems*, A. Kumar, A. Kumar, and A. Kumar (eds.), pp. 149–165. Cham: Springer. doi: 10.1007/978-3-031-58327-8_5.
59. Kumar, A., Gajjar, P., Gudadur, K. S., Kumar, A., Gupta, A. K., Choudhari, A., Chowdhury, N., and Kumar, A. (2024). Tribological hurdles in biomedical manufacturing: A comprehensive examination. In *Applications of biotribology in biomedical systems*, A. Kumar, A. Kumar, and A. Kumar (eds.), pp. 167–198. Cham: Springer. doi: 10.1007/978-3-031-58327-8_6.

60. Heidari, M., Thangavel, S., Rahmanivahid, P., Khashehchi, M., Kumar, A., Kumar, A., and Kumar, A. (2024). Navigating the landscape: Cutting-edge biomedical manufacturing techniques. In *Applications of biotribology in biomedical systems*, A. Kumar, A. Kumar, and A. Kumar (eds.), pp. 199–216. Cham: Springer. doi: 10.1007/978-3-031-58327-8_7.
61. Gour, R., Baliyan, N., Pal, A., Kumar, A., Kumar, A., and Kumar, A. (2024). Animal tribology. In *Applications of biotribology in biomedical systems*, A. Kumar, A. Kumar, and A. Kumar (eds.), pp. 217–234. Cham: Springer. https://doi.org/10.1007/978-3-031-58327-8_8.
62. Gupta, A. K., Choudhari, A., Kumar, A., Kumar, A., Gupta, A., Faisal, S., and Kumar, A. (2024). Composites for drug-eluting devices: Emerging biomedical applications. In *Applications of biotribology in biomedical systems*, A. Kumar, A. Kumar, and A. Kumar (eds.), pp. 251–311. Cham: Springer. doi: 10.1007/978-3-031-58327-8_10.
63. Choudhari, A., Gupta, A. K., Kumar, A., Kumar, A., Gupta, A., Chowdhury, N., and Kumar, A. (2024). Wear and friction mechanism study in knee and hip rehabilitation: A comprehensive review. In *Applications of biotribology in biomedical systems*, A. Kumar, A. Kumar, and A. Kumar (eds.), pp. 345–432. Cham: Springer. doi: 10.1007/978-3-031-58327-8_13.
64. Darrius, D., Bons, J., Kumar, A., Humaun Kabir, M., and Liang, H. (2024). Forever chemicals, per- and polyfluoroalkyl substances (PFAS), in lubrication. *Lubricants* 12 (4), 114.

Chapter 9

Biogreases derived from glycerides

Amit J. Agrawal, Pratibha S. Agrawal, and Vijay Y. Karadbhajne

9.1 INTRODUCTION

The ill effects of conventional fuels on climate and the easy availability of renewable energy sources have continuously increased their significance. Oils obtained from various nonedible seeds with a high percentage of oil content are increasingly utilized for creating biodegradable products because of two main reasons: (1) the least interest of the government in imported crude oil-based products and (2) increasing environment awareness among the public. To reduce the friction and increase the productivity and life of a machine, lubricating oil/grease is applied on its moving parts. A lubricant may be a fluid, a paste, or a fine powder present between two surfaces that are in relative movement to each other. Lubricating oils have specific deficiencies such as low consistency and even lower dropping point when they are applied in machines and exposed to high heat and working pressures. Lube oils are not fit to perform when a consistent increase in heat and working pressure in machines is noticed. This deficiency of lube oils leads an idea toward the synthesis and utilization of greases (Adhvaryu et al., 2004). Greases are semisolid paste-like structures and are defined as uniform dispersion of a suitable thickener (soap) in a liquid lubricating oil (Mang and Gosalia, 2017).

Greases are required in almost all moving parts of machinery to provide lubrication alongside the decrease in erosion between two complex components of the machine (Mendoza et al., 2011). They also confine the entry of any other foreign material in the framework, which may disturb the smooth functioning of the machine. During the operation, the consistency of grease permits them to go about as a sealant to forestall its spillage and dribbling (R. Kozdrach, 2020). For grease, it is necessary to keep up with its uniformity at the time of application and to go about as a viable sealant (Schneider, 2006).

Conventional grease is normally a blend of 60–95% base oil (mineral source oil), 5–25% thickener (the most commonly used thickener is soap),

DOI: 10.1201/9781003535447-9

and 0–10% additives to enhance its properties. Additives work on the efficiency of grease and increase its application span by further developing characteristics such as resistance to oxidation, protection from corrosion and wear, qualities against frothing, application under high pressures, and so forth (García-Zapateiro et al., 2014; Panchal et al., 2015).

Initially, crude-based oils were used as fluid for the synthesis of grease; however, lately, vegetable oils have acquired significance because of drastically increasing pollution issues and the need for the economic advancement of sustainable items. In this sense, it has been assumed that consistently a huge volume of oils is discarded in the climate, which can create soil and water pollution problems in the future for decades. Alongside this, oil spills are life threatening to aquatic animals and huge investment is required for its cleanup. Along these lines, the demand for green and biodegradable lubricants is continuously increasing day by day.

Plant-based nonedible oils are turning out to be more well-known these days to substitute conventional crude-based oils for manufacturing lubricants since they have comparable molecular structure and hydrocarbon chain length as that of mineral oils. Also, they have extra benefits of their sustainability, biodegradability, and the minimal expense of formulation alongside no damage to the climate (Mahadi et al., 2019; Gul et al., 2020). Recent advancements in the conservational norms of the environment and individual cognizance are the essential drivers for the production and manufacturing of such inexhaustible bioproducts on a large scale. Nonedible oil-based lubricants and greases are additionally easy to discard once their service life is over since they are non-hazardous and have no adverse effects on nature. Biofuels are as of now acquiring ubiquity all through the world in light of their endless benefits over mineral oil and an opportunity to lessen the reliance on traditional fuels (Casas-Godoy et al., 2020). Presently, an equal as well as practically identical development is normal for greases derived from nonedible vegetable oils where they specifically serve the purpose of lubrication (Rafał Kozdrach, 2021). Formulating biolubricants from nonedible oils and formulating bio-based greases from them have drawn in quite a large number of analysts to generate business opportunities for such bio-based new products having similar physical as well as chemical and tribological properties (McGuire, 2014; Vafaei et al., 2021). Such products can be utilized as a starting material, in blends or they may be utilized as added substances. Nonedible oil-based lubricant mixtures in addition to special property enhancers are capable of showing amazing results during their application. They may prove to be skeptical in reducing the reliance on conventional greases and their upgraded properties make them more significant modern products in commercial applications.

Some researchers carried out a detailed survey on properties, synthesis procedures, and utilization of biolubricants (Maleque et al., 2000, 2003; Masjuki et al., 1999). Porfiryev et al. claimed to prepare grease from base oils like mineral oil, synthetic poly-α-olefin oils, etc., and lithium soap of stearic acid as a thickener (Porfiryev et al., 2020). The dropping point of this grease was found to be 190°C. Hence, these greases are utilized for multipurpose functions since they have significant properties. Non-soap-based greases derived from silica are additionally utilized for multipurpose objectives because they can work under high heat (Skibińska, 2020).

Vegetable oil-based grease formulated from leftover jatropha oil as a base oil and lithium stearate/oleate soaps utilized as thickener along with zinc dialkyldithiophosphate as an additive gave a high dropping point in the range of 365°F–383°F and the NLGI grade of 2. These results comply with high-end greases. The biodegradable nature of these products is tested by changing the used products into CO_2 which was completely changed over into CO_2. The toxicity test of such products was also checked and found to be negative (Nagendramma and Kumar, 2015).

Additionally, two more formulations of greases were prepared by earlier laborers utilizing mineral and polyalphaolefin oil as base oils along with lithium complex and polyurea complex. The dropping points of the two prepared samples at various proportions, namely 80:20 and 85:15, were calculated. The most elevated dropping point for lithium complex grease was 315.33°C and for polyurea complex grease was found to be 345°C. Also, the penetration indices determined for both the greases and the highest values found for both the samples were reported as 28.64 for lithium complex grease and 42.64 for polyurea complex grease. Both the greases additionally showed other great properties like colloidal reliability, resistance to corrosion, and stability toward high temperatures, and these properties gave great outcomes when 80% crude oil-based oil was used for these two grease formulations (Ren et al., 2021).

Nonedible oil-based greases have many applications in diverse fields related to society, for example, ranger service apparatus used in forestry, construction vehicles, and machinery; railways and marine applications; hydroelectric fueled dams; farming; plant operations; mining; and digging (Japar et al., 2018). These products have gigantic demand in the market due to their user and environment friendliness. The biogrease formulated from palatable oil and their esters showed a higher cost of formulation, whereas if non-eatable oil is utilized, then the cost of biogrease falls. Also, greases are found to be biodegradable, equally efficient as mineral oil-based greases as well as environmentally acceptable (Sharma and Singh, 2017). The biogrease samples formulated from *Pongamia pinnata* and neem oil found applications as driers. The average weather conditions in India, basically Vidarbha district, are reasonable for *Madhuca indica* (mahua) and *Pongamia pinnata* (karanja) farming and harvesting (Banka and Parikh, 2019). Additionally, the non-harmfulness of mahua and karanja plants and their oils is low when compared to other non-eatable oils. Subsequently, both these oils were chosen for the current work.

9.2 MATERIALS AND METHODS

9.2.1 Materials

Refined *Madhuca indica* (mahua) oil and *Pongamia pinnata* (karanja) oil were bought from the nearby marketplace. The basic characteristic properties of both these oils were determined in the lab and are noted in Table 9.1. Other chemicals from Sigma-Aldrich were utilized in the examinations for carrying out experiments and grease formulation.

9.2.2 Methyl esters (ME) and polyol esters of oil

Both the oils were analyzed in the lab for their free fatty acid (FFA) content. The value was found on the higher side. Methyl esters (ME) of such oils with high FFA were ultimately synthesized utilizing two-stage reactions where esterification was carried out first and then was followed by a transesterification process to produce methyl esters (Baskar et al., 2017; Patel and Sankhavara, 2017). In transesterification, the catalyst used was sodium methoxide and methanol was used for ester synthesis. The upper layer separated after the total reaction was methyl ester which was finally purified to eliminate the remaining methanol (Sastry and Murthy, 2015).

Mahua oil methyl esters and karanja oil methyl esters were then independently reacted with trimethylolpropane (TMP) for the preparation of triesters (biolubricants) in 0.5 l batches. 1% w/w sodium methoxide was used as a catalyst for these reactions. The molar proportion of separate methyl esters to TMP taken for the reactions was 4:1. Oil bath was maintained at 120°C for 4 h and constant stirring was ensured during the due course of the reaction. The reaction was completed in the absence of air to ensure the formation of desired products (Heikal et al., 2017). The whole methanol generated as a byproduct was taken out of the system. Oleochemical esters based on trimethylolpropane (TMP) give great flow properties at low temperatures, and thus they are viewed as competent products for additional applications.

9.2.3 Biogrease

Formulation of biogrease consists of two steps, namely synthesis of soap and then blending of soap with base oil for the final formulation of grease. Karanja oil and mahua oil were separately used as the base oils in different weight ratios of oil and soap for the grease formulations. The best ratio of oil and soap was then selected for grease formulation from respective methyl esters and then triesters of the oils.

9.2.3.1 Thickener (soap)

Roughly 50 g of *Madhuca indica* oil was taken into a 250 ml container and then 40 ml of ethanol was added. At last, 25% (m/v) sodium hydroxide

Table 9.1 Characterization of *Madhuca indica* oil, *Pongamia pinnata* oil, and their methyl esters and biolubricants

Sr. No.	*Characteristics*	*Analytical method*	Madhuca indica *oil (mahua)*	Pongamia pinnata *oil (karanja)*	*Mahua—methyl ester*	*Karanja—methyl ester*	*Mahua polyol ester (biolubricant)*	*Karanja polyol ester (biolubricant)*
1	Density @ 20°C	ASTM-D1298	0.908 g/ml	0.936 g/ml	0.8528 g/ml	0.88 g/ml	0.879 g/ml	0.88 g/ml
2	Viscosity @ 40°C	ASTM-D445	37.181 cst	54.541 cst	4.49 cst	4.605 cst	69.34 cst	43.3 cst
3	Viscosity @ 100°C	ASTM-D445	11.65 cst	15.182 cst	3.028 cst	1.945 cst	8.56 cst	6.55 cst
4	Viscosity index	ASTM-D2270	325.61	304.93	447.52	205.67	103.56	101.21
5	Pour point	ASTM-D97	18°C	−1°C	0°C	−4°C	0°C	−2°C
6	Flash point	ASTM-D92	210°C	220°C	150°C	140°C	205°C	210°C
7	Fire point	ASTM-D92	220°C	240°C	165°C	148°C	218°C	219°C
8	Saponification value	ASTM-D94	185.16 mg KOH/g	188.54 mg KOH/g	—	—	—	—
9	Acid value	ASTM-D664	26.23 mg KOH/g	18.40 mg KOH/g	—	—	—	—

was added to the mixture. The blend was then heated by indirect means for around 1200 s, giving it steady and thorough mixing. Once the scent of liquor vanished, the reaction was considered to be overgiving a soap-like combination. The soap was then separated by adding 300 ml of a soaked sodium chloride solution to the mixture while continuous mixing was carried out and the container was kept in an ice-water bath. This helped in increasing the density of the watery arrangement by precipitating the soap and thus eliminating the unreacted sodium hydroxide from the soap. The soap was then filtered through filter paper and washed with super-cold water. Similarly, karanja oil soap was also prepared by replacing mahua oil with karanja oil. In this way, sodium, potassium, and calcium-based metallic soaps were prepared separately from three types of oils such as sodium hydroxide, potassium hydroxide, and calcium hydroxide.

9.2.3.2 *Blending of soap with base oil for biogrease formulation*

Mahua oil was utilized as a base oil for the synthesis of the first bio-based grease sample using metallic soaps as a thickener. Thickener was uniformly disseminated in the oil using a rapid stirrer and additionally applying thermomechanical arrangements. In the first step, the oil (80% by weight) was taken into the reactor and heated up to a temperature of 120°C. Then in the second step, after the temperature was achieved, metallic soap (20% by weight) was added to the reactor and blended vigorously with the oil at 200 rpm for uniform mixing. Constant mixing was ensured with the help of a stirrer and the temperature was slowly raised to 392°F. The mixture was constantly stirred till a uniform consistency was observed in the blending mixture. Once the consistency was observed, the temperature of the system was brought down continuously. With the decrease in temperature, the oil continuously became viscous and the soap tried to separate from the oil and settle down. Hence, stirring was continuously maintained during this span. By that time, the sample temperature came down to normal, and the thick oil transformed into a semisolid paste-like grease. Till this stage, no additives were added to this grease mixture (Buhlak et al., 2014; Fan et al., 2014; Nagendramma and Kumar, 2015).

Similarly, grease samples with 70:30 and 90:10 oil-to-soap ratios were also prepared and analyzed for their consistency properties. In the initial step, raw mahua oil was blended with soap for the formulation of the first type of biogrease. The second and third types of biogrease samples were prepared by replacing raw mahua oil with its methyl ester and TMP triester, respectively. Similarly, the fourth, fifth, and sixth types of biogrease samples were formulated by replacing manual oil with karanja oil, its methyl ester, and TMP triester. All the biogrease samples were then analyzed for the standard ASTM tests using the standard equipment as per the ASTM norms and the equipment was procured from Maharana Instruments Mfg. Co., Ajmer, India.

The dropping point of a grease is a significant characteristic property of the oil and it is only a mark of its resistance to high working temperatures. The dropping point of the prepared biogrease samples was performed by the standard test method ASTM-D2265. The dropping point was estimated by taking a measured amount of grease in a standard orifice cup and heating it till a temperature reached at which the grease became sufficiently liquid and fell from the opening of the test cup in the form of a liquid drop under the test conditions. Water washout resistance is nothing but the ability of grease to oppose the water impacts without critical losses in its core properties of lubrication. This property was determined by ASTM-D1264 for the prepared biogrease samples. The assembly consisted of a ball bearing that was coated with the biogrease samples and afterward, the bearing was infringed with a stream of water maintained at the predetermined flow rate and temperature. The percent loss of grease from the bearing on a weight basis in 1 h was determined as a water washout of the sample.

9.3 RESULTS AND DISCUSSION

9.3.1 Polyol ester synthesis from oils

Both *Madhuca indica* (mahua) oil and *Pongamia pinnata* (karanja) oils are nonconsumable oils. Nonetheless, mahua oil shows a few applications in the oleochemical industry like soap manufacturing, yet both these oils can prove to be a good choice to be the least expensive starting material for biodiesel (methyl ester) synthesis and its further conversion into triester (biolubricant). Subsequently, these oils can prove to be an effective and consistent source for the synthesis of biolubricants.

All the basic properties of the *Madhuca indica* oil and *Pongamia pinnata* oil carried out in the laboratory according to ASTM standards are shown in Table 9.1. Both the oils had higher free fatty acid content; hence methyl ester synthesis was consequently carried out in a strategy of esterification and then transesterification reaction (Patel and Sankhavara, 2017). Methyl esters of both oils were then independently reacted with trimethylol propane and thus triesters based on mahua oil and karanja oil methyl esters were synthesized.

9.3.2 Biogrease formulation using oils as base oils

Nonconsumable oils can be straightforwardly utilized as base oils for grease synthesis in the condition as it is they are present, yet they have specific detriments like poor thermal, oxidative, and hydrolytic strength; lower dropping point when converted into grease; and high pour points.

These disadvantages limit the immediate utilization of these nonedible oils as base oils. Consequently, the oils can be synthetically altered by various chemical reactions so that they can be efficiently utilized as biolubricants and greases. These reactions are esterification and transesterification, epoxidation, enzymatic reactions, and so on (Cecilia et al., 2020; Sharma and Singh, 2017). Tables 9.2 and 9.3 show the analysis of grease formulated utilizing *Madhuca indica* oil straightforwardly as a base oil with various proportions and types of soap on a weight basis, which was prepared in the lab as referenced previously. Three different types of greases based on different proportions of oil and soap were formulated as referenced in Table 9.3. All the significant physical characteristics of the greases as indicated by ASTM systems (ASTM Global, 2002) and described above were determined. Grease with 70% oil by weight and the remaining 30% soap brought about the most noteworthy characteristic values of the aforesaid standards, but the appearance of the product was found to be quite dry and the presence of lumps in the sample made it inconsistent. This type of appearance is not desired in greases. Then again, grease having 80% oil and 20% soap brought about grease with good consistency and had a smooth appearance. However, the proportion of 90% oil and 10% soap brought about an exceptionally oily, sleek, and extremely delicate final product.

Comparable outcomes were acquired when *Pongamia pinnata* oil (karanja oil) was straightforwardly utilized as a base oil for the synthesis of grease as displayed in Tables 9.2 and 9.3. Karanja oil-based biogrease looked more fibrous and uniform when compared with mahua oil-based biogrease. Likewise, the major property, i.e., dropping point for karanja oil biogrease was also found to be higher than mahua oil biogrease. The cone penetration index of this biogrease was also found to be higher as compared to *Madhuca indica*-based grease; however, it doesn't influence the NLGI grade of the grease. Many other researchers used different kinds of vegetable oils for biogrease formulation such as coconut oil, palm-based oil (Skibińska, 2020), olive oil (Sterpu et al., 2016), corn oil, and palm oil with calcium stearate soap (Vodounon et al., 2019). The ideal process parameters for the synthesis of these oil-based biogreases are kept at 212°F with 20–25% wt of soap content. The NLGI number is found to be 3–6 under ideal circumstances. As indicated by the rheological review, the palm oil-based biogrease was better than the corn oil-based biogrease. Looking toward the consistency of the grease, 80% oil and 20% soap by weight was viewed as reasonable.

9.3.3 Biogrease formulation using ME and polyol esters of oils

The ideal proportion of oil and soap for grease formulation was found by using raw oils as base oil. Subsequently, the same ratio was utilized for another type of grease formulation from methyl esters of *Madhuca indica*

Table 9.2 Analysis of *Madhuca indica* and *Pongamia pinnata* oil-based biogrease

Sr. No.	*Properties*	*Analytical method*	*Biogrease of* Madhuca indica *oil and sodium soap*					*Biogrease of* Pongamia pinnata *oil and sodium soap*				
			Oil:soap ratio			ME:soap ratio	Triester:soap ratio	Oil:soap ratio			ME:soap ratio	Triester:soap ratio
			70:30	80:20	90:10	80:20	80:20	70:30	80:20	90:10	80:20	80:20
1	Appearance	—	Dry and lumpy	Smooth and consistent	Very soft and oily	Smooth and consistent	Smooth and consistent	Dry and lumpy	Smooth, fibrous	Very soft and oily	Smooth, fibrous	Smooth, fibrous
2	Cone penetration index (worked) @ 25°C	ASTM-D217	264	280	316	275	270	268	294	326	287	288
3	NLGI grade	—	2	2	1	2	2	2	2	1	2	2
4	Dropping point	ASTM-D2265	149°C	136°C	120°C	139°C	173°C	170°C	146°C	133°C	156°C	177°C
5	Water washout resistance @ 38°C	ASTM-D1264	7%	10%	<14%	8%	7%	6%	8%	11%	8%	7%

Table 9.3 Analysis of *Madhuca indica* oil-based biogrease using different soaps

Sr. No.	*Properties*	*Analytical method*	*Potassium-based biogrease*					*Calcium soap-based biogrease*				
			Oil:soap ratio			ME:soap ratio	Triester:soap ratio	Oil:soap ratio			FAME:soap ratio	Triester:soap ratio
			70:30	80:20	90:10	80:20	80:20	70:30	80:20	90:10	80:20	80:20
1	Appearance	—	Dry and lumpy	Smooth and consistent	Very soft and oily	Smooth and consistent	Smooth and consistent	Dry and lumpy	Smooth, fibrous	Very soft and oily	Smooth, fibrous	Smooth, fibrous
2	Cone penetration index (worked) @ 25°C	ASTM-D217	264	278	316	281	285	268	290	312	285	285
3	NLGI grade	—	2	2	1	2	2	2	2	1	2	2
4	Dropping point	ASTM-D2265	140°C	136°C	117°C	148°C	173°C	210°C	186°C	173°C	186°C	195°C
5	Water washout resistance @ 38°C	ASTM-D1264	7%	10%	<14%	8%	7%	<5%	6%	7%	5%	5%

oil (ME), and afterward, polyol esters of TMP were utilized as a base oil for yet another type of biogrease formulation. Examination of both types of greases is given in Table 9.2. From the results, it is obvious that the grease type with polyol esters of TMP utilized as base oil gave the best outcomes. Its dropping point was obtained to be the highest among the other types of greases 173°C. The dropping point is a major property for any grease since it defines the maximum temperature limits where the grease can work efficiently under the applied load and genuine working conditions (Suetsugu et al., 2013).

In this study, biogrease was formulated using calcium and magnesium-based metallic soaps. El-Adly et al. (2015) formulated the biogrease from used cooking oil and biochar in varying concentrations. The dropping point of this product is claimed to be 374°F, which is excellent. The advantage of such type of product is that it is environment friendly during and after use. ME and polyol esters of TMP based on karanja oil were also utilized in the same way to prepare karanja-based biogrease samples and analyzed them in the lab. From the results, it has been seen that biogrease synthesized from polyol esters of TMP based on karanja oil gave surprisingly better outcomes with regard to the dropping point which was observed to be 177°C. It is obvious from the results that the biogreases gave higher dropping points due to the high viscosity of the polyol esters from which they were formulated. Similarly, polyol esters also have higher oxidation stability, which is an added advantage for biogreases. Similar types of greases were formulated using potassium and calcium-based soap, where it was observed that biogrease formulated from calcium soap yielded the best results as compared to sodium and potassium-based greases. On the other hand, the toxicity of karanja oil toward aquatic life restricted its use where the biogrease is directly utilized in marine applications.

9.4 CONCLUSION

The oils, i.e., karanja oil and mahua oil, selected for biogrease formulation are safe and the biogreases thus obtained are biodegradable. These biogreases can be utilized altogether in all applications and they are capable of easily replacing mineral-based greases. Because of their good water washout resistance, which is 7%, these biogreases can easily prove to be more efficient than conventional greases, especially in direct applications like agribusiness, railways, and marine applications where these products are directly released into the earth or in water bodies after their application. Biogrease samples formulated with 80% oil and 20% soap by weight showed the best results and more consistency. Other important properties like dropping point and NLGI grade for both the biogreases are comparable with conventional mineral oil-based greases. *Pongamia pinnata* oil (karanja oil)-based grease showed a dropping point of 177°C, whereas the value for *Madhuca indica* (mahua oil)-based grease is 173°C. But the NLGI grade for both products falls in the range of 2. Biogrease formulated from crude *Pongamia pinnata* oil and *Madhuca indica* oil showed poor oxidation stability and thermal resistance when it was compared with the biogrease formulated from respective methyl esters and polyol esters, yet the product formulated from non-eatable sources is both prudent and biodegradable. Hence, the base oil and the additives should be selected properly for the sustainable production of biogreases. The development of such biogreases could be significant in the modern era. It is accepted that such kind of exploration study could help analysts to address the difficulties ahead given sustainable and eco-friendly products.

REFERENCES

Adhvaryu A, Erhan SZ, Perez JM (2004) Preparation of soybean oil-based greases: Effect of composition and structure on physical properties. *J Agric Food Chem*. https://doi.org/10.1021/jf049888r

Banka S, Parikh SP (2019) Nonedible oil biodiesels: The cutting-edge future of renewable energy in India. Asia-*Pac J Chem Eng*. https://doi.org/10.1002/apj.2310

Bashiri S, Ghobadian B, Dehghani Soufi M, Gorjian S (2021) Chemical modification of sunflower waste cooking oil for biolubricant production through epoxidation reaction. *Mater Sci Energy Technol*. https://doi.org/10.1016/j.mset.2021.03.001

Baskar G, Gurugulladevi A, Nishanthini T, Aiswarya R, Tamilarasan K (2017) Optimization and kinetics of biodiesel production from Mahua oil using manganese doped zinc oxide nanocatalyst. *Renew Energy*. https://doi.org/10.1016/j.renene.2016.10.077

Buhlak S, Ibrahim B, Alhamoui M (2014) Manufacturing (lithium-sodium) lubricant grease based on Syrian base oil and studying its physical, chemical, and rheological properties. *Int J ChemTech Res*. Vol. 6, No. 4, pp. 2247–2254.

Casas-Godoy L, Barrera-Martínez I, Ayala-Mendivil N, Aguilar-Juárez O, Arellano-García L, Reyes AL, Méndez-Zamora A, Sandoval G (2020) Biofuels. In *Biobased Products and Industries*. https://doi.org/10.1016/B978-0-12-818493-6.00004-X

Cecilia JA, Plata DB, Saboya RMA, de Luna FMT, Cavalcante CL, Rodríguez-Castellón E (2020) An overview of the biolubricant production process: Challenges and future perspectives. In *Processes*. https://doi.org/10.3390/pr8030257

El-Adly RA, Yossef MA, Modather FH, Ismail EA, Abbas DM (2015) Biogrease based on biochar from rice straw and waste cooking oil. *Int J Adv Pharm, Biol Chem*, Vol. 4, No. 4, pp. 1–5.

Fan X, Xia Y, Wang L (2014) Tribological properties of conductive lubricating greases. *Friction*. https://doi.org/10.1007/s40544-014-0062-2

García-Zapateiro LA, Valencia C, Franco JM (2014) Formulation of lubricating greases from renewable basestocks and thickener agents: A rheological approach. *Ind Crops Prod.* https://doi.org/10.1016/j.indcrop.2014.01.020

Gul M, Masjuki HH, Kalam MA, Zulkifli NWM, Mujtaba MA (2020) A review: Role of fatty acids composition in characterizing potential feedstock for sustainable green lubricants by advance transesterification process and its global as well as Pakistani prospective. In *Bioenergy Research.* https://doi.org/10.1007/s12155-019-10040-7

Heikal EK, Elmelawy MS, Khalil SA, Elbasuny NM (2017) Manufacturing of environment friendly biolubricants from vegetable oils. *Egyptian J Petroleum.* https://doi.org/10.1016/j.ejpe.2016.03.003

Japar A, Aziz NSA, Razali MN (2019) Formulation of fumed silica grease from waste transformer oil as base oil. *Egyptian J Petroleum.* https://doi.org/10.1016/j.ejpe.2018.12.001

Japar NSA, Aziz MAA, Razali MN, Rahman NWA (2018) Grease and its application on electrical equipment: A review. *Int J Eng Technology (UAE),* p. 7/12. https://doi.org/10.14419/ijet.v7i3.26.17455

Kozdrach R (2020) The tribological, physico-chemical and rheological properties of lubricating grease produced on epoxy resin base. *Tribol Ind.* https://doi.org/10.24874/ti.760.09.19.04

Kozdrach R (2021) The influence of base oil type on the rheological properties of ecological lubricating greases. *Nafta - Gaz.* https://doi.org/10.18668/NG.2021.02.07

Mahadi MA, Choudhury IA, Azuddin M, Yusoff N, Yazid AA, Norhafizan A (2019) Vegetable oil-based lubrication in machining: Issues and challenges. IOP Conference Series: Materials Science and Engineering. https://doi.org/10.1088/1757-899X/530/1/012003

Maleque MA, Masjuki HH, Haseeb ASMA (2000) Effect of mechanical factors on tribological properties of palm oil methyl ester blended lubricant. *Wear.* https://doi.org/10.1016/S0043-1648(00)00319-7

Maleque MA, Masjuki HH, Sapuan SM (2003) Vegetable-based biodegradable lubricating oil additives. In *Industrial Lubrication and Tribology.* https://doi.org/10.1108/00368790310470976

Mang T, Gosalia A (2017) Lubricants and their market. In *Lubricants and Lubrication.* https://doi.org/10.1002/9783527645565.ch1

Masjuki HH, Maleque MA, Kubo A, Nonaka T (1999) Palm oil and mineral oil based lubricants - their tribological and emission performance. *Tribol Int.* https://doi.org/10.1016/S0301-679X(99)00052-3

McGuire N (2014) *Bio-based greases.* Tribology and Lubrication Technology.

Mendoza G, Igartua A, Fernandez-Diaz B, Urquiola F, Vivanco S, Arguizoniz R (2011) Vegetable oils as hydraulic fluids for agricultural applications. *Grasas Aceites.* https://doi.org/10.3989/gya.056210

Nagendramma P, Kumar P (2015) Eco-friendly multipurpose lubricating greases from vegetable residual oils. *Lubricants.* https://doi.org/10.3390/lubricants3040628

Panchal T, Chauhan D, Thomas M, Patel J (2015) Bio based grease A value added product from renewable resources. *Ind Crops Prod.* https://doi.org/10.1016/j.indcrop.2014.09.030

Patel RL, Sankhavara CD (2017) Biodiesel production from karanja oil and its use in diesel engine: A review. In *Renewable and Sustainable Energy Reviews*. https://doi.org/10.1016/j.rser.2016.12.075

Porfiryev Y, Shuvalov S, Popov P, Kolybelsky D, Petrova D, Ivanov E, Tonkonogov B, Vinokurov V (2020) Effect of base oil nature on the operational properties of low-temperature greases. *ACS Omega*. https://doi.org/10.1021/acsomega.9b04087

Ren G, Zhang P, Ye X, Li W, Fan X, Zhu M (2021) Comparative study on corrosion resistance and lubrication function of lithium complex grease and polyurea grease. *Friction*. https://doi.org/10.1007/s40544-019-0325-z

Sánchez R, Franco JM, Delgado MA, Valencia C, Gallegos C (2008) Effect of thermo-mechanical processing on the rheology of oleogels potentially applicable as biodegradable lubricating greases. *Chem Eng Res Des*. https://doi.org/10.1016/j.cherd.2008.05.002

Sastry S, Murthy C (2015) Synthesis of biodiesel by In-situ transesterification of karanja oil. *Bangladesh J Sci Industrial Res*. https://doi.org/10.3329/bjsir.v49i4.22623

Schneider MP (2006) Plant-oil-based lubricants and hydraulic fluids. *J Sci Food Agric*. https://doi.org/10.1002/jsfa.2559

Sharma UC, Singh N (2017) Biogreases for environment friendly lubrication. *Environmental Science and Engineering Vol. 1: Sustainable Development*. Publisher: Studium Press LLC USA, pp. 305–317.

Skibińska A (2020) Multifunctional silica lubricating greases. *Nafta - Gaz*. https://doi.org/10.18668/NG.2020.04.07

Sterpu AE, Prodan G, Teodorescu N, Prodea IM, Dumitru AI, Koncsag CI (2016) Lubricating greases from olive oil, corn oil and palm oil. *Revista de Chimie*, Vol. 67, No. 8, pp. 1575–1582.

Suetsugu Y, Sekiguchi H, Nakanishi Y, Fujinami Y, Ohno T (2013) Basic study of grease rheology and correlation with grease properties. *Tribology Online*. https://doi.org/10.2474/trol.8.83

Vafaei S, Fischer D, Jopen M, Jacobs G, König F, Weberskirch R (2021) Investigation of tribological behavior of lubricating greases composed of different bio-based polymer thickeners. *Lubricants*, pp. 8/12. https://doi.org/10.3390/lubricants9080080

Vodounon CA, Sterpu AE, Prodea IM, Koncsag CI (2019) Influence of manufacturing temperature on the rheological behavior of some vegetable lubricating greases. *Ovidius Univ Annals Chem*. https://doi.org/10.2478/auoc-2019-0010

Zhang E, Li W, Zhao G, Wang Z, Wang X (2021) A study on microstructure, friction and rheology of four lithium greases formulated with four different base oils. *Tribol Lett*. https://doi.org/10.1007/s11249-021-01469-z

Chapter 10

Lubricant

Soft materials (hydrogel)

Sakib Faisal, Nusrat Chowdhury, Abhishek Kumar, Avinash Kumar, Ashwani Kumar, Amit Choudhari, and Ashish Kumar Gupta

10.1 INTRODUCTION

Natural tribological systems, such as those found in the mouth and joint cartilage, provide exceptional, enduring smooth life [1]. These systems are characterized by a biphasic makeup including mucins, polysaccharides, phospholipids, and glycoproteins, complemented by a water-based element. Since 1960, hydrogels, which mimic the composition of these biological systems and are known for their water affinity, compatibility with biological tissues, and adaptable structures, have been extensively utilized to explore the principles of biological smoothness and for uses like replacing cartilage [2–5]. Creating water-based lubricants with minimal friction is key for technological and biological uses, leveraging hydrogels and microgels to enhance the properties of materials and offer efficient hydration lubrication for elastic surfaces [6]. Hydration of lubrication mechanisms, encompassing elements like hydrated ions, surfactants, biological and biomimetic polymers, and phosphatidylcholine vesicles, leverage highly hydrated surface groups for reduced friction, with PC lipids in joints being notably effective lubricants, suggesting that hydrogel studies could enhance understanding of cartilage lubrication and inspire advanced treatments for osteoarthritis and the development of lubricious hydrogels for biomedical uses [7].

In tribology research, incorporating stimuli-responsive characteristics into hydrogels allows their surface friction to be adjusted through external factors such as pH levels, temperature, electrolytes, and solvents. Hydrogels exhibit properties of both solids and liquids; they can undergo deformation under applied force and return to their original shape once the force is removed, akin to solids. Similarly, they facilitate the movement and diffusion of small solutes through their network, much like liquids. Their soft and moist nature makes hydrogels comparable to biological soft tissues [8]. The other significant sections of polymer-based lubrication material include microgels, polymer brushes, powdered polymers, etc. Though hydrogels are mostly used in soft material tribology, the other polymer-based lubrication other than hydrogels can be significantly used in solid-metal lubrication systems. Developed polymers such as polytetrafluoroethylene (PTFE), polyether

DOI: 10.1201/9781003535447-10

ether ketone (PEEK), and polyamide (nylon) are commonly utilized in solid material lubrication applications due to their excellent tribological properties, including high wear resistance and low friction coefficients. These polymers are preferred in various industrial, mechanical, and aerospace applications for reducing wear and extending the lifespan of components. On the other hand, soft materials with high water content, often referred to as hydrogels, are extensively used in biomedical and tissue engineering applications. These hydrogels mimic the natural extracellular matrix found in living tissues, making them ideal for tissue replacement, wound healing, and drug delivery systems. Their high water content ensures compatibility with biological environments, allowing for applications in artificial organs, soft tissue replacements, and other bio-related fields. This distinct use of soft, water-rich polymers in the biomedical sector contrasts with the more mechanically focused applications of PTFE, PEEK, and nylon in solid lubrication [19].

Furthermore, fillers or composites within hydrogels are also significant ways to improve lubrication, strength, porosity, and other tribological characteristics of soft materials. The integration of micron-sized fibers and composite materials into hydrogels significantly improves their mechanical and tribological attributes, rendering them ideal for biolubrication in demanding environments. Composites such as GelMA–PAM–PMPC and MXene/polyvinyl alcohol (PVA) exhibit enhanced lubrication capabilities due to the formation of hydration layers and the effects of multi-crosslinking, respectively, showcasing their suitability for uses like lubrication in water-lubricated stern tube bearings. Furthermore, the creation of photoresponsive hydrogels through the inclusion of α-cyclodextrins, polyethylene glycol, and azobenzene compounds offers an innovative approach to dynamically adjusting lubrication properties in response to varying light exposures. Such advancements herald a new era in the development of hydrogel-based lubricants for diverse applications across the biomedical and engineering sectors [30–37].

Additionally, the chapter will address the intricate interplay between the hydrogel's physical properties and its lubrication efficiency, highlighting how the viscoelastic nature of hydrogels contributes to their unique lubricating capabilities. Special attention is given to the advancements in hydrogel technology that enhance their functionality, such as the incorporation of nanofillers and the development of composite hydrogels, which further improve their tribological performance. The discussion extends to the challenges faced in optimizing hydrogel formulations for specific applications, including the need for balance between hydration levels and mechanical integrity. The potential of hydrogels to replicate the natural lubrication found in human joints positions them at the forefront of materials science research, with implications for the design of more effective prosthetics and bioinspired lubrication systems. Through an interdisciplinary approach, this chapter aims to bridge the gap between material science and biomedical engineering, offering a comprehensive perspective on the current state and future directions of hydrogel-based lubrication solutions.

Hydrogels have become fundamental in cell culture due to their ability to mimic the natural extracellular matrix, offering a supportive scaffold for cell growth, differentiation, and migration. Their high water content and customizable mechanical properties make them suitable for a wide range of cell types, enhancing cell function and enabling the creation of complex tissue models. Functionalized with bioactive compounds, hydrogels facilitate advanced cell signaling and tissue formation, revolutionizing tissue engineering and regenerative medicine. This progress supports deeper insights into tissue biology and opens new avenues for drug discovery and personalized medicine [89–91].

The illustration provides a comprehensive overview of hydrogels' role as lubricative agents in applications such as cell culture and articular cartilage. It further delves into the utilization of hydrogel composites and polymer-infused hydrogels, highlighting their versatility beyond conventional hydrogels in these domains. Additionally, it underscores their contribution to enhancing the properties and performance of biological substitutes.

10.2 POLYMER-BASED MATERIALS AS LUBRICATION

Polymer-based lubricants offer distinct benefits, primarily through their easy customization and adaptable functionality. For instance, polymeric hydrogels that respond to stimuli like pH changes, temperature shifts, and light enable the precise management of surface friction, making them ideal for crafting intelligent devices and adjustable mechanical systems [8].

In the field, researchers delve into the unique tribological aspects of polymers, such as their time-sensitive viscoelastic nature, lubricant absorption causing swelling, and the effects of brittleness, surface properties, and magnetic fields on their wear and frictional behavior. Effects of composition, pretreatment, and strain hardening are available on the tribological properties of the polymer-based materials [9]. The concept of hydration lubrication, which involves adding hydratable molecules to reduce interfacial friction, serves as a foundational approach to developing aqueous lubricants. Yet, for complex applications like smart friction management, arthritis treatment, and bioinspired lubrication, traditional methods may fall short, necessitating the emergence of functional hydration lubrication that integrates additional capabilities into the lubricating materials for enhanced performance [8]. In polymer-based lubrication, there are stimuli-responsive polymer-based lubricating materials, polymer brushes, microgels, and hydrogels. In this chapter, we will discuss hydrogels in detail, and we will also provide some background for polymer-based other materials in this section (Figure 10.1).

Materials that are sensitive to stimuli undergo property alterations, such as changes in hydrophilicity or hydrophobicity, interfacial adhesion, surface

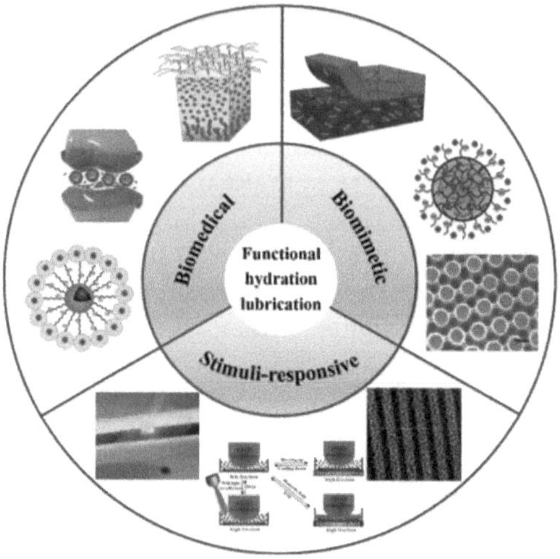

Figure 10.1 An illustration depicting how functional hydration lubrication is achieved through three categories of polymer-based lubricants: those that are responsive to stimuli, those designed for medical applications, and those that mimic biological systems [8].

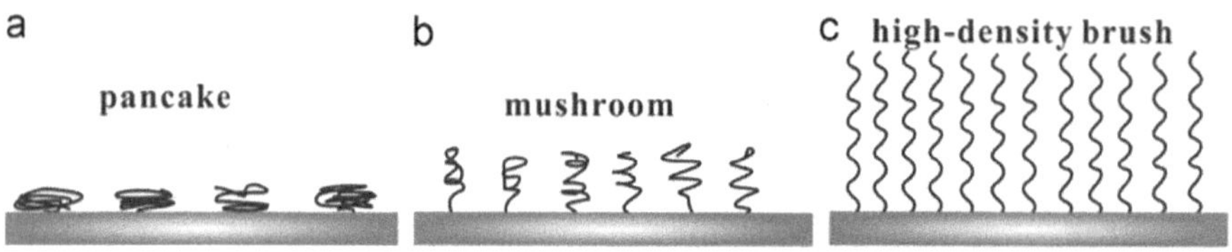

Figure 10.2 Illustration of polymer-brushed lubrication and their reaction to the lubricant [21].

texture, and interfacial organization, in response to variations in environmental factors like temperature, pH, presence of electrolytes, magnetic and electric fields, light, and solvents. These changes are crucial in influencing interfacial friction [10–17]. Measurements of force and friction demonstrated that the polymer brush exhibited effective lubrication properties under acidic conditions, with a noticeable increase in friction at alkaline pH levels. Additionally, temperature adjustments influenced the hydrophobic interactions between the polymer chains and the solvent, altering the brush's density and structure and affecting its frictional behavior as shown in Figure 10.2.

Investigations into polymer brush tribology reveal that the interaction between polymer-coated surfaces in a solvent prevents the brushes from intertwining due to entropic forces, creating a thin lubricating fluid layer that significantly reduces friction, even under high pressure. This phenomenon contrasts with the high friction observed between hard surfaces, where direct asperity contact occurs. The tribological properties of polymer brushes can be finely adjusted by manipulating factors like grafting density, chain length, stiffness, solvent interaction, and crosslinking, allowing for tailored frictional responses based on diverse brush architectures [18]. Some advancements in polymer-based lubrication include (1) enhanced natural polymer coatings, newly developed for their lubricating and wear-resistant capabilities; (2) artificial polymer coatings, designed with biomimetic structures to achieve superior lubrication and durability; and (3) specially formulated hydrogels and elastomers, soft yet robust materials engineered with tailored chemical compositions and structures to improve their friction-reducing and wear resistance attributes [20]. The advancement of polymer brushes can be composed with electrochemical or photochemical processes, the polymer brushes' features and involvement with the solvent can significantly reduce the friction behavior as shown in Figure 10.3 [21]. Polymer brushes, due to their unique properties, have been widely applied across diverse fields like nanotechnology, material science, biomedical engineering, catalysis, photonics, and electronics, leading to the development of various functionalized surfaces and interfaces, as extensively reviewed in the literature [22–32].

Biological interfaces, such as those found in joints, typically exhibit remarkably low friction coefficients (ranging from 0.001 to 0.03), largely due to the role of biological lubricants. The high water content within biological tissues contributes to their effective biolubrication through mechanisms of aqueous lubrication. However, water's inherently low viscosity and high fluidity can lead to suboptimal lubrication performance under heavy loads, potentially compromising the protective hydration layer at the boundary. The effectiveness of aqueous lubrication is enhanced by strong interactions, such as electrostatic and dipole–dipole interactions, between the biological lubricant and water molecules, creating a more resilient hydrated barrier at the friction interface. This barrier makes it energetically

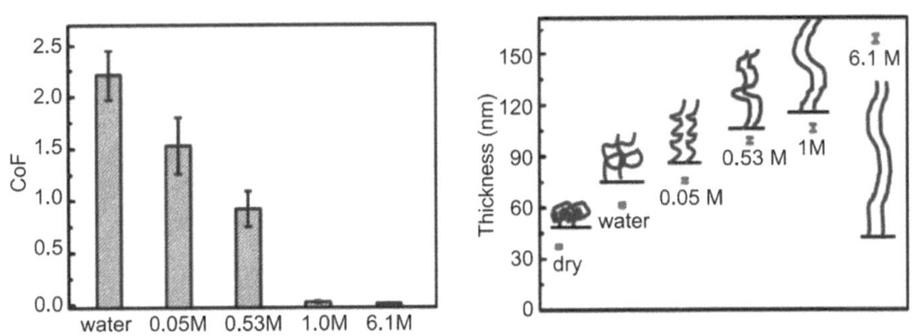

Figure 10.3 Changes in the length of polymer-based brush chains and their lubricating characteristics as influenced by the concentration of NaCl [20].

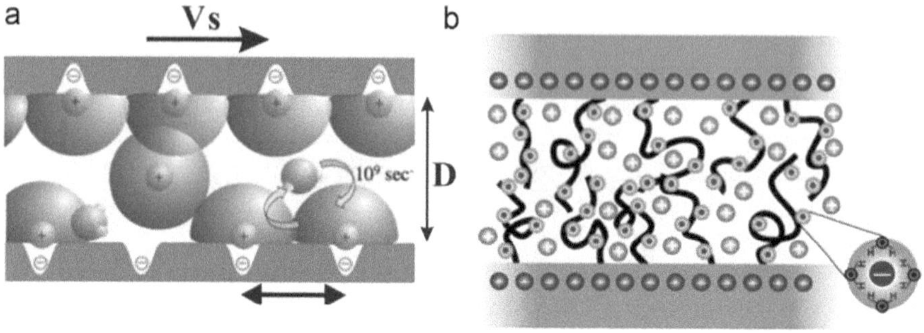

Figure 10.4 (a) It is shown that the moisture layer is capable of supporting significant weight without being expelled, and the rapid exchange rate of water molecules within this layer with those in the surrounding environment or other moisture layers maintains its fluid nature and (b) and the charged polymer brushes [21].

and entropically unfavorable to displace water molecules, allowing them to absorb and dissipate shear forces dynamically. These unique characteristics afford biological interfaces exceptional lubrication, even under significant pressure [21]. Figure 10.4 depicts the behavior of lubrication in the presence of charged brushes and neutral brushes and how the charge affects the brush behavior of polymers.

10.3 HYDROGEL-BASED LUBRICATED COMPOSITE AS LUBRICATION

Conventional hydrogels typically demonstrate inadequate tribological properties under conditions of high stress, attributable to their deficient mechanical robustness and resilience. By incorporating distinct types of micron-scale fibers into purely chemically cross-linked hydrogels, it is possible to fulfill the requisite criteria for mechanical strength and toughness, thereby markedly ameliorating their tribological efficacy as biolubricants [33]. Figure 10.5 shows that fiber-reinforced hydrogels within PVP-PHEMA have rigid structures that have been shown to increase the lubrication behavior by decreasing friction. Fiber-reinforced composite hydrogels significantly outperform pure hydrogels in mechanical strength, with the embedded fibers extending fractures to absorb energy during shearing and tearing, thereby improving durability and toughness, as demonstrated by spun lace fiber composites maintaining low friction and high wear resistance even after 10,000 sliding cycles with a polydimethylsiloxane (PDMS) ball. Other composites like gelatin-based composite hydrogels, namely GelMA–PAM–PMPC, MXene/polyvinyl alcohol (PVA) (MS-MP) hydrogel composite, high hydration potential PAAm-*co*-PAMPS microgel, bioactive

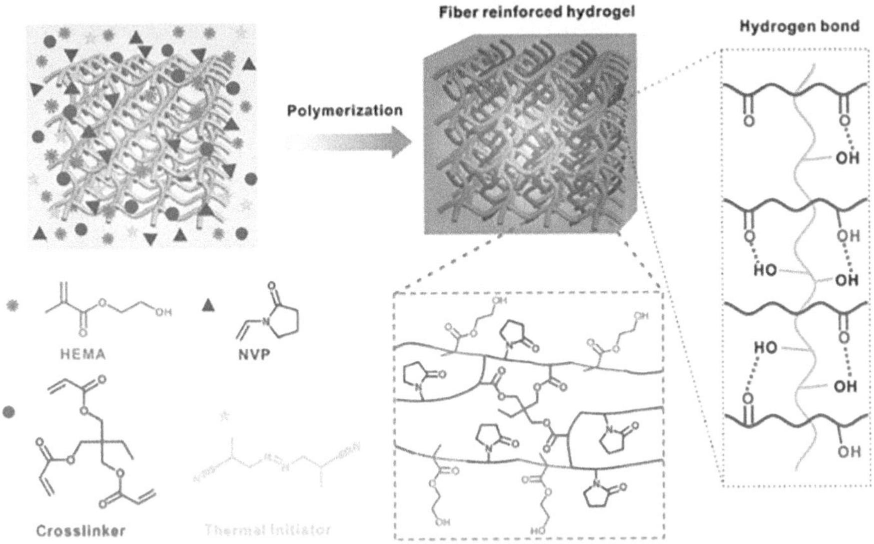

Figure 10.5 A schematic depiction illustrates the fabrication method of the fiber-reinforced PVP-PHEMA composite hydrogels, encompassing the associated molecular and network configurations, as well as the hydrogen bonding interactions between the surface of the fibers and the hydrogel matrix [33].

glass (BAG) particles embedded in gellan gum (GG) hydrogel composite, and PAAm hydrogel microspheres were fabricated through emulsion polymerization, and subsequently, composites of these microspheres with ultra-high-molecular-weight polyethylene (UHMWPE) were produced using the melt blending technique [34, 36, 37]. These composites are a blend of soft materials for biological application as inorganic materials won't be amenable to biological tissue cells. Incorporating zinc ions into MXene/PVA hydrogels enhances their structure and lubrication through multi-crosslinking and salting out effects, offering a novel approach to creating solid–liquid lubricating materials with excellent load capacity and lubrication [36]. Research on how particle size, crosslinking density, and PAAm microsphere filler levels affect the tribological properties of composites highlighted their effectiveness in creating a lubricating hydration layer, thus improving functionality and presenting an innovative strategy for developing sophisticated materials for water-lubricated stern tube bearings [37].

The coefficient of friction (CoF) for GelMA–PAM–PMPC composites demonstrates a decline from 0.018 to 0.011 as the rotational speed increases from 0.5 to 5 Hz, a trend also observed in GelMA, attributed to the formation of an effective hydration layer between contact surfaces at higher speeds, leading to lower CoF values. The enhanced lubrication of GelMA–PAM–PMPC hydrogels is due to their superior water retention, facilitating hydration lubrication. Durability tests conducted with hard steel and soft polydimethylsiloxane (PDMS) as counter surfaces sliding against these hydrogels under a 0.5 N load and at 5 Hz for 10,000 cycles further evaluate their performance [34].

A hydrogel exhibiting reversible photo-induced lubrication was developed by incorporating a responsive system of α-cyclodextrins and polyethylene glycol (α-CD/PEG), along with the competitive guest molecule 1-[*p*-(phenylazo)benzyl]pyridinium bromide (AzoPB), into a poly(vinyl alcohol) (PVA) and polyacrylamide (PAAm) base. This semiconvertible mechanism allows for the alteration of lubrication properties and mechanical toughness upon exposure to different light conditions, specifically under ultraviolet (UV) and visible (Vis) light. The hydrogel's molecular composition, including acrylamide, PVA, α-CD, PEG, and AzoPB, facilitates this unique photoresponsive behavior. Friction tests conducted using a ball-on-disk tribometer under varying humidity levels and with a water film on the hydrogel surface demonstrated enhanced lubrication, with significantly reduced friction coefficients after UV and Vis light exposure compared to a standard PVA and PAAm hydrogel mix. These findings suggest the critical role of supramolecular components in achieving photoresponsive lubrication as shown in Figure 10.6 [38].

Figure 10.7 depicts that hydrogels embedded with lipids achieve self-lubrication through the strategic incorporation of lipid molecules within

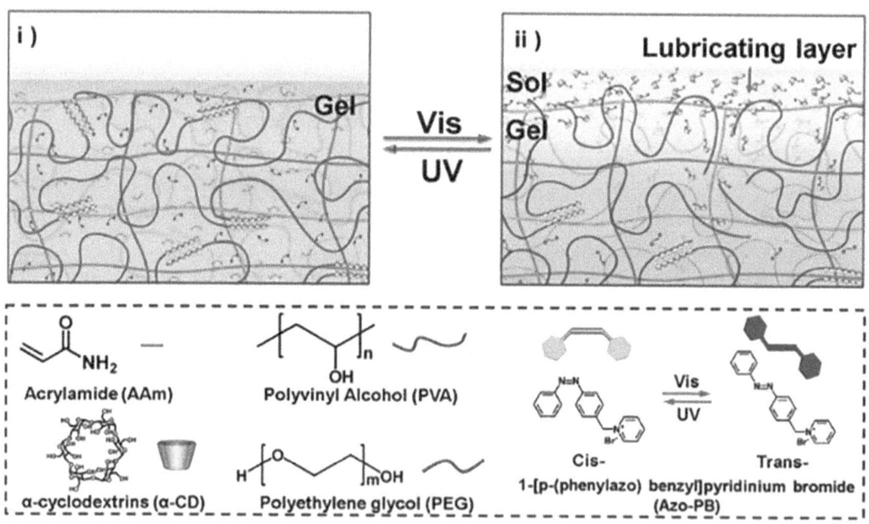

Figure 10.6 Illustration of the photoresponsive lubrication mechanism in the engineered semiconvertible supramolecular hydrogel upon exposure to (i) ultraviolet (UV) and (ii) visible (Vis) light, including the molecular structures of acrylamide, poly(vinyl alcohol), α-cyclodextrins, polyethylene glycol, and 1-[*p*-(phenylazo)benzyl]pyridinium bromide [38].

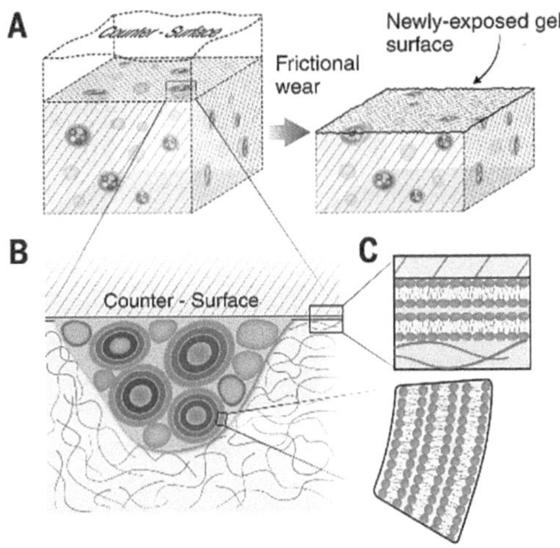

Figure 10.7 Diagram depicting how hydrogels embedded with lipids achieve self-lubrication [55].

their polymeric networks, creating micro-reservoirs that store lubricating substances. Upon the application of mechanical stress or friction, these lipid micro-reservoirs are exposed, releasing the lipids to the hydrogel surface. This release forms a boundary lubrication layer, significantly reducing friction by creating a slippery interface between the hydrogel and contacting surfaces. The lipid layer also facilitates hydration lubrication, where the presence of water molecules further enhances the lubricity by forming a lubricating film. This innovative approach mimics biological lubrication systems, offering a biomimetic solution to reduce wear and friction in various applications [55].

10.4 HYDROGEL USE IN CARTILAGE REPLACEMENT

Methods for addressing cartilage damage due to osteoarthritis have involved the use of hydrogel-based materials as a substitute for damaged cartilage, filling the defects, or as a scaffold for cartilage tissue engineering, where scaffolds containing cells are secured within the damaged areas. In both scenarios, ensuring that the hydrogel surfaces are adequately lubricated is crucial to prevent the layers from being torn away (delamination) under significant pressures in the joints through continuous interaction, compression, and movement [7]. Double network (DN) hydrogels, which combine a stiff polyelectrolyte matrix with a flexible, lightly cross-linked secondary polymer, exhibit remarkably low coefficients of friction (CoF), enhancing their suitability for applications requiring high durability under sliding conditions. To replicate the complex, porous structure of natural cartilage, researchers have developed hydrogels with embedded polymer chains in macroporous structures, significantly reducing CoF through enhanced lubrication. Further innovations include the creation of hydrogels with surface-grafted polymer brushes to simulate the layered lubrication mechanism of articular cartilage, combining a softer surface layer with a firmer underlying matrix. Additionally, hydrogels infused with liposomes have been designed to mimic the self-lubricating lipid layers found in living cartilage, offering a bioinspired approach to achieving low friction surfaces in synthetic materials [7].

Articular cartilage lubrication primarily operates through three modes: hydrodynamic, boundary, and weeping lubrication. Hydrodynamic lubrication occurs when the cartilage surfaces in motion generate a pressurized fluid film that separates the surfaces, reducing direct contact and friction. This model is dependent on the relative velocity of the surfaces and the viscosity of the synovial fluid. In boundary lubrication, molecularly thin layers of lubricating molecules, such as lubricin and hyaluronic acid, adsorb onto the cartilage surfaces, creating a slippery interface that prevents wear even when the joint is under high load and the fluid film is minimal. Weeping lubrication, on the other hand, involves the expulsion of interstitial fluid from the cartilage under compression, which then seeps back, carrying away debris and providing a lubricating layer [39–46]. The structural and chemical properties direct the frictional behavior of hydrogels like articular cartilage, which plays a significant role in moderating their tribological properties [89, 91]. Figure 10.8 illustrates the intricate lubrication system within a synovial joint, emphasizing the layered composition of the boundary layer with hyaluronic acid, lubricin, and phospholipids, all resting on a collagen foundation. It further details how fluid film lubrication maintains a thin liquid barrier between cartilage surfaces to reduce friction and wear, governed by the Newtonian relationship between shear stress, shear rate, and fluid viscosity. Additionally, it delves into boundary lubrication, where friction reduction is primarily dependent on the molecular interactions at the surface interface, independent of the deeper structural layers. This depiction underscores the complexity and efficiency of joint lubrication, which is crucial for smooth and pain-free movement [7].

The lubrication of hydrogels, materials widely used in biomedical engineering due to their high water content and biocompatibility, shares similarities with cartilage lubrication. Hydrogels can mimic the hydrodynamic and boundary lubrication mechanisms of articular cartilage. In hydrodynamic lubrication, the movement of hydrogel surfaces relative to each other can generate a fluid film derived from the water within the hydrogel, reducing friction. For boundary lubrication, hydrogels can be functionalized with molecules similar to those in synovial fluid, such as hyaluronic acid or phospholipid analogs, to create a lubricious surface layer. Moreover, the inherent porosity and water retention capabilities of hydrogels allow for a weeping-like lubrication mechanism, where water

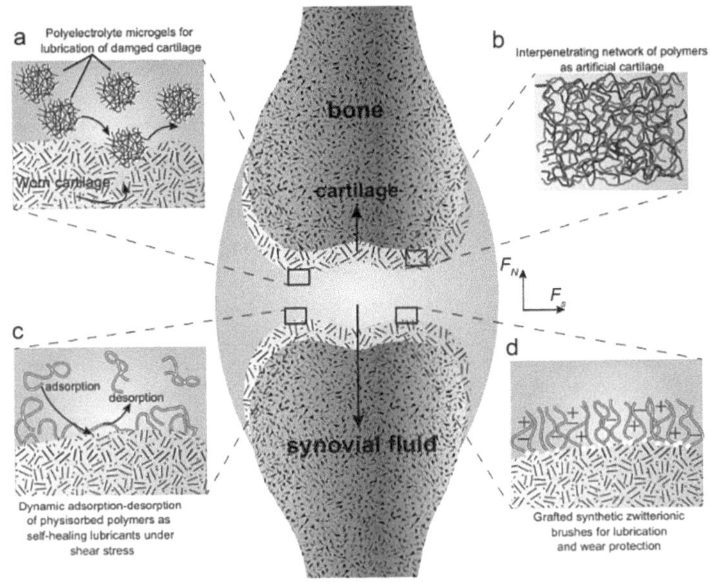

Figure 10.8 Synthetic polymers are utilized in biolubrication and enhancing wear resistance through various applications: (a) utilizing microgels not only for lubrication but also for delivering medication to damaged cartilage areas, (b) creating interpenetrating polymer networks that form hydrogels with superior wear resistance and mechanical characteristics suitable for synthetic cartilage, (c) developing polymeric coatings that are physically adsorbed onto surfaces, offering lubrication along with self-repair capabilities, and (d) employing densely packed synthetic brushes that are firmly attached to surfaces to provide both lubrication and increased resistance to wear [20].

is expelled and reabsorbed, facilitating the removal of wear particles and reducing surface friction [46–48]. The coefficient of friction for hydrogels is influenced by velocity rather than hardness, with the Stribeck curve indicating hydrodynamic lubrication at high speeds and low loads, and *N*,*N*-dimethylacrylamide (DMAA)) hydrogels demonstrating lower friction coefficients than AAm hydrogels despite similar hardness. Hydrogels achieve reduced friction through the development of a fluid film, maintained by the attractive forces between water molecules, which are crucial for minimizing friction. [49] (Figure 10.9).

Hasan et al.'s research has shown that hydrogels with gradient-stiffness layers, mimicking the structure of biological tissues like cartilage, exhibit unique contact mechanics and adhesion properties due to their internal polymer distribution, as revealed by nanoindentation tests using atomic force microscopy. These findings, particularly the increased adhesion and elasticity observed in hydrogels molded with different materials, underscore the potential of gradient-stiffness hydrogels in diverse applications within health science and environmental fields [50]. Hydrogels designed to mimic

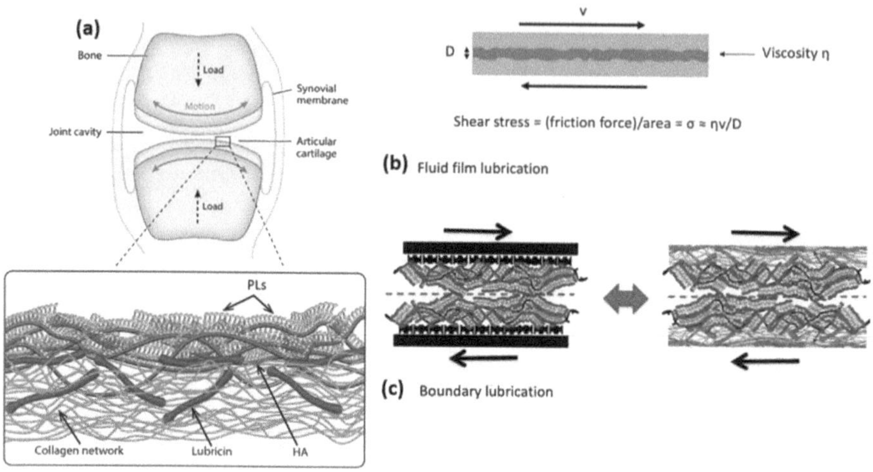

Figure 10.9 (a) Depicts a synovial joint, highlighting its boundary layer composed of hyaluronic acid (HA), the lubricating mucinous glycoprotein known as lubricin, and phospholipids (PLs) atop a collagen matrix. (b) Describes the fluid film lubrication mechanism where a slender fluid layer keeps the cartilage surfaces apart during motion, minimizing direct contact and significantly reducing friction and wear, with the shear stress σ_s being expressible through the Newtonian equation, σ_s = (shear rate) × (effective fluid viscosity). (c) Explains that in the boundary lubrication mode, frictional energy loss happens at the contact point between opposing boundary layers, heavily reliant on the precise molecular arrangement at this junction and largely unaffected by the underlying substrates of these layers [7].

the lubrication and wear resistance of natural cartilage facilitate smooth, pain-free movement across numerous interaction cycles, with their tribological properties shaped by factors like polymer composition and cross-linking. Experimental studies on hydrogels such as gelatin and PAAm have highlighted their load and speed-dependent frictional behaviors, suggesting a range of lubrication mechanisms influenced by the hydrogels' viscoelastic nature and interfacial characteristics [51]. The thermosensitive hydroxypropyl chitin/hyaluronic acid (HPCH/HA) hydrogel presented here offers a dual advantage of effective joint lubrication and cartilage protection, showcasing the potential for synergistic osteoarthritis therapy. Although HPCH alone lacks the optimal lubricating qualities of HA, its gel form provides some lubrication in knee joints and supports cartilage matrix synthesis and cartilage cell repair, potentially offering superior osteoarthritis treatment compared to HA alone. Adding HA to HPCH hydrogels improves their mechanical strength and slows degradation, while HA's gradual release extends its effectiveness, reducing the need for frequent intra-articular injections. This released HA forms a lubricating film on joint surfaces, lowering friction, and HPCH also aids in converting stem cells to cartilage, enhancing chondrocyte matrix production, and repairing damaged cartilage cells [52] (Figure 10.10).

Freeman et al. delved into the tribological characteristics of poly-2-hydroxyethyl methacrylate (poly-HEMA) hydrogels using a device tailored for tribology studies, especially articular cartilage replacement, structured around a two-level, four-variable factorial experiment. The setup featured a 6-mm diameter stainless steel sphere in contact with poly-HEMA hydrogel disks, examining the effects of variables like applied load, lubrication, the crosslink density of the hydrogel, and its hydration level. Conducting linear oscillatory sliding assessments over 30 min, significant primary effects and variable interactions were identified. Results showed that escalating the load from 6 to 20 N led to a 125% increase in hydrogel wear, while enhancing crosslink density decreased wear by more than 60%, and higher hydration levels caused a 130% rise in wear. The friction coefficient varied widely from 0.02 to 1.7, and wear differed by a factor greater than 60. Notably, the interplay between hydration and lubrication, as well as hydration and crosslinking, significantly influenced wear outcomes. The absence of a direct link between friction and wear in hydrogels, a frequently discussed topic in hydrogel studies, was a pivotal finding. The data is primarily split into two groups: one exhibiting low wear and friction, and the other with high wear but still low friction, providing crucial insights for enhancing hydrogel tribological performance to minimize wear and friction [53, 56–60].

10.5 HYDROGEL-BASED LUBRICATION CELL CULTURE

Hydrogels, with their high water content and tunable physical properties, have emerged as a cornerstone in the realm of cell culture, providing a biomimetic environment that closely replicates the natural extracellular matrix (ECM) of tissues. Their three-dimensional network offers an ideal scaffold that supports cell attachment, proliferation, differentiation, and migration, essential for the development of tissue-like structures [62, 63]. The inherent porosity of hydrogels facilitates efficient nutrient and waste exchange, while their mechanical properties can be finely adjusted to match those of native tissues, ranging from soft brain tissue to more rigid cartilaginous structures. This versatility allows for the cultivation of a diverse array of cell types, including sensitive primary cells and stem cells, under conditions that promote optimal cell function and phenotype expression [64, 65]. Moreover, hydrogels can be functionalized with bioactive molecules such as growth factors, peptides, and adhesion proteins to further enhance cell signaling and tissue development. The integration of hydrogels in 3D cell culture systems has revolutionized tissue engineering and regenerative medicine by enabling the construction of complex, multicellular architectures that better mimic in vivo environments. This advancement not only

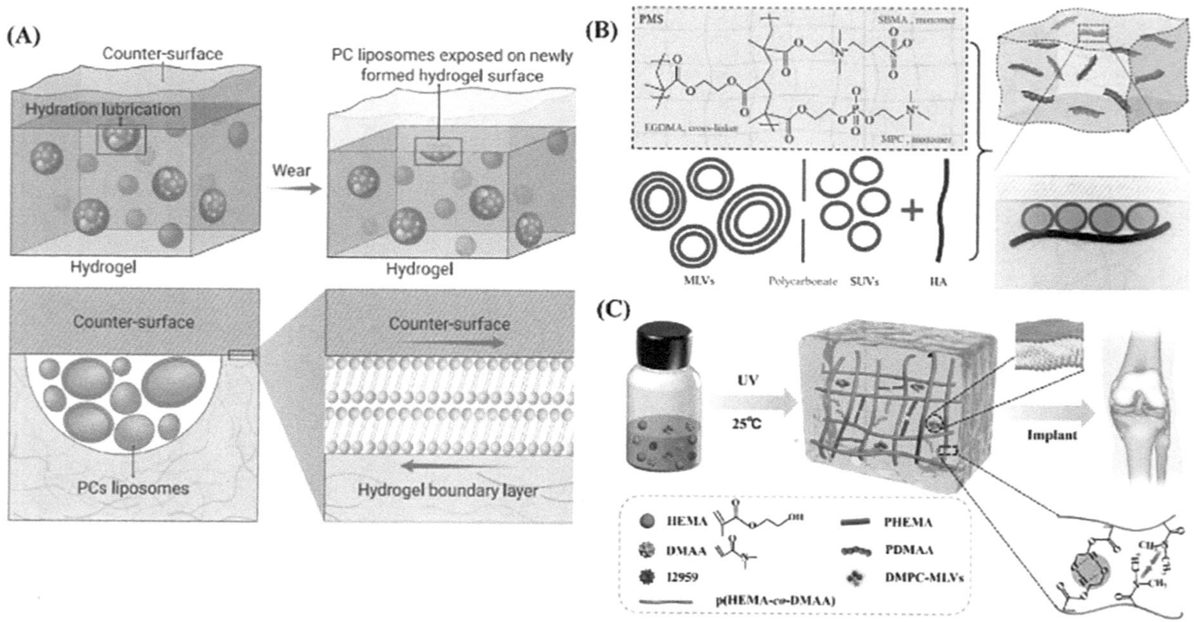

Figure 10.10 (A) Depicts a hydrogel integrated with lipids that create micro-reservoirs within its structure, with additional reservoirs becoming accessible under frictional forces, facilitating the formation of a lipid boundary lubrication layer on the surface, thereby enhancing lubrication through hydration mechanisms. (B) Describes the creation of the PMS-HSPC(SUV)-HA hydrogel and its combined lubrication effect, where the inclusion of lipid SUVs and HA results in a highly lubricated state due to the cooperative interaction between lipids and HA, with lipid liposomes forming around the HA, leading to improved lubrication. (C) Illustrates the development of lipid-lubricated hydrogels designed for biocompatibility and high-performance lubrication, leveraging lipid-based lubrication mechanisms [54].

improves our understanding of tissue development and disease progression but also holds significant promise for drug screening, toxicology studies, and the development of personalized medicine approaches. Figure 10.11 represents the hydration lubrication mechanism of interactions between biological organisms [66, 67].

Synthetic hydrogels, created from artificial polymers like PVA and PEG, offer structural support to cells but lack biological activity, necessitating modifications with bioactive molecules for enhanced cellular interactions. Incorporating elements such as arginine-glycine-aspartate (RGD) peptides into these hydrogels has been shown to facilitate fibroblast activity and stem cell differentiation, vital for 3D cultures. Hyaluronic acid (HA), a key component in tissue regeneration, has been integrated into hydrogels to simulate the tumor environment for drug testing and improve cell culture outcomes by supporting cell proliferation and matrix interaction. Recent advancements include HA-based bio-inks for 3D bioprinting and their combination with other polymers to strengthen hydrogel matrices for applications like cartilage regeneration [61, 68, 69, 91]. Figures 10.12 and 10.13 represent the adhesion and proliferation of cells within a hydrogel matrix and make a 3D cell culture environment rather than 2D, which is a more realistic representation of the human body.

Peptide-based hydrogels, capable of self-assembling into highly hydrated networks, offer a biocompatible, biodegradable, and functional three-dimensional framework for studying cellular dynamics, facilitated by alternating hydrophobic and hydrophilic amino acid sequences under specific pH and ionic conditions [70–76]. These hydrogels support cell adhesion through interactions between cell surface receptors like integrins, which connect to the cell cytoskeleton, and hydrogel-bound ligands, enhancing cell attachment, migration, and matrix remodeling. Additionally, other receptors such as syndecans and cadherins complement integrin function, promoting adhesion to the matrix and neighboring cells, thereby influencing cell behavior and matrix synthesis [77–83]. Stem cell differentiation is

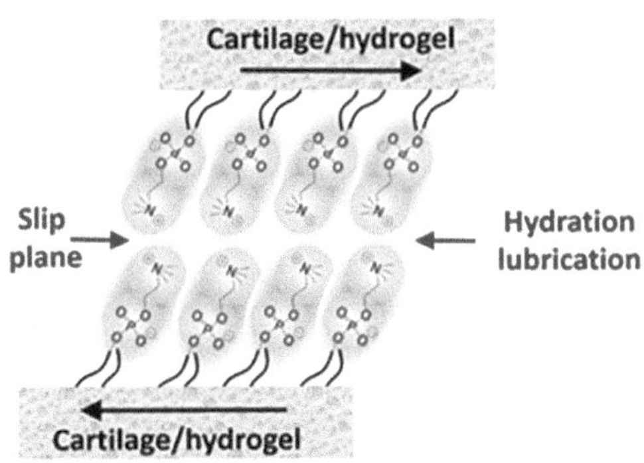

Figure 10.11 The hydration lubrication visible within the hydrogel/cartilage surfaces [54].

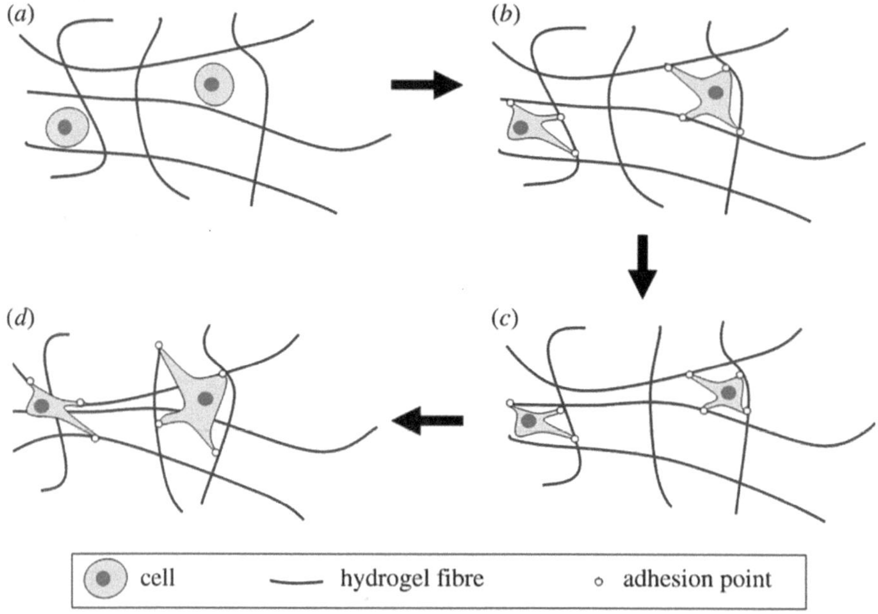

Figure 10.12 Diagram illustrating the contraction mechanism within a hydrogel containing cells: (a) the initial state where cells are dispersed throughout the hydrogel network, (b) the phase where cells extend and attach to the network's fibers, (c) a stage where the cells' traction forces lead to fiber deformation, and (d) the final process where cells detach and rebind to different fibers, culminating in the hydrogel's overall contraction [62]

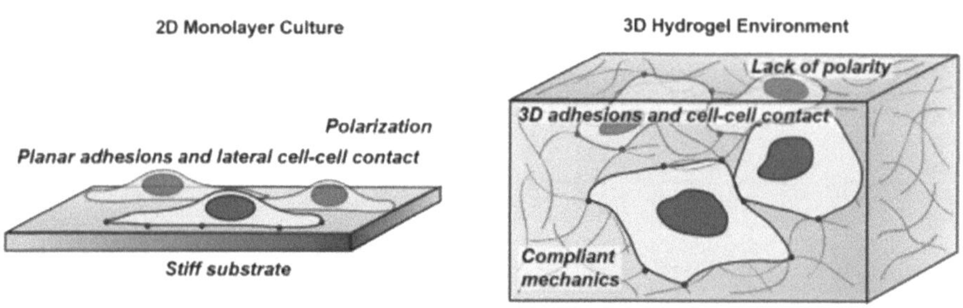

Figure 10.13 In a two-dimensional monolayer setup (on the left), stem cells experience polarization due to anchoring on one side to a rigid plastic base, resulting in restricted engagements with neighboring cells and three-dimensional matrix elements. Conversely, within a 3D hydrogel context (on the right), the environment furnishes multidimensional signals and fosters extensive interactions among cells and with the matrix in a three-dimensional space [63].

influenced by the stiffness of the substrates, with material rigidity in three-dimensional hydrogels affecting cell behavior and remodeling through polymer concentration, arrangement, and crosslink density. Specifically, mesenchymal stem cells exhibit enhanced proliferation on stiffer hydrogels, whereas neuronal stem cells show reduced proliferation with increasing hydrogel stiffness [62, 66–69]. Micro-friction experiments conducted between HEMA- and silicone-based hydrogels and a confluent HCE-T layer revealed a friction coefficient of around 0.03, with increased sliding cycles leading to more extensive damage to the cell layer. The detachment of cells during sliding was directly linked to the friction coefficient, indicating that the hydrogel-on-cell setup resulted in minimal friction [90].

Hydrogels can enhance lubrication in cell culture by providing a highly hydrated, three-dimensional matrix that mimics the natural extracellular environment, reducing friction between cells and culture substrates [84]. Their water-rich composition facilitates the formation of a lubricating layer, minimizing mechanical stress and shear forces on cells during movements or manipulations. The viscoelastic properties of hydrogels allow for the absorption and dissipation of applied forces, further contributing to a lubricative effect. By closely replicating the lubrication mechanisms found in biological tissues, hydrogels can improve cell viability and function, making them ideal for dynamic cell culture systems that require low friction and high wear resistance as shown in Figure 10.14 [85–90, 93-95].

10.6 CONCLUSION

The comprehensive investigation into hydrogels and polymer-based materials as lubricants underscores their transformative potential in reducing friction and serving as biological replacements within the human body. These materials, characterized by their high water content, biocompatibility, and tunable physical properties, mimic the lubrication mechanisms inherent in natural biological systems, offering a promising avenue for enhancing tribological performance in various applications. Hydrogels, in particular, have demonstrated remarkable versatility in replicating the extracellular matrix's structure and function, facilitating a conducive environment for cell growth, proliferation, and differentiation. This makes them invaluable in tissue engineering and regenerative medicine, where they can act as scaffolds for tissue regeneration, cartilage replacement, and even in the development of synthetic tissues that closely emulate natural counterparts. The ability of hydrogels to reduce friction and wear in biological systems is further enhanced by incorporating bioactive molecules and leveraging stimuli-responsive properties, which enable precise control over their mechanical and tribological characteristics. Polymer-based materials extend this functionality by offering customizable solutions for specific lubrication needs, whether in industrial applications or within the human

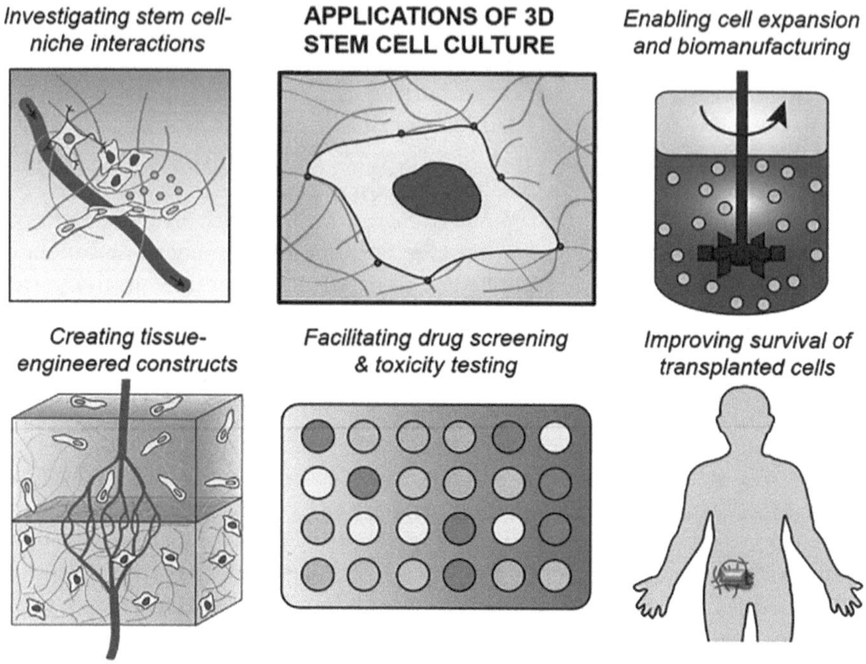

Figure 10.14 Growing stem cells within three-dimensional hydrogels facilitates various applications such as exploring stem cell environments, developing tissue-engineered models, enhancing drug testing platforms, advancing bioproduction processes, and boosting cell viability and performance after transplantation [63].

body. The integration of polymers like PTFE, PEEK, and nylon, known for their excellent wear resistance and low friction coefficients, with hydrogels, introduces composite materials that combine the best of both worlds: the mechanical robustness of polymers and the biological mimicry of hydrogels.

Advancements in hydrogel technology, such as the development of fiber-reinforced composites, lipid-incorporated hydrogels, and photoresponsive hydrogels, have opened new horizons for creating lubrication systems that not only reduce friction but also actively contribute to cellular and tissue health. These innovations have shown potential in creating more effective prosthetic devices, enhancing the longevity and functionality of implants, and providing novel solutions for drug delivery and wound healing. In conclusion, the strategic use of hydrogels and polymer-based materials as lubricants represents a significant leap forward in materials science and biomedical engineering. By closely mimicking the natural lubrication found in the human body, these materials offer a pathway to reducing friction and wear in biological systems, improving the quality of life for individuals through enhanced prosthetic devices and regenerative therapies, and paving the way for the next generation of biomedical innovations [92]. Their continued development and integration into medical applications promise not only to advance our understanding of tribological interactions at the biological interface but also to revolutionize the design and functionality of biomedical devices and systems.

REFERENCES

1. Shoaib, T. & Espinosa-Marzal, R. M. (2020). *Colloids Interfaces*, *4*(4), 54.
2. Mantelli, F. & Argueso, P. (2008). *Curr. Opin. Allergy Clin. Immunol.*, *8*, 477–483.
3. Sterner, O., et al. (2017). *ACS Appl. Mater. Inter.*, *9*, 20150–20160.
4. Swann, D. A., et al. (1974). *Ann. Rheum. Dis.*, *33*, 318–326.
5. Schmidt, T. A., et al. (2007). *Arthritis Rheum.*, *56*, 882–891.
6. Hu, J., et al. (2020). *ACS Macro Lett.*, *9*(12), 1726–1731.
7. Lin, W. & Klein, J. (2022). *Acc. Mater. Res.*, *3*(2), 213–223.
8. Liu, G., et al. (2022). *Che. Eng. J.*, *429*, 132324.
9. Brostow, W., et al. (2010). *J. Mater. Edu.*, *32*(5), 273.
10. Wu, Y., et al. (2015). *Adv. Mater. Interfaces*, 2(2), 1400392.
11. Datta, A., et al. (2022). Advanced materials in biological implants and surgical tools. In: *Advanced Materials for Biomedical Applications.* CRC Press, pp 21–43.
12. Kumar, A., et al. (2022a). Recent advancements and future trends in next-generation materials for biomedical applications. In: *Advanced Materials for Biomedical Applications.* CRC Press, pp 1–19.
13. Kumar, A., et al. (2016a). Liquid slippage in confined flows: Effect of periodic micropatterns of arbitrary pitch and amplitude. In: *ASME 2016 5th International Conference on Micro/Nanoscale Heat and Mass Transfer, MNHMT* 2016.
14. Kumar, A., et al. (2016b). *Phys Fluids,* 28.
15. Kumar, A., et al. (2018a). Reduction of hydraulic friction in confined flows by laser texturing: Experiments and theoretical validation. In: *ASME 2018 16th International Conference on Nanochannels, Microchannels, and Minichannels.*
16. Xia, F., et al. (2006). *Adv. Mater.*, *18*(4), 432–436.
17. Zhang, Y., et al. (2021). *Chem. Mater.*, *33*(8), 2785–2795.
18. Singh, M. K. (2020). Polymer brush based tribology. In: *Tribology in Materials and Applications* (pp. 15–32). Cham: Springer International Publishing.
19. Shiao, S. J., et al. (2001). *J. Appl. Polym. Sci. J. App. Pol. Sci.*, *80*(9), 1514–1519.
20. Adibnia, V., et al. (2020). *Prog. Polym. Sci.*, *110*, 101298.
21. Yang, W. & Zhou, F. (2017). *Biosurf. Biotribol.*, *3*(3), 97–114.
22. Park, J. T., et al. (2013). *Adv. Funct. Mater.*, *23*, 26–33.
23. Hwang, D. -H., et al. (2012). *J. Nanosci. Nanotechnol.*, *12*, 4137–4141.
24. Cayre, O. J., et al. (2011). *Soft Matter.*, *7*, 2211–2234.
25. Chen, T., et al. (2010). *Prog. Polym. Sci.*, *35*, 94–112.
26. Kumar, A., et al. (2018). *J. Heat Transfer,* 140.

27. Kumar, A., et al. (2016). Liquid slippage in con fined flows: Effect of periodic micropatterns of arbitrary pitch and amplitude. In: *ASME 2016 5th Micro/Nanoscale Heat and Mass Transfer International Conference (MNHMT2016)*, Biopolis, Singapore.
28. Kumar, A., et al. (2022). Biomedical study of femur bone fracture and healing. In: *Advanced Materials for Biomedical Applications*. CRC Press, pp. 235–250.
29. Kumar, A., et al. (2013). *J. Micro./Nanolithogr.*, MEMS, MOEMS 12.
30. Kumar, A., et al. (2018). *Tribol. Trans.*, *61*, 951–959.
31. Zhao, B. & Zhu, L. (2009). *Macromolecules*, *42*, 9369–9383.
32. Weir, M. P. & Parnell, A. J. (2011). *Polymers*, *3*, 2107–2132.
33. Li, J., Gao, et al. (2022). *Friction*, *10*, 54–67.
34. Zhang, K., et al. (2022). *Friction*, 1–15.
35. Qu, M., et al. (2020). *Chem. Mat.*, *32*(18), 7805–7813.
36. Miao, X., et al. (2023). *Chem. Eng. J.*, *476*, 146848.
37. Wang, C., et al. (2021). *Wear*, *477*, 203841.
38. Wang, J., et al. (2021). *Matter*, *4*(2), 675–687.
39. Lin, W. & Klein, J. (2022). *Acc. Mater. Res.*, *3*(2), 213–223.
40. Ren, N. (2009). Advanced modeling of mixed lubrication and its mechanical and biomedical applications (Doctoral dissertation, Northwestern University).
41. Kumar, A. & Panda, U. (2022). Microfluidics-based devices and their role on point-of-care testing. In: *Biosensor-based Advanced Cancer Diagnostics*. Elsevier, pp. 197–224.
42. Kumar, A., et al. (2022). *Trends Fabr. Polym. Polym. Compos.*, 1–3.
43. Kumar, A., et al. (2022). Clinically available/under trial drugs and vaccines for treatment of SARS-CoV-2. In: *Computational Approaches for Novel Therapeutic and Diagnostic Designing to Mitigate SARS-CoV2 Infection*. Elsevier, pp. 451–488.
44. Kumar, A., et al. (2022). *ACS Appl. Bio. Mater.*, *5*, 2046–2068.
45. Kumar, A., et al. (2023). Physics of laser—matter interaction in laser-based manufacturing. In: *Laser-Based Technologies for Sustainable Manufacturing*. CRC Press, pp. 45–54.
46. Freeman, M. E., et al. (2000). *Wear*, *241*(2), 129–135.
47. Murakami, T., et al. (2015). *Proc. Inst. Mech. Eng. Part H: J. Eng. Med.*, *229*(12), 864–878.
48. Blum, M. M. & Ovaert, T. C. (2013). *Wear*, *301*(1–2), 201–209.
49. Ishikawa, Y., Hiratsuka, K. I., & Sasada, T. (2006). *Wear*, *261*(5–6), 500–504.
50. Hasan, M. M. & Dunn, A. C. (2023). *J. Chem. Phys.*, *159*(18).
51. Geurds, L., Xu, Y., & Stokes, J. R. (2021). *Biotribology*, *25*, 100162.
52. Liu, W., et al. (2022). *Mater. Des.*, *217*, 110579.
53. Freeman, M. E., et al. (2000). *Wear*, *241*(2), 129–135.
54. Qiu, F., et al. (2023). *Gels*, *9*(2), 144.
55. Lin, W., et al. (2020). *Science*, *370*(6514), 335–338.
56. Kumar, A., et al. (2023). *Lubricants*, *11*, 64.
57. Kumar, A., et al. (2022). Influence of hexagonal boron nitride additive nanocutting fluid on the machining of AA6061-T6 alloy using minimum quality lubrication. *Proc. Inst. Mech. Eng. Part E J. Process. Mech. Eng.* 09544089221110980.

58. Kumar, S., et al. (2022g). *Trends Fabr. Polym. Polym. Compos.*, 10–11.
59. Leventini, S. D., et al. (2022). *Tribol. Lett.*, *70*, 127.
60. Mehra, A. K., et al. (2021). *Compos. Commun.*, *25*, 100732.
61. Park, Y., et al. (2021). *Int. J. Mol. Sci.*, *22*(5), 2491.
62. Ahearne, M. (2014). *Interface Focus*, *4*(2), 20130038.
63. Wilson, J. L. & McDevitt, T. C. (2017). Biofunctional hydrogels for Three-dimensional stem cell culture. In: *Biology and Engineering of Stem Cell Niches*. Academic Press, pp. 345–362.
64. Brown, R. A., et al. (1998). *J. Cell. Physiol.*, *175*(3), 323–332.
65. Kharkar, P. M., et al. (2013). *Chem. Soc. Rev.*, *42*(17), 7335–7372.
66. Mauch, C., et al. (1989). *FEBS Lett.*, *250*(2), 301–305.
67. Lutolf, M. P., et al. (2003). *Proc. Nat. Acad. Sci.*, *100*(9), 5413–5418.
68. Ahearne, M., et al. (2010). *Tissue Eng. Part C: Methods*, *16*(2), 319–327.
69. Chung, C., et al. (2012). *J. Soft Matter.*, *8*(39), 10141–10148.
70. Miller, J. S., et al. (2010). *Biomaterials*, *31*(13), 3736–3743.
71. Parihar, A., et al. (2023). Cryopreservation: A comprehensive overview, challenges, and future perspectives. *Adv. Biol.*, 2200285.
72. Parihar, A., et al. (2021). *Regen. Eng. Transl. Med.*, 1–27.
73. Parmar, V., et al. (2018). *J. Laser. Appl.*, 30.
74. Parmar, V., et al. (2019). *Mech. Mater.*, 137.
75. Pathak, A., Kumar, A., Kumar, A., & Kumar, A. (2023). Application of laser technology in the mechanical and machine manufacturing industry. In: *Laser-based Technologies for Sustainable Manufacturing*. CRC Press, pp. 107–155.
76. Hadjipanayi, E., et al. (2009). *Cell Mot. Cytoskel.*, *66*(3), 121–128.
77. Gupta, S., et al. (2020). Performance analysis of gudgeon pin of various cross sections by FEM. *Int. J. Recent. Technol. Eng.*, *8*, 4569–4573.
78. Shrivastava Prateek, S. R., et al. (2020). Investigation of torsional rod to minimize vibration in automobile using ANSYS. In: *2020 1st International Conference on Innovative Research in Applied Science, Engineering and Technology (IRASET)*, pp. 1–6.
79. Singh, M., Kumar, A., & Khan, A. R. (2020). *Phys. Rev. Fluids*, *5*, 102101.
80. Singh R. K., et al. (2014b). *Microsyst. Technol.*, *20*, 101–111.
81. Subramanian, Y., et al. (2023). *J. Electron. Mater.*, 1–11.
82. Vats, P., Gajrani, K. K., & Kumar, A. (2023). Laser-based additive manufacturing. In: *Laser-based Technologies for Sustainable Manufacturing*. CRC Press, pp. 67–83.
83. Dias, D., et al. (2024). *Lubricants*, *12*(4), 114.
84. Cao, Y., et al. (2017). *J. Mech. Behav. Biomed. Mat.*, *68*, 163–172.
85. Han, Y., et al. (2021). *Bioact. Mater.*, *6*(10), 3596–3607.
86. Pan, M., et al. (2023). *Chem. Mater.*, *35*(13), 4998–5008.
87. Rafid, M., et al. (2021). Effects of surface discontinuity on droplet dynamics in a microfluidic channe (2021), 15th International conference on heat transfer, fluid mechanics and thermodynamics proceedings, Vol-15, pp. 754–759.
88. Dolan, G. K., et al. (2017). *Soft Matter*, *13*(19), 3592–3601.
89. Chowdhury, N. (2023). Structural property relation in improving chemically and physically crosslinked hydrogel surface properties (Doctoral dissertation, University of Illinois at Urbana-Champaign).

90. Dunn, A. C., et al. (2008). *Tribol. Lett.*, *30*, 13–19.
91. Chowdhury, N., et al. (2024). Characterization of Hydrogel Properties in the Advancement of Bio-Tribology, In *Application of Biotribology in Biomedical Systems*, pp 27–78. Cham: Springer Nature Switzerland.
92. Pal, A., Gour, R., Kumar, A., Kumar, A., Kumar, A., & Chowdhury, N. (2024). Tribological Measurements of Human Skin. In *Applications of Biotribology in Biomedical Systems*. pp. 149–165.
93. Choudhari, A., Gupta, A. K., Kumar, A., Kumar, A., Gupta, A., Chowdhury, N., & Kumar, A. (2024). Wear and Friction Mechanism Study in Knee and Hip Rehabilitation: A Comprehensive Review. In *Applications of Biotribology in Biomedical Systems*, 345–432.
94. Kumar, A., Kumar, A., Kumar, A., Choudhari, A., Gupta, A. K., & Faisal, S. (2024). Introduction to Biotribology: A Science of Surface Interaction. In *Applications of Biotribology in Biomedical Systems,* pp. 1–26. Cham: Springer Nature Switzerland.
95. Kumar, A., Gajjar, P., Gudadur, K. S., Kumar, A., Gupta, A. K., Choudhari, A., ... & Kumar, A. (2024). Tribological Hurdles in Biomedical Manufacturing: A Comprehensive Examination. In *Applications of Biotribology in Biomedical Systems*. pp. 167–198. Cham: Springer Nature Switzerland.

Chapter 11

Recent advances, current challenges, and future perspectives of viscosity index improvers

Yatika Gori, Chandra Kishore, Pravin P. Patil, and M. R. Tyagi

11.1 INTRODUCTION

In recent years, problems such as global warming, the depletion of natural resources, and air pollution caused by emissions are emerging on a global scale. These problems call for efforts directed toward the development of fuel-efficient engines and exhaust gas reduction measures. As a solution to these issues, performance improvements should be achieved on the oil that lubricates the sliding sections of engines. The human need for energy is insatiable. Man's use of conventional and unconventional hydrocarbon sources has served human energy requirements, but not without associated challenges of renewability, toxicity, depletion, degradability, and environmental hazards. Since a complete substitute for fossilized energy sources is yet to be realized, one viable option available to man is alternative, renewable, ecofriendly, and biodegradable energy sources – thus, minimizing complete reliance on mineral-based energy sources [1–3].

Lubricants are finding relevance in automobiles, product processing, transportation, agriculture, and other key sectors of industrial development as a result of the use of machines to aid human operations. While more than 70% of commercially available lubricants are from crude oil/hydrocarbon, commercial formulations of lubricants using vegetable base oils are gaining global interest. Lubricants obtained from bio-based oils are eco-compatible but are widely reported to be thermally and oxidatively unstable, especially when such biolubricants are produced from the crude form of vegetable oils. Lubricants are composed of 70–100% base oil and 1–30% additives and surfactants incorporated to enhance their performances and provide dispersion stability to the lube oil. The base oil could be mineral base oil, synthetic base oils, re-refined base oils, biomass/vegetable base oils, or a blend of them [4–6].

Multiple approaches are being pursued to reduce dependence on petroleum HEV, PHEV, non-petroleum-based fuels, advanced combustion concepts, lightweight materials, regeneration, waste heat recovery, and

DOI: 10.1201/9781003535447-11

advanced technological vehicles. But Figure 11.1 shows increasing demands and dependency on petroleum. Fuel-efficient lubricants offer a pathway to improve the fuel economy of conventional vehicles. The following are three ways to control friction and wear to increase fuel economy:

Development, material, and production costs are always important factors to consider when the market potential for any of these approaches is assessed. This report focuses on the third option, which is lubricants for increasing fuel economy.

In a machine, the friction between metal to metal parts arises due to moving surfaces and the machine experienced resistance that retards its movement. More energy is lost to friction than delivered to the wheels. Approximately 10% in the engine and 5% in the drivetrain. Due to friction, a large amount of energy is liberated in the form of heat (Figure 11.2), which reduces the efficiency of the I.C. Engine. "Substances which apply between two moving and sliding surfaces to reduce friction between them are known as lubricants", and the process by which the friction between sliding surfaces is reduced is known as lubrication.

11.1.1 Classification of lubricants

Figure 11.3 shows the classification of lubricants in the following manner:

A. **Liquid lubricants:** These include animal oils, vegetable oils, petroleum oils, and synthetic lubricants. *Animal oils*: tallow oil, whale oil, etc. *Vegetable oils*: castor oil, palm oil, etc. *Petroleum oils*: petroleum fractions. *Synthetic lubricants*: polyglycol, silicones, etc.

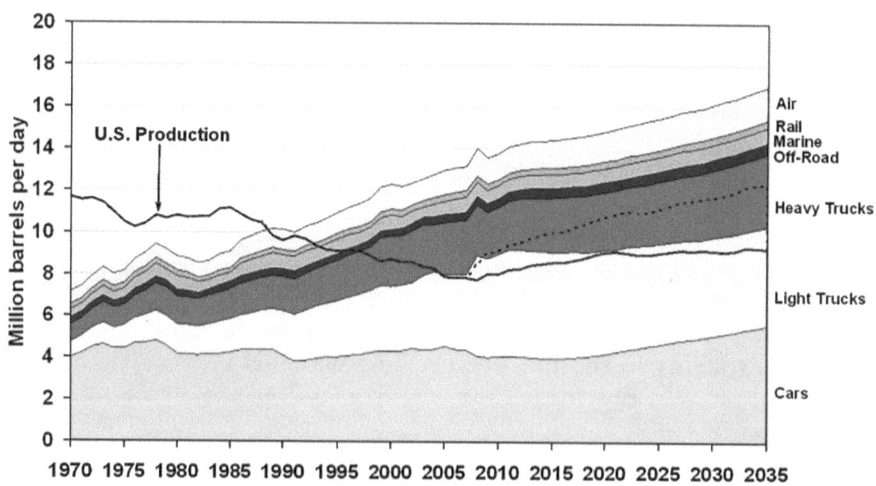

Figure 11.1 Historical use of petroleum for transportation [18]

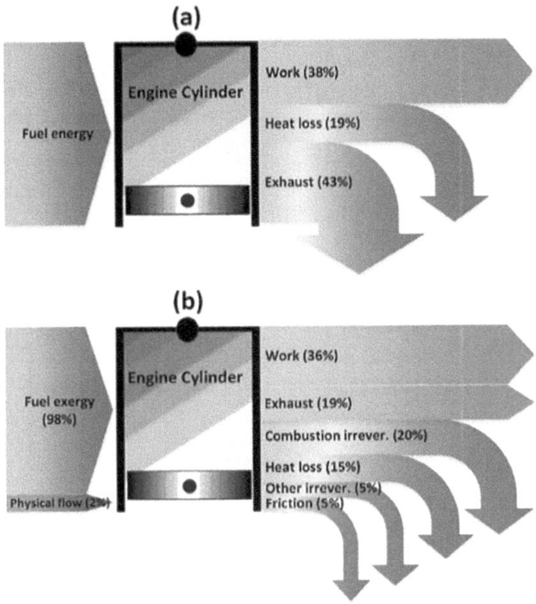

Figure 11.2 Energy losses in internal combustion engine [19].

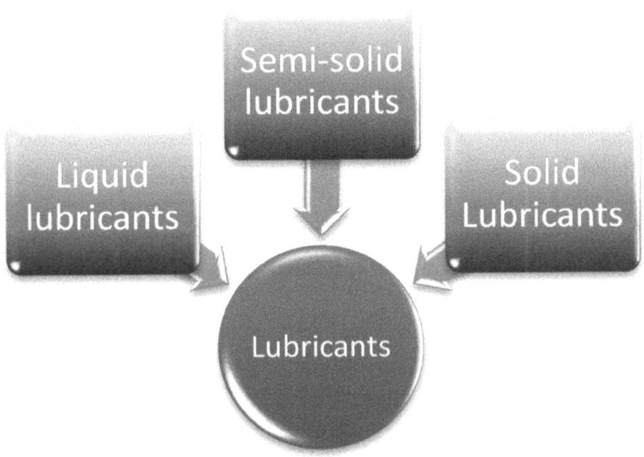

Figure 11.3 Classification of lubricants.

B. **Semi-solid lubricants** (**grease**): Semi-solid lubricants are formed by emulsifying oil and fat with thickening agents like soap of sodium, calcium, lithium, aluminum at higher temperature. *Soda based*: In this case, sodium soaps are used as a thickening agent in mineral or petroleum oil. They are slightly soluble in water. They can be used up to 175°C. *Lithium based*: In this case, lithium soaps are emulsifying

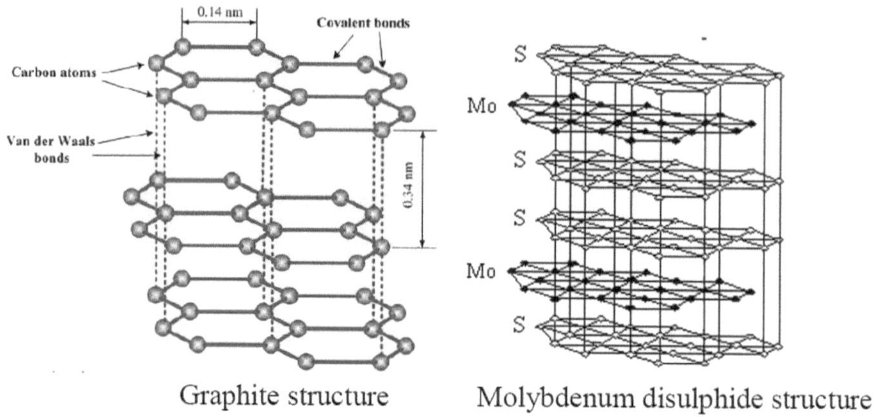

Figure 11.4 Graphite and molybdenum disulfide structure.

with petroleum oil. They are water resistant and used up to 15°C. *Calcium based*: In this case, calcium soaps are emulsified with petroleum oil. They are also water resistant and used up to 80°C. At higher temperatures, soap and petroleum oil are separate from each other.

C. **Solid lubricants:**

Graphite, molybdenum disulfide (MoS_2), and boron nitride $(BN)_x$ are predominantly used as solid lubricants. They are used under high temperature and high load (pressure).

Graphite: It is most widely used as a solid lubricant. Graphite has a layered structure; layers are held together with the help of weak van der Waals' forces, which facilitate the easy sliding of one layer onto the other layer. It is very soapy to touch and non-inflammable. It is used at higher temperature (around 450°C) conditions. They are either used in powder form or mixed with oil or water.

Molybdenum disulfide (MoS_2): It is a sandwich-like structure in which a hexagonal layer of molybdenum (Mo) lies between two hexagonal layers of sulfur (S) atom (Figure 11.4). Like graphite, each of the layers are held together with weak van der Waals' forces. It differs from graphite because it is used in high vacuum unlike graphite (graphite is mixed with water or oil). It adheres even more strongly to the metal or other surface.

11.2 RECENT DEVELOPMENTS

Formulation of engine oils and additives using conventional fossil energy sources is associated with numerous challenges such as toxicity, non-biodegradability, and rising cost as a result of their global shortages [1–3].

Alang et al. [4] and Owuna et al. [3] discovered that mineral-based oil consumption rate exceeds the rate of natural formation of its deposit, and for this reason, the world is at risk of harsh energy challenges if alternative energy solutions are not found. In response to this energy demand, lubricating oil producers and consumers are adopting bio-based materials for the production of lubricants and additives—this is also gaining renewed interest from the research community [5–8]. Bio-based sources, unlike conventional mineral energy sources, are renewable, ecofriendly, and biodegradable [9–14].

The choice of base oil is important because the qualities and performances of the formulated lubricant are dependent on the type of base oils, the refining method adopted, and the type and quality of additive incorporated [14]. For sustainability of development, with the less net negative effect on the environment, the world needs environmentally friendly lubricating oils [15].

Compared to oil lubrication, solid lubrication offers a superior solution for reducing friction, conserving energy and materials, and avoiding the problems encountered with oil-based lubricants [15–17]. PTFE (polytetrafluoroethylene) is known for imparting very low friction and enhancing wear resistance and hence is used as a very efficient solid lubricant (SL) for engineering polymers such as PEEK (polyetheretherketone), PAEK (polyaryletherketone), PIs (polyimides), etc. [18–22]. The reason behind such improvement was studied to a great extent revealing the counterpart interaction with PTFE particles under shear. PTFE is likely to transfer an entire chain by C–C bond scission to the metal surface during sliding. There are excellent discussion of how current and new lubricant additive technology affects friction and thus fuel efficiency [23–26].

By understanding how friction varies under different lubrication regimes (boundary, mixed, and hydrodynamic), engineers can select the optimal lubricant viscosity and operating conditions to minimize friction and wear in machines. This reduces energy consumption and extends machine life. The Stribeck curve is important in the field of tribology, which deals with friction, lubrication, and wear. At low speeds or high loads, a thin lubricant film separates the contacting surfaces imperfectly. This is the boundary lubrication regime, where friction is high due to direct contact of asperities on the surfaces. As speed or load decreases, a thicker lubricant film forms, entering the mixed lubrication regime. Here, friction reduces as the lubricant carries some of the load, but asperity contact still occurs. Finally, at high speeds or low loads, a complete lubricant film separates the surfaces in the hydrodynamic lubrication regime. This film is strong enough to support the load entirely, leading to the lowest friction. The Stribeck curve depicts this transition between regimes, showing a drop in friction as lubrication improves [27].

Optimizing thermal contact between gear teeth and lubricant film minimizes frictional heating. Analyzing thermal contact resistance can help identify areas for improvement in gearbox design [28–34]. Gearbox design considerations like material selection, gear geometry, and lubrication system layout significantly impact thermal performance. A holistic approach that integrates lubricant selection with gearbox design is vital [35–41].

Utilizing finite element analysis (FEA) allows for simulating gear contacts under various operating conditions. This helps predict wear patterns, stress distribution, and thermal behavior within the gearbox [42–46]. Artificial Neural Network (ANN) offer a powerful tool for modeling complex relationships between lubricant properties, operating conditions, and gearbox performance. By training an ANN with experimental data, researchers can predict lubricant performance under various scenarios. By combining these approaches, researchers and engineers can achieve a comprehensive understanding of lubricant performance in gearboxes [47–51].

The interest in the subject is clearly reflected in the information available through the literature. Research as reported in the literature does indicate the effectiveness of the role of VIIs in getting benefits in improving viscometric properties [52–53]. However, the main problem with their use is their non-Newtonian behaviors. This results in uncertainty in the prediction of lubricant film thickness even for the well-established elastohydrodynamic lubrication (EHD) theories. It is needless to comment on the more complex friction behavior. So, there is a definite need to understand the tribology behavior of VIIs, particularly, in improving the viscometric properties of the blend with an aim of energy saving by the reduction in friction. Friction modifier (molybdenum phosphothioates), extreme pressure antiwear (EP/AW) additives, boundary-lubricity additives (fatty amides, esters), and slippery surface deposits (graphite, Teflon, MoS_2) are used for energy-efficient lubricants [54–57] (Figure 11.5).

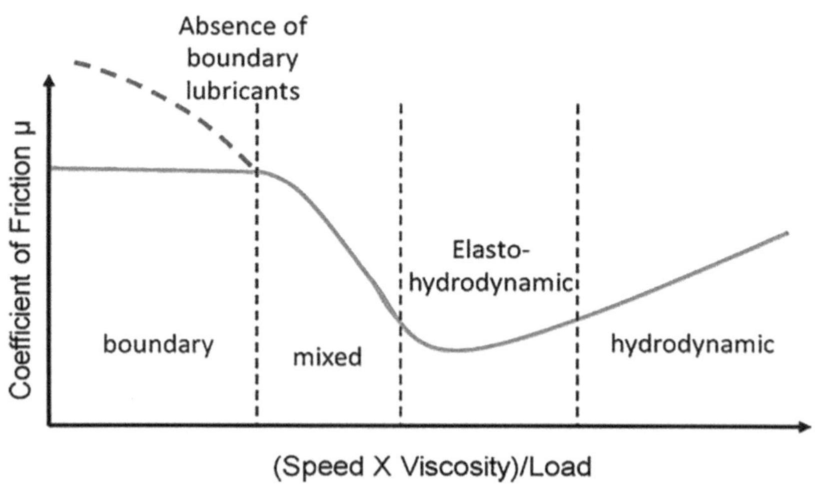

Figure 11.5 Stribeck curve [27].

11.3 ROLE OF LUBRICANT ADDITIVES ON FUEL EFFICIENCY

Asperity friction is a surface phenomenon and friction modifiers compete with other functional properties like viscosity index, pour point, antiwear, extreme pressure, detergents, dispersants, antioxidants, corrosion inhibitors, and anti-foaming. Therefore, it is often necessary to reformulate the total additive package to optimize lubricant performance. Additives (friction modifiers) that reduce boundary/asperity friction are required to enable the use of low-viscosity fluids. Under typical driving cycles, asperity friction accounts for approximately 10% of total engine friction, yet lowering asperity friction by the use of additives that form low-friction boundary films mitigates mixed and boundary friction and enables the use of low-viscosity fluids. Figure 11.6 shows a comparative study of additive and non-additive lubricants. Graphene was used in lubricant to increase fuel economy by reducing friction. It was observed that using graphene an additive reduces boundary friction by 70–77%. Superior results with graphene are due to the consistent formation of tribofilms under all conditions.

11.4 EXPERIMENTAL METHODS FOR PERFORMANCE EVALUATION

The purpose of this test is to determine the wear-preventive characteristics of a lubricant (Figure 11.7). In the four-ball wear test, a steel ball is rotated against three lubricated stationary steel balls under a specified load, speed, temperature, and time. The better the lubricant is at preventing wear, the smaller the wear scar will be on the three stationary balls. Four-ball data is useful to differentiate between lubricating greases having different levels of load-bearing capabilities, wear protection, and friction reduction. Although these results do not necessarily correlate with real-life

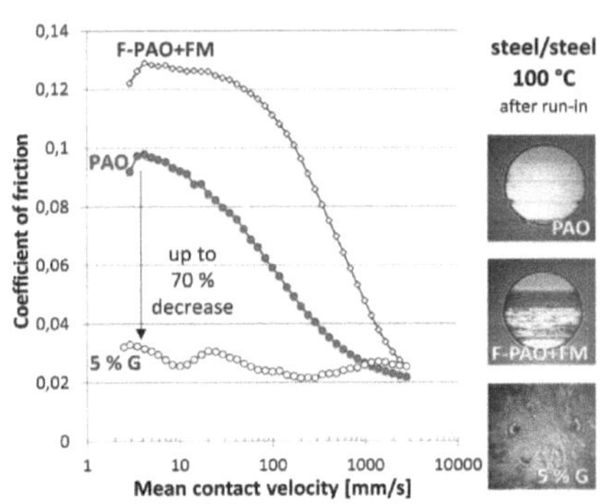

Figure 11.6 Comparative study of lubricant using additives and non-additives [20].

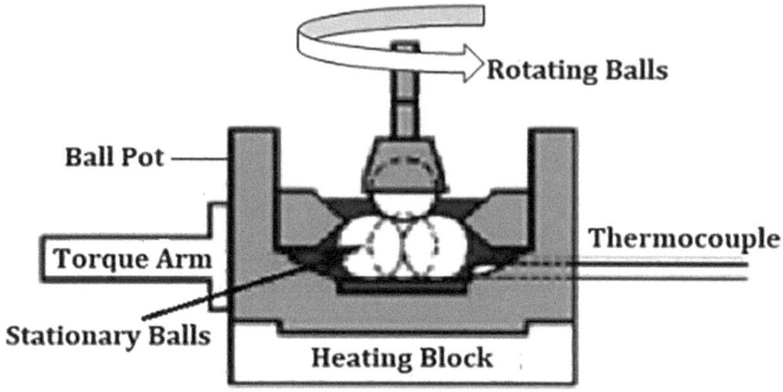

Figure 11.7 Schematic diagram of four-ball geometry experiment (point contact).

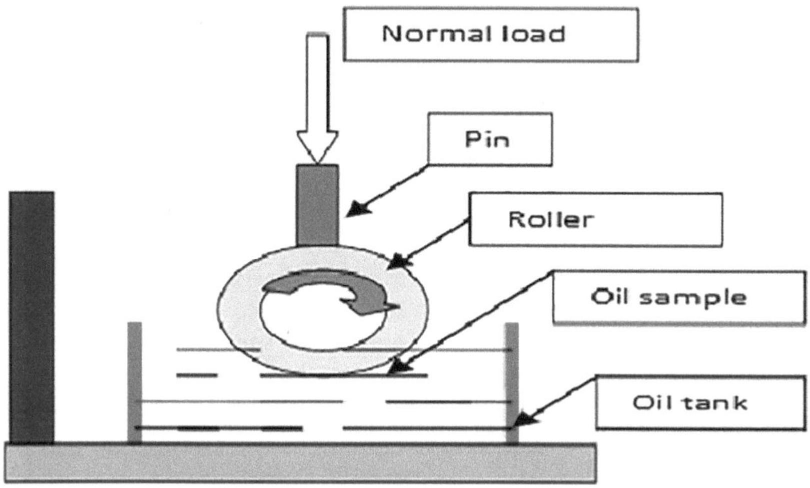

Figure 11.8 Layout of pin on disk testing rig.

applications, the method has become an industry standard in comparing EP, wear, and frictional functionality

The schematic diagram of pin on disk testing setup is shown in Figure 11.8. The stationary pin is pressed against the rotating disk under the given load. Steel pin and roller are mounted on the rotating shaft. There was a lubricated line contact between the stationary steel pin and steel roller. The roller is immersed in an oil tank for testing the oil sample. The shaft can be rotated to a desired speed, which generates tangential friction force between the pin and the roller that force can be sensed by the load cell attached to the roller shaft assembly. During the test, the friction force, wear, and

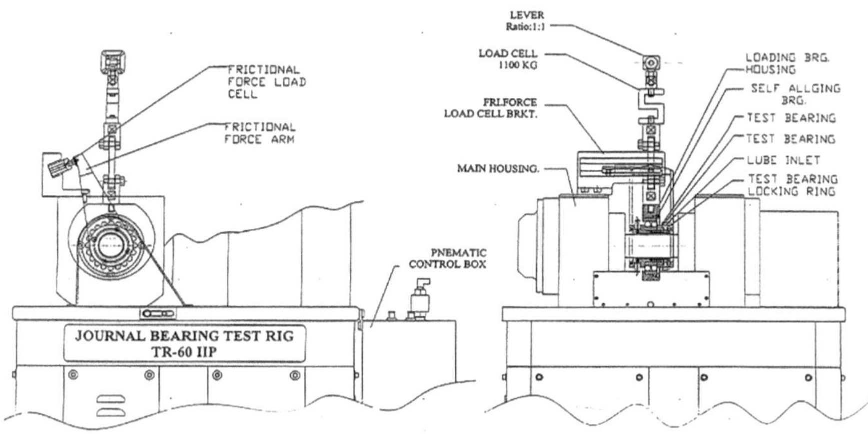

Figure 11.9 Assembly drawing journal bearing test rig.

temperature are continuously monitored. The shaft can be rotated to a desired speed, which generates tangential friction force between the pin and the roller that force can be sensed by the load cell attached to the roller shaft assembly. The bearing assembly has a highly polished journal size of φ 50 × 1000 mm. The journal is made of case-hardened steel having 60 HRC and supported on either end of bearings. The bearings are housed inside spindle housing and in turn in housing. This housing is firmly fixed on the base plate and has relief at the middle to accommodate test bearing and sump for lubricant. The journal is driven by an AC motor through a 1:1 pulley and teeth belt at a maximum speed of 3000 rpm (Figure 11.9). The maximum desired load of 2000 N was applied using pneumatic loading. The lubrication system consists of an oil filter, valves, temperature controller, reservoir, and oil pump. The lubrication system is used for supplying heated oil. Load cell is used for sensing friction force and all data is transferred for recording and processing. Similarly, the film thickness is measured with the help of linear variable differential transformer (LVDT). Load, speed, and coefficient of friction are shown on the computer screen.

11.5 DISCUSSION

When it comes to lubricants for high-pressure applications, maintaining viscosity across a wide range of temperatures is crucial. This is where high-performance viscosity index improvers (VIIs) come in. These additives help lubricants thicken at high temperatures and thin out at low temperatures, ensuring proper lubrication and reducing friction. But for environmentally conscious applications, sustainable options are becoming increasingly important. Examples of high-performance, sustainable VIIs for high-pressure applications are as follows:

Polyalphaolefins (PAO): These synthetic VIIs are derived from long-chain hydrocarbons and offer excellent thermal and shear stability. They are particularly compatible with synthetic base oils, making them ideal for high-performance lubricants. PAOs are known for their long lifespans and are generally considered biodegradable, making them a sustainable choice.

Ethylene propylene copolymers (EPM): These VIIs are based on copolymers of ethylene and propylene. They offer good viscosity improvement and compatibility with a wide range of mineral and synthetic base oils. EPMs are generally considered recyclable and have a lower environmental impact compared to traditional VIIs.

Bio-based VIIs: This is a developing area with ongoing research into VIIs derived from renewable resources like vegetable oils or natural polymers. These bio-based VIIs offer the advantage of being readily biodegradable and reducing dependence on petroleum-derived products.

In high-pressure applications, the VII needs to maintain its thickening ability even under shearing forces. PAOs excel in this area. Effective VIIs should help lubricants reduce friction across a wide temperature range, minimizing wear and tear on components.

As mentioned earlier, the ideal sustainable VII should be readily biodegradable to minimize environmental impact after the lubricant's lifespan. It is suggested to use VIIs derived from renewable resources to reduce reliance on fossil fuels. Choose VIIs with low toxicity profiles to minimize environmental and human health risks.

In conclusion, while significant strides have been made in developing bio-based and high-performance VIIs, challenges regarding compatibility, performance consistency, and extreme temperature resistance persist. Future research efforts directed toward material science advancements and a deeper understanding of VII–base oil interactions are essential to achieving high-performance, sustainable VIIs that meet the evolving demands of modern lubrication technology.

REFERENCES

1. W.L.Y. Hsien (2015) Utilization of vegetable oil as biolubricant and additives, in: *Towards Green Lubrication in Machining, Spring Briefs Green Chemistry for Sustainability*, W.L.Y. Hsien (ed.), Springer: Berlin, pp. 7–17.
2. S.C. Mat, M.Y. Idroas, Y.H. Teoh, and M.F. Hamid (2018) Physical, performance, combustion and emission characteristics of melaleuca cajuputi oil-refined palm oil hybrid biofuel blend, *Energie*s. 11, 1–20.
3. F.J. Owuna, M.U. Dabai, M.A. Sokoto, C. Muhammad, and A.L. Abubakar (2019) Analysis of blends of vegetable oils with mineral based lube oil, *Journal of Scientific and Engineering Research*. 6, 41–48.

4. M.B. Alang, M.K. Ndikontar, Y.M. Sani, and P.T. Ndifon (2018) Synthesis and characterisation of a biolubricant from cameroon palm kernel seed oil using a locally produced base catalyst from plantain peelings, *Green and Sustainable Chemistry*. 8, 275–287.
5. E.S. De Souza, G. Belinato, R.L.S. Otero, É.C.A. Simêncio, S.C.M. Augustinho, W. Cupelupi, C. Conconi, L.C.F. Canale, and G.E. Toten (2011) Thermal oxidation stability of vegetable oils as metal heat treatment quenchants, *Journal of ASTM International*. 9, 1–30.
6. F.J. Owuna, M.U. Dabai, M.A. Sokoto, C. Muhammad, and A.L. Abubakar (2018) Use of Lagenaria siceraria seed oil for the production of environmentally friendly biolubricant, *American Journal of Applied and Industrial Chemistry*. 2, 1–7.
7. H.M. Mobarak, M.E. Niza, H.H. Masjuki, M.A. Kalam, K.A.H. Al Mahmud, M. Habibullah, and A.M. Ashraful (2014) The prospects of biolubricants as alternative in automotive applications, *Renewable and Sustainable Energy Reviews*. 33, 34–43.
8. C. Muhammad, Z. Usman, and F. Agada (2018) Biodiesel production from Ceiba pentandra seed oil using CaO derived from snail shell as catalyst, *Journal of Petroleum Science and Engineering*. 2, 7–16.
9. M.U. Dabai, F.J. Owuna, M.A. Sokoto, and A.L. Abubakar (2018) Assessment of quality parameters of ecofriendly biolubricant from waste cooking palm oil, *Asian Journal of Applied Chemistry Research*. 1, 1–11.
10. K.M. Jagadeesh, V.K. Satish, K. Venkatesh, and N. Kathyayini (2018) Evironmentally friendly functional fluids from renewable and sustainable sources—a review, *Renewable and Sustainable Energy Reviews*. 18, 1787–1801.
11. O. Ocholi, M. Menkiti, M. Auta, and I. Ezemagu (2017) Optimization of the operational parameters for the extractive synthesis of biolubricant from sesame seed oil via response surface methodology, *Egyptian Journal of Petroleum*. 27, 265–275.
12. K. Naima, and A. Liazid (2013) Waste oils as alternative fuel for diesel engine: A review, *The Journal of Petroleum Technology and Alternative Fuels*. 4, 30–43.
13. N.A. Arianti, and W. Widayat (2018) A review of bio-lubricant production from vegetable oils using esterification and transesterification process, *MATEC Web of Conferences*. 156, 2–7.
14. F.J. Owuna, M.U. Dabai, M.A. Sokoto, and A.L. Abubakar (2019) Analysis of properties of biolubricant obtained from blends of used palm oil, theory and applications of chemistry published by book publisher international (a part of Science Domain International). 2, 123–135.
15. T. Onodera, K. Kawasaki, T. Nakakawaji, Y. Higuchi, N. Ozawa, K. Kurihara, and M. Kubo (2014) Effect of tribochemical reaction on transfer-film formation by poly (tetrafluoroethylene), *Journal of Physical Chemistry C*. 118 (22), 11820–11826.
16. B. Bhushan (2001) *Modern Tribology Handbook. 1. Principles of Tribology*, CRC Press, Boca Raton.
17. S.K. Sinha, and B.J. Briscoe (2009) *Polymer Tribology*, World Scientific, London.

18. Historical Use of Petroleum for Transportation in the US (Accessed February, 2024) https://www.energy.gov/eere/vehicles/articles/fotw-1000-october-23-2017-us-petroleum-production-met-demand-transportation
19. Z. Wang, S. Shuai, Z. Li, and W. Yu (2021) A review of energy loss reduction technologies for internal combustion engines to improve brake thermal efficiency, *Energies*. 14 (20), 6656. doi: 10.3390/en14206656.
20. J. Kogovšek, and M. Kalin (2023) Comparison of graphene as an oil additive with conventional automotive additives for the lubrication of steel and DLC-coated surfaces, *Tribology International*. 180, 108220. doi: 10.1016/j.triboint.2023.108220.
21. B. Zhmud, and M. Roegiers (2009) Base oils pose a challenge for solubility and lubricity, *TLT*. 65 (7), 34–39.
22. M. Roegiers and B. Zhmud (2009) Tribological performance of ionized vegetable oils as lubricity and fatty oiliness additives in lubricants and fuels, *Lubrication Science*. 21, 169–174.
23. A. Somers, P. Howlet, J. Sun, D. MacFarlane, and M. Forsyth (2010) Transition in wear performance for ionic liquid lubricants under increasing load, *Tribology Letters*. 40 (2), 279–284.
24. V. Totolin, I. Minami, C. Gabler, and N. Doerr (2013) Halogen-free borate ionic liquids as novel lubricants for tribological applications, *Tribology International*. 67 (11), 191–198.
25. L. Zhu, L. Chen, S. Xiang, G. Chen, and X. Yang Xin (2012) Tribological properties of functionalized ionic liquids containing ester-group as lubricants for steel-steel system, *China Petroleum Processing and Petrochemical Technology*. 14 (2), 60–65.
26. D. Li, M. Cai, D. Feng, F. Zhou, and W. Liu (2011) Excellent lubrication performance and superior corrosion resistance of vinyl functionalized ionic liquid lubricants at elevated temperature, *Tribology International*. 44 (10), 1111–1117.
27. D. Puhan (2020) Lubricant and lubricant additives, in: *Tribology in Materials and Manufacturing—Wear, Friction and Lubrication*, Amar Patnaik, Tej Singh, and Vikas Kukshal (eds.). doi: 10.5772/intechopen.93830.
28. R.K. Upadhyay, V.P. Singh, and A. Kumar (2024) Sustainable transportion: Policy, planning and implementation, in: *Energy Efficient Vehicles: Technologies and Challenges*, Varun Pratap Singh, Ashwani Kumar, Chandan Swaroop Meena, and Gaurav Dwivedi, CRC Press, Boca Raton, FL, pp. 1–22.
29. V.P. Singh, and A. Kumar (2024) Techno-economic and future aspects of the HEV-EV-FCV, in: *Energy Efficient Vehicles: Technologies and Challenges*, Varun Pratap Singh, Ashwani Kumar, Chandan Swaroop Meena, and Gaurav Dwivedi (eds.), CRC Press, Boca Raton, FL, pp 1–20.
30. A. Kumar, V.P. Singh, C.S. Meena, and N. Dutt (eds.) (2023) *Thermal Energy Systems: Design, Computational Techniques and Applications*. doi: 10.1201/9781003395768.
31. A. Kumar, and P.P. Patil (2016) FEA simulation based performance study of multi-speed transmission gearbox, *IJMMME*. 6 (1), 57–67. doi: 10.4018/IJMMME.2016010103.
32. A. Kumar, R. Jain, and P.P. Patil (2016) Dynamic analysis of heavy vehicle medium duty drive shaft using conventional and composite material, *IOP: Materials Science and Engineering*. 149, 1–11. doi: 10.1088/1757-899X/149/1/012156.

33. A. Kumar, and P.P. Patil (2015) FEA based dynamic vibration response analysis of loose mounted heavy vehicle transmission gearbox casing, *IJAMMC*. 5 (1), 136–141. http://dx.doi.org/10.11127/ijammc.2015.03.02.
34. A. Kumar, and P.P. Patil (2014) Dynamic vibration analysis of heavy vehicle truck transmission gearbox housing using FEA, *JESTR*. 7 (4), 66–72.
35. A. Kumar, H. Jaiswal, R. Jain, and P.P. Patil (2014) Free vibration and material mechanical properties influence based frequency and mode shape analysis of transmission gearbox casing, *Procedia Engineering*. 97, 1097–1106. doi: 10.1016/j.proeng.2014.12.388.
36. A. Kumar, Y. Gori, C.S. Meena, V.P. Singh, and V.K. Sharma (2022) *Design and Analysis of Heavy Vehicle Medium Duty Transmission Gearbox System*, CRC Press, Boca Raton, FL, pp. 67–86. doi: 10.1201/9781003360001-4.
37. A. Kumar, S. Rana, Y. Gori, and N.K. Sharma (2021) Thermal contact conductance prediction using FEM based computational techniques, in: *Advanced Computational Methods in Mechanical and Materials Engineering*, A. Kumar, Y. Gori, N. Dutt, Y.K. Singla, and A. Maurya (eds.), CRC Press, Boca Raton, FL, pp. 183–220. doi: 10.1201/9781003202233-13.
38. A. Kumar, Y. Gori, N. Dutt, Y.K. Singla, and A. Maurya (eds.) (2021) Advanced computational methods, in: *Mechanical and Materials Engineering*. doi: 10.1201/9781003202233.
39. K.K. Gajrani, A. Prasad, and A. Kumar (eds.) (2022) *Advances in Sustainable Machining and Manufacturing Processes*. doi: 10.1201/9781003284574.
40. S. Singh, S.K. Behura, A. Kumar, and K. Verma (eds.) (2022) *Nanomanufacturing and Nanomaterials Design: Principles and Applications*. doi: 10.1201/9781003220602.
41. V.K. Sharma, A. Kumar, M. Gupta, V. Kumar, D.K. Sharma, and S. Sharma (eds.) (2022) *Additive Manufacturing in Industry 4.0: Methods, Techniques, Modeling, and Nano Aspects*. doi: 10.1201/9781003360001.
42. R. Kumar, A. Prasad, and A. Kumar (eds.) (2023) *Sustainable Smart Manufacturing Processes in Industry 4.0*. doi: 10.1201/9781003436072.
43. A. Kumar, and P.P. Patil (2017) Free vibration and connecting bolt constraint based FEA analysis of heavy vehicle medium duty transmission gearbox housing made from AISI 4130 alloy, in: *Mathematics Applied to Engineering*, pp. 133–146. doi: 10.1016/B978-0-12-810998-4.00007-7.
44. A. Kumar, A. Dwivedi, H. Jaiswal, and P.P. Patil (2015) Material-based vibration characteristic analysis of heavy vehicle transmission gearbox casing using finite element analysis, *Advances in Intelligent Systems and Computing*. 308, 527–533. doi: 10.1007/978-81-322-2012-1_56.
45. F. Ahmad, V. Tomer, A. Kumar, and P.P. Patil (2016) FEA simulation based thermo-mechanical analysis of tractor exhaust manifold, 173–181. doi: 10.1007/978-81-322-2740-3_18.
46. V.P. Singh, A. Kumar, C.S. Meena, and G. Dwivedi (eds.) (2024) *Energy Efficient Vehicles: Technologies and Challenges* (1st ed.), CRC Press, Boca Raton, FL. doi: 10.1201/9781003464556.
47. V.P. Singh, A. Kumar, C.S. Meena, and G. Dwivedi (2024) Introduction to energy-efficient vehicles for sustainable transportion, in: *Energy Efficient Vehicles: Technologies and Challenges*, Varun Pratap Singh, Ashwani Kumar, Chandan Swaroop Meena, and Gaurav Dwivedi (eds.), CRC Press, Boca Raton, FL, pp. 1–18.

48. P. Patil, Y. Gori, A. Kumar, and M.R. Tyagi (2021) Experimental analysis of tribological properties of polyisobutylene thickened oil in lubricated contacts, *Tribology International*, 1–7. doi: 10.1016/j.triboint.2021.106983.
49. P.P. Patil, and A. Kumar (2017) Dynamic structural and thermal characteristics analysis of oil lubricated multi speed transmission gearbox: Variation of load, rotational speed and convection heat transfer, *ISTM*. 41, 281–291. doi: 10.1007/s40997-016-0063-z.
50. A. Kumar (2018) Low to high speed transient structural and thermal temperature measurement of oil-lubricated multi-speed heavy vehicle transmission gearbox system based on FEA, in: *Advanced Numerical Simulations in Mechanical Engineering*, pp. 1–21. doi: 10.4018/978-1-5225-3722-9.ch001.
51. A. Shukla, and A. Kumar (2021) Design and thermal characteristic analysis of gearbox system based on finite element analysis (FEA), *AIP Conference Proceedings*. 2317, 030002. doi: 10.1063/5.0036235
52. A. Kumar, Y. Gori, and P.P. Patil (2020) Finite element analysis based thermo-mechanical performance study of heavy vehicle medium duty transmission gearbox, in: *Advancements in Manufacturing Materials and Mechanical Engineering*, IGI Global. doi: 10.4018/978-1-7998-4939-1.
53. S. Rana, and A. Kumar (2018) FEA based design and thermal contact conductance analysis of steel and Al rough surfaces, *IJAER*. 13 (16): 12715–12724.
54. A. Kumar, H. Jaiswal and P.P. Patil (2017) Parametric optimization of vibrating heavy vehicle medium duty transmission gearbox housing using response surface method, *International Journal of Vehicle Structures and Systems*. 9 (3), 149–153. doi: 10.4273/ijvss.9.3.04.
55. S. Bhoi, A. Prasad, A. Kumar, R.B. Sarkar, B. Mahto, C.S. Meena, and C. Pandey (2022) Experimental study to evaluate the wear performance of UHMWPE and XLPE material for orthopedics application, *Bioengineering*. 9, 676. doi: 10.3390/bioengineering9110676.
56. A.K. Paul, A. Prasad, and A. Kumar (2022) Review on artificial neural network and its application in the field of engineering, *Journal of Mechanical Engineering*. 1(1), 53–61. doi: 10.56697/JMEP.2022.1107.
57. A.G. Pandey, A. Kumar, P.P. Patil, and V. Kumar (2022) FEA based contact analysis of railway track and wheel, *AIP Conference Proceedings*. 2481, 050010-1–050010-8. doi: 10.1063/5.0103726.

Chapter 12

State of the art in lubricants for extreme cold weather

Sandeep Kumar, Avinash Kumar, Abhishek Kumar, Ashwani Kumar, Amit Choudhari, Ashish Kumar Gupta, and Sakib Faisal

12.1 INTRODUCTION

12.1.1 Background and rationale

Lubrication is a crucial component in the effective running of modern machinery and equipment, providing essential tasks such as friction reduction, wear minimization, and overall operating efficiency. However, the performance of traditional lubricants faces daunting hurdles in very cold weather situations, where low temperatures have a substantial impact on their physical and chemical properties. This chapter investigates the background and rationale for the critical need for specialized lubricants capable of withstanding extreme cold conditions, particularly as industries and applications expand into harsh climate regions such as the Arctic, high-altitude regions, and even outer space.

The demand for such specialist lubricants has increased as industries expand into these tough regions. Research by Huang et al. [1] highlights the key aspects of lubricants, such as extreme pressure and lubricating performance, which can limit the application range of machinery and equipment as illustrated in Figures 12.1 and 12.2. John and Menezes [2] actively contribute to the development of novel lubricant technologies and prospective lubricating materials to address friction and wear challenges. Furthermore, research by Dai et al. [3] underlines the necessity of increasing liquid droplet mobility on rough surfaces, which can be used for everything from condensation heat transfer to icing and frosting prevention.

Furthermore, studies by Petersson et al. [4] and Lin et al. [5] highlight the problems posed by extreme weather events and urge for novel lubricant additives to improve performance in such circumstances. Adera et al. [6] investigated nanostructured oil-infused surfaces while by Freschi et al. [7] concentrated on micro- and nano-WS_2 structures, demonstrating ongoing efforts to improve tribological properties and reduce energy consumption in a variety of industries using advanced lubrication technology.

Finally, the listed references provide a detailed summary of the issues that lubricants face in extremely cold temperatures. The current research efforts

DOI: 10.1201/9781003535447-12

Figure 12.1 The bite of winter: machinery struggles in extreme cold.

Figure 12.2 Frozen gear: lubrication challenges at low temperatures.

exhibited in these studies represent a determined pursuit of developing specialized lubricants that can thrive in such severe settings, ensuring not only effective machinery functioning but also reduced wear and friction, ultimately contributing to enhanced longevity and performance.

12.1.2 Significance of extreme low temperatures in lubrication

Extremely low temperatures have a significant impact on lubricant performance, causing increased friction, accelerated wear, and the risk of equipment failure. This impact is especially noticeable in a wide range of industries, including automotive, aerospace, construction, mining, oil and gas, and scientific research missions navigating the harsh conditions of the Arctic areas, high-altitude landscapes, and the perilous vastness of outer space. The chapter examines the crucial importance of keeping correct lubrication in these severe cold circumstances, as it is critical to guaranteeing the durability, dependability, and efficient operation of machinery and equipment. The long-term benefits include increased productivity, lower maintenance costs, and an overall improvement in safety standards.

Extremely low temperatures present a variety of issues that influence a wide range of industries. Research by Yan et al. [8] provides valuable insights by investigating the rheological and flow behavior of lubricating grease under extremely cold conditions, revealing a significant increase in yield stress at low shear rates and lower temperatures. This study emphasizes the complex job of maintaining adequate lubricating qualities in the face of severe cold temperatures. Furthermore, research by Su et al. [9] describes a novel method for synthesizing self-lubricating ceramic composites with temperature-sensitive solid lubricants, giving insight on possible materials capable of withstanding severe cold ambient temperatures.

Furthermore, the repercussions of very cold weather go beyond mechanical performance. Research by Lin et al. [10] explores the impact of very low ambient temperatures on out-of-hospital cardiac arrest, focusing on potential health consequences such as causing coronary artery spasm or producing blood clotting, which can result in deadly myocardial infarction. This comprehensive approach emphasizes the interdependence of extreme cold weather concerns, recognizing the necessity for sophisticated lubrication methods and materials to enable optimal performance and safety in locations where machinery runs in extreme cold temperatures.

12.1.3 Objectives of the chapter

The goal of this chapter is to thoroughly examine the most recent developments in lubricants designed for extremely cold climates. It explores the many kinds, characteristics, and difficulties associated with low-temperature lubricants, as well as formulation techniques and additives. Trends and potential future developments are also covered. It also highlights how important specialty lubricants are to a variety of applications, thus making sure machinery runs dependably under challenging conditions. This chapter provides industry professionals and researchers with vital insights to help them navigate the complexity of extreme weather lubrication. Figure 12.3 illustrates the functions of various lubricants.

12.2 LUBRICANTS

As essential parts of mechanical systems, lubricants are crucial for promoting smooth functioning and reducing wear caused by friction. This section

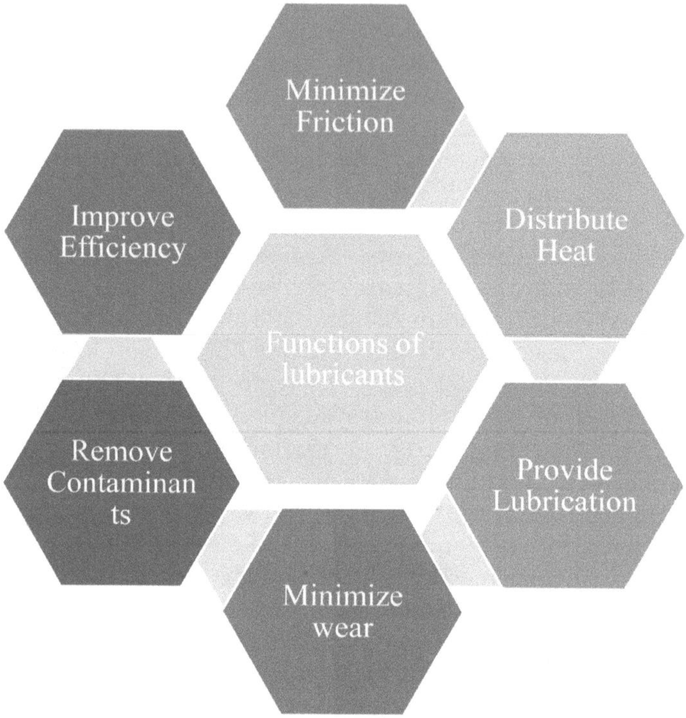

Figure 12.3 Functions of lubricants.

begins a thorough investigation into the complex world of lubricants by looking at their various types and characteristics.

12.2.1 Types

Lubricants serve an important role in many applications and are divided into distinct categories based on their composition and physical state. These classifications include:

1. **Mineral oils:** Mineral oils, derived from crude oil, are frequently used due to their availability and low cost. Nonetheless, they may have limits in terms of low-temperature performance and oxidative stability.
2. **Synthetic lubricants:** Chemically engineered, synthetic lubricants outperform mineral oils, especially at high temperatures. Examples include polyalphaolefins (PAOs), esters, polyalkylene glycols (PAGs), and perfluoropolyethers (PFPEs).
3. **Bio-based lubricants:** Bio-based lubricants, derived from renewable sources such as vegetable oils and animal fats, provide environmentally

favorable options. They are biodegradable and have low toxicity, making them ideal for environmentally sensitive habitats.
4. **Greases:** Semi-solid lubricants having a thickening agent dispersed in a lubricating fluid, greases are useful in situations that require the lubricant to maintain its position and protect against contamination.
5. **Solid lubricants:** Solid lubricants, which include materials such as molybdenum disulfide (MoS_2), graphite, and polytetrafluoroethylene (PTFE), are useful in situations where conventional liquid or semi-solid lubricants fail, such as in vacuum or at high temperatures.

Recent research has focused on new lubrication technologies, such as exploration of bioinspired lubricant-infused surfaces [11] and analysis of ionic liquids for green molecular lubrication [12]. These studies highlight the emergence of lubricant solutions with unique features. Furthermore, research work by Noble et al. [13] on quantifying lubricant droplet spreading using molecular dynamics provides insights into the underlying physical principles that drive lubricant behavior on solid substrates.

12.2.2 Properties

Lubricants' performance is inextricably linked to their physical and chemical qualities, which are critical in evaluating their suitability for various applications and operating situations. Several key properties have a substantial influence on lubricating performance.

1. **Viscosity:** Viscosity, or a lubricant's resistance to flow, is a critical feature that influences lubrication performance. Low-viscosity lubricants are commonly used in low-temperature applications to guarantee optimum flow and distribution.
2. **Pour point:** The pour point is the lowest temperature at which a lubricant can flow under certain conditions. Lubricants with lower pour points are more suited for extreme cold weather applications.
3. **Flash point:** The flash point is the lowest temperature at which a lubricant's vapors will ignite when exposed to an ignition source. A greater flash point is desirable for safety and fire resistance.
4. **Oxidation stability:** Oxidation stability refers to a lubricant's capacity to withstand chemical breakdown caused by oxygen exposure and high temperatures. Improved oxidation stability can increase the lubricant's service life.
5. **Thermal conductivity:** The thermal conductivity of a lubricant is critical in applications involving high-speed or high-load circumstances, since it allows for effective heat transmission.
6. **Compatibility:** Lubricants must be compatible with the materials they encounter, such as seals, gaskets, and metals, in order to avoid degradation or unpleasant reactions.

Gasni et al. [14] investigated the wear mechanism of ball bearings lubricated with biolubricants while Renzis et al. [15] applied a one-dimensional fuel-oil dilution model to predict in-cylinder oil effects in a direct injection engine, thus shedding light on the impact of physical properties on lubricant performance and behavior. These studies highlight the need to understand lubricant characteristics in order to optimize performance in a variety of applications. Furthermore, Blanck et al.'s [16] study on lubricant adhesion on aluminum via additive head-group adsorption and Waltman et al.'s [17] study on the role of configuration entropy in boundary lubricants based on the *n*-perfluoro propylene oxide main chain structure shed light on the physical interactions that influence lubricant behavior.

12.2.3 Low-temperature effects on lubrication

12.2.3.1 Challenges presented by extremely low temperatures

Extreme cold weather poses considerable challenges to lubricants, resulting in increased viscosity, poor pumpability, and the possibility of solidification or wax precipitation. These difficulties, if not handled, might result in insufficient lubrication, increased friction, wear, and the possibility of equipment failure or shutdown. Yan et al. [8] studied the rheological and flow behavior of lubricating grease in extremely cold weather, and Liu and Wang [18] studied biolubricants derived from waste cooking oil with improved oxidation stability and low-temperature properties. These studies shed light on the impact of cold temperatures on lubricant behavior and explore alternative lubricant solutions.

Furthermore, the research by Ravensteijn et al. [19] on triple-function lubricant additives based on organic–inorganic hybrid star polymers, as well as by Park et al. [20] on the different characteristics of cold day and cold surge frequency over East Asia in a global warming scenario, provides insights into improving lubricant performance and understanding the climatic factors influencing extreme cold events. These research studies help to advance lubrication technologies and techniques for reducing the effects of very cold weather on lubricant performance, ultimately ensuring the reliability and longevity of machinery and equipment operating in harsh cold climates.

12.12.3.2 Impact on lubricant viscosity and flow properties

Low temperatures have a significant impact on lubricant viscosity and flow characteristics. According to the well-known Arrhenius equation, as temperature drops, lubricant viscosity increases exponentially. This increased viscosity presents a significant obstacle, impeding the lubricant's ability to flow and distribute effectively throughout the lubricated system. As a result, this phenomenon increases wear and reduces overall efficiency.

The temperature–viscosity relationship is frequently examined using the viscosity index (VI), which quantifies the rate at which viscosity changes with temperature. Lubricants having a high VI show low-viscosity variation across a wide temperature range, making them ideal for applications involving extreme temperature fluctuations. Research by Covitch and Trickett [21] studying the role of polymers as viscosity index improvers in lubricating oils, as well as research by Paredes et al. [22] investigating into the impact of pressure on the dynamic viscosities of poly(propylene glycol)dimethyl ether lubricants, provides valuable insights into lubricant temperature-dependent behavior and highlights the importance of the viscosity index in lubricant design. Furthermore, research by Fan et al. [23] studying high-performance lubricant base stocks derived from biorenewable gallic acid, as well as Shah et al. [24] investigating grease performance requirements for electric and hybrid vehicle applications, provides important insights into improving lubricant properties for specific applications. Understanding and regulating the temperature–viscosity connection is critical for developing lubricants that can survive severe cold temperatures while maintaining maximum functioning and longevity in a variety of applications.

12.2.3.3 Cold start issues and bearing failures

The higher viscosity of lubricants in extremely cold temperatures makes it difficult to properly lubricate during cold starts, especially in applications like industrial machinery and automobile engines. Excessive viscosity restricts flow, which results in inadequate lubrication of vital parts like bearings, increasing wear and friction and possibly causing equipment failure.

Yan et al. [8] investigated the rheological and flow behavior of lubricating grease in extremely cold weather, and Pohrer et al. [25] investigated the practical application of CO_2 as a viscosity index improver for lubricants. These findings highlight the need to understand and regulate lubricant viscosity for optimal performance in extremely cold settings.

Furthermore, the research work by Sarpal et al. [26] on the correlation of structure and properties of base oils, as well as Bonnevie et al.'s [27] investigation into the impact of sub-critical impact on the lubricating mechanisms of articular cartilage, provides insights into lubricant properties and behavior under different conditions. These researches help to improve lubricant performance and understand the elements that influence lubrication efficacy and are vital for ensuring consistent and efficient performance, particularly during key startup periods in a wide range of industrial applications.

12.2.4 Properties of ideal low-temperature lubricants

12.2.4.1 Viscosity and viscosity index

Ideal low-temperature lubricants are characterized by their ability to maintain low viscosity at cold temperatures, ensuring proper flow and distribution, while simultaneously exhibiting adequate viscosity at higher temperatures for effective lubrication and wear protection. Achieving a high viscosity index (VI), which measures the rate of viscosity change with temperature, is crucial for consistent performance across a broad temperature range.

Wang et al. [28] conducted research on the synthesis of a high viscosity index base stock and its lubricating properties, while Li et al. [29] investigated extreme pressure lubrication with the high thermal diffusivity of liquid metal. These studies underline the importance of viscosity control in improving lubrication performance.

Research study by Lin et al. [5] on water lubrication additives for silicon nitride achieves superlubricity under extreme conditions, while the study by Peña-Parás et al. [30] on extreme pressure properties of nanolubricants for metal-forming applications provides insights into innovative lubricant formulations for better performance in challenging environments. These investigations help to advance lubricant technologies that are tailored to satisfy specific operational needs.

12.2.4.2 Pour point and cloud point

Pour point and cloud point are important features of low-temperature lubricants. The pour point is the lowest temperature at which a lubricant flows under certain conditions, whereas the cloud point is the temperature at which wax crystals begin to form, making the lubricant seem cloudy or hazy.

Lubricants with low pour points and cloud points are favored for use in very cold weather because they maintain appropriate fluidity and prevent wax precipitation, ensuring reliable lubrication even at extremely low temperatures. Strategic formulation approaches, such as the use of pour point depressants and cold flow improvers, are vital for lowering these critical temperatures.

Dandan et al. [31] investigated the effect of varying percentages of pour point depressant (PPD) on the tribological properties of palm kernel oil, while Kamal et al. [32] investigated high-density polyethylene waste-modified lube oil nanocomposites as pour point depressants.

Furthermore, research by Müller et al.'s [33] study on bismuth (III) sulfide as an additive for enhanced lubricity without toxicity, as well as Sharma et al.'s [34] investigation into the influence of flow improvers on the rheological and microscopic properties of Indian waxy crude oil, provides valuable information on novel additives and their effects on lubricant properties. These researches help to advance lubricant technologies, with the goal of achieving optimal performance under demanding settings.

12.2.4.3 Antiwear and extreme pressure properties

To improve wear protection and protect against severe pressure circumstances, low-temperature lubricants must have effective antiwear and extreme pressure (EP) capabilities. These critical qualities are often obtained with the use of specialist additives such as zinc dialkyldithiophosphates (ZDDPs), organic friction modifiers, and molybdenum-based compounds.

Antiwear chemicals serve an important function by producing protective layers on metal surfaces, reducing friction and avoiding adhesive wear. Extreme pressure additives, on the other hand, are critical for preventing severe wear and surface damage, particularly when exposed to high loads or shock loading circumstances. These additives are especially important in low-temperature applications, where factors such as increased viscosity and poor lubrication can exacerbate wear difficulties.

Research study by Huang et al. [1] on oil-soluble ionic liquids as antiwear and extreme pressure additives in poly-α-olefin for steel/steel contacts and Li et al.'s [29] study on extreme pressure lubrication with high thermal diffusivity of liquid metal highlight the importance of additives in improving lubricant performance under extreme conditions. Furthermore, Zhao et al.'s [35] study on using talc nanosheets as lubricant additives in harsh conditions and Guan et al.'s [36] study on aeronautical steel under oil–air lubrication with extreme pressure and antiwear additives provide valuable insights into innovative lubricant additives and their impact on tribological performance. These research studies help to progress lubricant technologies by focusing on improving wear prevention and lubrication effectiveness [37, 38].

12.2.4.4 Seal and material compatibility

Extreme cold weather conditions can cause brittleness in seals and other materials, reducing flexibility and increasing the risk of leakage and failure. Low-temperature lubricants must be compatible with the seals and materials used in the system to avoid deterioration, swelling, or shrinkage.

Fluoroelastomers, nitrile rubber, and polyacrylate rubber are common seal materials used in low-temperature applications. Ensuring lubricant compatibility with these materials is critical for proper sealing and preventing leakage, which could result in lubricant loss, contamination, or equipment damage. Notably, research study by Bulut et al. [39] on the lubricant compatibility of FKM seals in synthetic oils and Zhao et al.'s [40] investigation into the friction and wear behavior of various seal materials under water-lubricated conditions shed light on the importance of seal material compatibility with lubricants.

Furthermore, Meng et al. [41] used a quasi-3D model to conduct a thermohydrodynamic analysis of herringbone-grooved mechanical face seals and Zhang et al. [42] investigated coupled heat transfer characteristics in a spiral groove mechanical seal lubricated by magnetic fluid, providing insights into seal behavior and lubricant–seal interactions under a variety of conditions. Such research contributes to a better understanding of the complex dynamics that regulate seal functioning and lubricant–seal compatibility.

12.2.5 Common lubricant additives for low temperatures

12.2.5.1 Pour point depressants

Pour point depressants are essential for decreasing lubricant pour points and improving flow at low temperatures. These additives work by altering the crystal structure of wax, preventing the production of big crystals that obstruct flow. They produce smaller, scattered crystals, which improve lubricant fluidity. They are often made of polymeric compounds such as polymethacrylates and ethylene-vinyl acetate copolymers.

The importance of pour point depressants is highlighted in research done by Zhao et al. [40], which focuses on the lubricant compatibility of FKM seals in synthetic oils, emphasizing the importance of selecting lubricants compatible with seal materials to ensure optimal sealing performance.

12.2.5.2 Antiwear additives

Antiwear additives are critical for preventing wear and minimizing friction, especially in low-temperature conditions where lubricant viscosity increases. Zinc dialkyldithiophosphates (ZDDPs) and organic friction modifiers are two widely used antiwear additives. ZDDPs work by generating protective layers on metal surfaces, which effectively prevents metal-to-metal contact and reduces adhesive wear. Research by Zhang and Spikes [43] into the mechanism of ZDDP antiwear film formation sheds light on the usefulness of this additive in lubricants.

Organic friction modifiers, such as fatty acids and esters, help to build protective coatings and improve the lubricant's load-carrying ability. Wang et al. [44] investigated the tribological performance of organic molybdenum in the presence of an organic friction modifier, providing useful insights into the synergistic impacts of several antiwear additives. Furthermore, Dzięgielewski et al. [45] explored tribochemical interactions between carbon nanotubes and ZDDP antiwear additives, providing insights into creative techniques to improve lubricant performance.

12.2.5.3 Viscosity modifiers

Viscosity modifiers are essential lubricant additives that provide constant viscosity across a wide temperature range and are important in composition of lubricants as shown in Figure 12.4. They are especially important for low-temperature applications because they preserve viscosity at higher temperatures while promoting ideal flow and dispersion at lower ones. Mostly composed of polymeric materials, viscosity modifiers efficiently control the viscosity of lubricants by adjusting their molecular connections and structure to temperature fluctuations. Polymethacrylates, olefin copolymers, and polyisobutylenes are typical examples.

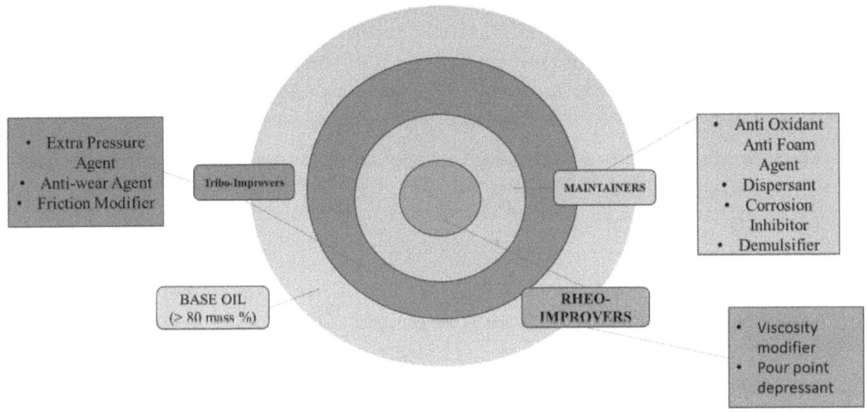

Figure 12.4 Typical composition of lubricants.

Research by Martini et al. [46] stressed the importance of viscosity modifiers as lubricant additives in a comprehensive review, highlighting their role in optimizing viscosity for a wide range of temperature conditions. Furthermore, Robinson et al.'s study [47] investigated the use of highly branched polyethylenes as both viscosity and friction modifiers, demonstrating the adaptability of viscosity modifiers in improving lubricant performance. Also Marx et al. [48] provided valuable insights into shear thinning and hydrodynamic friction of oils containing viscosity modifiers, while Meheust et al. [49] studied bio-based polyricinoleate and polyhydroxystearate as viscosity modifiers.

12.2.5.4 Cold flow improvers

Cold flow improvers are specialty additives that increase lubricant flow at low temperatures and inhibit the development of big, obstructive networks by altering the crystal structure of wax. They co-crystallize with wax molecules, usually polymeric compounds, resulting in smaller, dispersed crystals. This ensures ideal flow and distribution within the lubricated system by maintaining fluidity at low temperatures.

12.3 CHALLENGES IN EXTREME WEATHER LUBRICATION

Despite advancements in low-temperature lubricant formulations, several challenges persist in extreme weather lubrication. These challenges include:

1. **Maintaining consistent performance across a wide temperature range:** While low-temperature performance is critical, lubricants used

in adverse weather situations must also work well at higher temperatures. Achieving a wide operating temperature range while maintaining consistent lubrication and wear prevention might be difficult.

2. **Compatibility with various materials and seals:** Low-temperature lubricants must be compatible with a diverse range of materials and seals used in a variety of applications. Ensuring compatibility to minimize degradation, swelling, or shrinkage is critical for avoiding leaks and equipment failure.
3. **Addressing environmental concerns:** The environmental impact of certain lubricant additives or base oils is an increasing concern, especially in vulnerable environments like the Arctic. Creating ecologically friendly and biodegradable lubricant compositions that preserve performance is a continuous problem.
4. **Balancing performance and cost:** Although high-performance low-temperature lubricants have several advantages, their production costs can be greater than conventional lubricants. Manufacturers and end-users alike must strike the correct balance between performance, cost, and economic sustainability.
5. **Extreme conditions and harsh environments:** Some applications, such as deep space exploration or operations in isolated and inhospitable locations, involve extreme conditions and harsh environments that test lubricant performance. Creating lubricants that can endure these environments while remaining trustworthy is a huge task.

Impact of the above challenges and their possible solutions along with their respective research areas can be seen in Table 12.1 and Figure 12.5 shows the mechanisms that get affected if right lubricants are not selected.

12.4 CURRENT TRENDS, RECENT INNOVATIONS, AND TECHNOLOGICAL ADVANCEMENTS

It's critical to keep up with the latest developments in technology and trends in the ever-changing world of lubricants designed for extremely cold settings. This section explores current developments and developing trends. First, it examines how nanotechnology is revolutionizing low-temperature lubrication.

12.4.1 Nanotechnology in low-temperature lubrication

The use of nanotechnology in lubricant formulations has emerged as a viable area of research for extreme cold weather applications. The addition of nanoparticles, such as metal oxides, carbon nanotubes, graphene, and

Table 12.1 Challenges in extreme weather condition

Challenge	*Description*	*Impact*	*Possible solutions*	*Research areas*
Maintaining consistent performance over wide temperature range	Lubricants must maintain adequate performance from extremely low to high temperatures, requiring a wide operating range.	Reduced lubrication efficiency, increased wear, and potential failures at temperature extremes.	Advanced additive packages, tailored base oil selection, viscosity modifiers.	Development of multifunctional additives, novel base oils, and nanocomposites.
Compatibility with materials and seals	Low-temperature lubricants must be compatible with various materials and seals used in different applications.	Material degradation, swelling, shrinkage, and leakage leading to equipment failures.	Rigorous compatibility testing, development of specialized lubricant formulations for specific materials.	Exploration of new materials and surface treatments for improved compatibility.
Addressing environmental concerns	Certain lubricant additives or base oils may have adverse environmental impacts, particularly in sensitive ecosystems.	Environmental contamination, bioaccumulation, and potential regulatory non-compliance.	Development of biodegradable and environmentally friendly lubricants, use of renewable feedstocks.	Research on bio-based additives, sustainable base oils, and life cycle assessments.
Balancing performance and cost	High-performance low-temperature lubricants may have higher production costs compared to conventional lubricants.	Economic viability and competitiveness challenges for manufacturers and end-users.	Optimization of formulations, process improvements, and economies of scale.	Cost-effective alternative base oils and additives, process efficiency studies.
Extreme conditions and harsh environments	Some applications involve extreme conditions and harsh environments that push the limits of lubricant performance.	Potential equipment failures, safety risks, and operational disruptions.	Development of specialized lubricant formulations, advanced condition monitoring systems.	Exploration of novel materials, advanced manufacturing techniques, and intelligent lubricant systems.

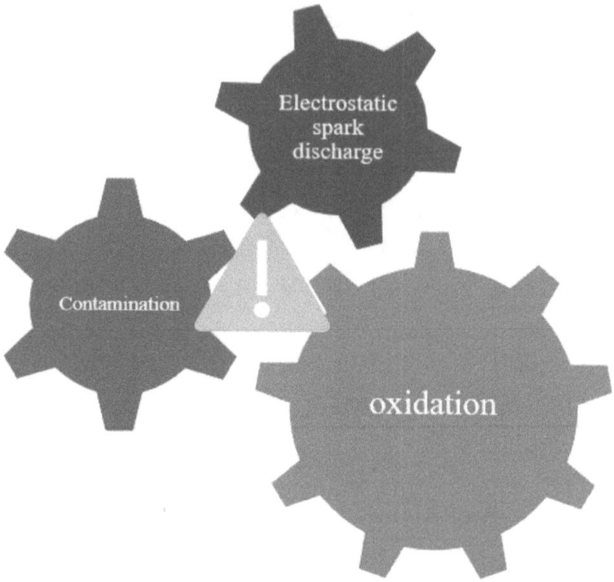

Figure 12.5 Mechanisms that get affected with wrong lubricant selection.

other nanostructured materials, can improve the characteristics of lubricants, including:

1. **Viscosity behavior:** Nanoparticles can change the viscosity of lubricants, increasing low-temperature fluidity while preserving appropriate viscosity at higher temperatures. This is accomplished by techniques such as changing the wax crystal structure or adjusting the lubricant's rheological characteristics.
2. **Wear resistance:** Nanoparticles can create protective tribofilms on metal surfaces, reducing friction and avoiding wear even at high pressures and low temperatures.
3. **Friction reduction:** The use of nanoparticles can reduce friction coefficients in lubricated systems, boosting energy efficiency and lowering wear.
4. **Thermal conductivity**: Certain nanoparticles, such as carbon nanotubes and graphene, can improve lubricant thermal conductivity, resulting in better heat transfer and dissipation in high-load or high-speed applications.
5. **Oxidation stability**: Some nanoparticles, such as metal oxides, can act as antioxidants, increasing the oxidation stability and service life of low-temperature lubricants.

Research on the tribological properties of MoS_2 nanoparticles interacting with ZDDP tribofilms demonstrates the ability of nanoparticles such as

WS_2 to dramatically reduce friction when paired with traditional antiwear additives [50–54]. Furthermore, research on the tribological behavior of Cu nanoparticles in palm olein-based grease lubricants shows that adding nanoparticles improves dropping point temperature and penetration depth [55]. Furthermore, studies on the creation of robust interfacial tribofilms using borate- and polymer-coated ZnO nanoparticles highlight the improved wear protection achieved under a boundary lubrication regime [56]. Studies on the tribological effects of Cu, Fe, and Zn nanoparticles in mineral and bio-based lubricants demonstrate nanoparticles' ability to improve lubricity and performance in a variety of lubricating oils [57].

12.4.2 Biodegradable lubricants for cold environments

The development of biodegradable and ecologically friendly lubricants for cold environments has received a lot of attention as environmental concerns and regulations have grown. These lubricants, generated from renewable sources such as vegetable oils, animal fats, and synthetic esters, are intended to have a minimal detrimental impact on the environment. Biodegradable lubricants have various advantages over mineral oil-based and synthetic lubricants.

12.4.2.1 Lowered environmental impact

Biodegradable lubricants are easily broken down by microorganisms, limiting long-term environmental contamination and the possibility of bioaccumulation. Research on the tribological performance of biodegradable lubricants demonstrates their ability to offer adequate lubrication while remaining environmentally benign.

12.4.2.2 Low toxicity

Biodegradable lubricants are naturally less hazardous to marine and terrestrial species. This feature makes them appropriate for usage in environmentally sensitive places. Studies on the rheological and flow behavior of lubricating grease in extremely cold conditions highlight the relevance of understanding how temperature affects lubricant performance and dynamics [58].

12.4.2.3 Renewable and sustainable

Biodegradable lubricants, which are derived from renewable feedstocks, help to create a more sustainable and environmentally friendly lubricating system. Research into the effects of extreme weather events on hydrological processes emphasizes the importance of considering the indirect responses to cold weather-induced disturbances, particularly in wooded areas [59].

12.4.2.4 Potential for low-temperature performance

Biodegradable lubricants can achieve comparable or even superior low-temperature performance to traditional lubricants when carefully formulated and by using additives. The economic value of climate change-induced mortality emphasizes the necessity of understanding the long-term effects of temperature extremes [60].

Oxidation stability, compatibility with materials and seals, and cost-effectiveness are all issues that must be solved before biodegradable lubricants may be widely used in extreme cold weather applications. The research of the hydrological impact of extreme weather-induced forest disturbances in tropical watersheds emphasizes the interconnectedness of ecological and hydrological processes influenced by cold waves [59]. To summarize, the development and acceptance of biodegradable lubricants for cold conditions represent a substantial progress toward sustainable and environmentally conscious lubrication practices.

12.4.3 Smart lubricants and condition monitoring systems

The incorporation of smart technology and condition monitoring systems into low-temperature lubricants is a new trend in extreme weather lubrication. These systems provide real-time monitoring of lubricant performance, wear conditions, and prospective equipment failures, allowing for proactive maintenance and lubricant optimization.

1. **Sensor integration:** Smart lubricants use embedded sensors or sensor systems to monitor factors such as temperature, pressure, viscosity, and contamination levels [11]. These sensors provide significant information on the lubricant's state and function, allowing for predictive maintenance and prompt interventions.
2. **Internet of Things (IoT) connection:** Smart lubrication systems can be connected with IoT technologies to provide remote monitoring, data analysis, and real-time decision-making [13]. This enables for centralized asset and location monitoring, resulting in more efficient maintenance planning and resource allocation.
3. **Condition monitoring algorithms:** Advanced algorithms and machine learning techniques can use smart lubricant data to reliably forecast lubricant deterioration, wear patterns, and probable problems [12]. This information aids in optimizing lubricant replacement intervals and reducing unnecessary downtime.

4. **Additive replenishment:** In some advanced systems, smart lubricants can be designed to self-replenish critical additives as they are depleted, maintaining optimal performance and extending the lubricant's service life.
5. **Integration with maintenance systems:** Smart lubricant systems can be integrated with existing maintenance management systems, providing a comprehensive view of equipment health and enabling coordinated maintenance activities across multiple systems and components.

The implementation of smart lubricants and condition monitoring systems offers several benefits, including increased equipment reliability, reduced maintenance costs, optimized lubricant usage, and improved operational efficiency. However, challenges such as system integration, data security, and cost-effectiveness need to be addressed for widespread adoption in extreme weather lubrication applications.

12.5 FUTURE SCOPE IN EXTREME WEATHER LUBRICATION

The subject of extreme weather lubrication is constantly expanding, with ongoing research and development activities aimed at identifying new methods and improved materials to overcome the challenges faced by extreme cold temperatures. The following are some upcoming technologies and fields of research.

12.5.1 Emerging technologies

Lubrication science is being developed to improve performance, sustainability, and efficiency in severe settings. In this section, we'll look at how developing technologies like improved additives, AI integration, alternative base oils, nanocomposites, and innovative production methods are driving breakthroughs in lubricating solutions.

12.5.1.1 Advanced additive packages

Advanced additive packages are being developed to provide multifunctional additives with desirable qualities such as low-temperature fluidity, wear protection, and oxidation stability all in one package. This method seeks to simplify formulas while improving overall performance. Ravensteijn et al. [19] investigated triple-function lubricant additives based on organic–inorganic hybrid star polymers, highlighting the possibility for combining several functionalities in a single additive package to streamline formulations and increase lubricant performance. The study highlights the benefits of minimizing formulation complexity through multifunctional additives.

Furthermore, Jiang et al.'s [61] study on the tribological properties of tungsten disulfide nanoparticles surface-capped by oleylamine and maleic anhydride dodecyl ester as additives in diisooctylsebacate demonstrates the importance of multifunctional additives for synthetic ester base oils. This study emphasizes the importance of integrating additives that provide numerous benefits for improved lubricant performance.

12.5.1.2 Machine learning and artificial intelligence

These technologies are being used in lubricant formulation and condition monitoring to optimize additive packages, predict degradation, and design intelligent maintenance plans. These technologies present opportunities to improve lubricant performance and efficiency in a variety of applications. Research studies by Jordon and Mitchell [62] on machine learning trends and prospects, as well as research by Jiang et al. [63] on artificial intelligence in healthcare, provide insights into how machine learning and artificial intelligence techniques are applied in many fields. These studies demonstrate the potential of these technologies for optimizing processes and decision-making [64–67].

12.5.1.3 Alternative base oils

While synthetic base oils such as polyalphaolefins (PAOs) and esters are commonly used in low-temperature lubricants, researchers are looking into alternative base oils derived from renewable sources or novel chemical structures in order to improve performance and environmental sustainability. These alternative base oils seek to improve lubricant characteristics while being consistent with sustainability aims. Researchers Holmberg and Erdemir's [68] study on the impact of tribology on global energy consumption, costs, and emissions demonstrates the promise of polyalkylene glycol (PAG)-based lubricants as an alternative to mineral and synthetic hydrocarbon oils. The study emphasizes PAG-based lubricants' decreased viscosity and enhanced environmental compatibility.

Furthermore, the study of [69] on smart-responsive colloidal capsules as a multifunctional lubricant additive reveals the difficulties encountered in microcapsule breakage under extreme stressed-shearing circumstances. This study sheds light on the difficulty of developing multifunctional lubricant additives to increase performance.

12.5.1.4 Nanocomposites

The creation of nanocomposites, which incorporate nanoparticles into polymeric matrix or lubricant formulations, is a potential field of study. These nanocomposites have unique features such as increased wear resistance, higher thermal conductivity, and customized rheological behavior, making them useful in a variety of applications. Wen et al. [70] investigated polymeric nanocomposite membranes for water treatment, while Fischer et al. [71] investigated totally miscible nanocomposites, highlighting the potential benefits of integrating nanoparticles into polymeric matrix. These studies provide insights into the problems and benefits of using nanocomposites in many fields [72, 73].

Furthermore, the work of Tehfe et al. [74] on the production of silver/polymer nanocomposites, as well as [75] on highly refractive barium-titanate-incorporated polyimide nanocomposite films, highlights nanocomposites' various uses and properties in materials research. These researches help us comprehend the versatility and promise of nanocomposite materials. Finally, the references provide significant insights into the creation and implementation of nanocomposites, demonstrating their ability to improve material characteristics and performance in a variety of sectors [76–81].

12.5.1.5 Advanced manufacturing techniques

The use of sophisticated manufacturing processes, such as additive manufacturing (3D printing) and micro/nanostructuring, can result in the creation of novel lubricant delivery systems or surface treatments that increase lubrication performance under severe cold environments. Etsion [82] conducted research on the state of the art in laser surface texturing, while Na et al. [83] investigated the tribological performance of SiC surfaces with squamous groove microtexture, demonstrating the possibility of new manufacturing processes in improving lubrication performance. These researches provide important insights into the use of micro/nanostructuring to improve tribological qualities [65, 84–86].

Furthermore, Wang et al. [87] on durable nanofluid-infused hierarchical surfaces and Taha-Tijerina [88] on tribological and thermal transport of Ag-vegetable nanofluids prepared by laser ablation provide insights into the development of advanced manufacturing techniques for lubricant applications. These studies help to understand the possibility of innovative production processes for improving lubrication performance. Finally, the references offer useful insights into the use of innovative manufacturing techniques to improve lubrication performance in extremely cold circumstances [89].

12.5.2 Future developments in low-temperature lubrication

As research and development activities continue, numerous projected advances in low-temperature lubrication are expected to meet current issues and push performance boundaries.

12.5.2.1 Extended operating temperature range

Future lubricants are anticipated to possess a broader operating temperature range, ensuring consistent performance from extremely low to high temperatures. This advancement will be facilitated through sophisticated formulations and the integration of innovative additives and base oils. Research conducted by Erhan et al. [90] on the lubricant base stock potential of chemically modified vegetable oils and by Sharma et al. [91] on the preparation and properties evaluation of biolubricants derived from canola oil and canola biodiesel demonstrates the promise of alternative base oils in broadening the operating temperature range of lubricants. These researches provide useful information about the use of plant oils to improve lubricant performance. Furthermore, the studies by Hernández-Sierra et al. [92] on friction and wear enhancement using curcumin as a natural additive in green lubricants and Woma et al. [93] on vegetable oil-based lubricants highlight the need for novel additives and base oils to improve lubrication performance across a wider temperature spectrum. These researches provide substantial contributions to our understanding of developments in lubricant compositions for a wide range of operating circumstances. Finally, the references provide useful insights into the creation of lubricants with a wider operating temperature range using creative formulations and the incorporation of different base oils and additives [94–100].

12.5.2.2 Improved biodegradability and environmental compatibility

With growing environmental concerns and harsher restrictions, developing highly biodegradable and environmentally friendly low-temperature lubricants has become a top goal. This development may include the use of sustainable feedstocks, bio-based additives, and novel formulation methodologies. The incorporation of clay nanoparticles into polymeric matrices for regenerative medicine and biomaterial design [101], as well as the wear-resistant nanocomposites based on biomedical-grade UHMWPE paraffin oil and carbon nanofillers [102], highlights the potential of nanoparticles in enhancing polymeric materials' properties and biocompatibility. Research study by Öncü and E. Durak [103] on using waste vegetable oil methyl esters as lubricant oil, and Hou et al.'s study [104] on reducing and polymerizing graphene oxide/ styrene mixtures to create polymer nanocomposites with tuneable dielectric constants, highlight the need for alternative feedstocks and innovative formulations in sustainable lubricant development. These works provide substantial contributions to our understanding of advances in biodegradable and environmentally friendly lubricants. Finally, the references offer useful insights into the creation of lubricants with improved biodegradability and environmental friendliness through the use of innovative additives and base oils [105–109].

12.5.2.3 Improved efficiency and energy savings

Future low-temperature lubricants are expected to be more energy efficient by lowering friction and limiting viscous losses, resulting in significant energy savings and lower operating costs. Research studies by Greiner et al.

[110] on solids under extreme shear and the energy consumed in overcoming friction in the transportation sector, as well as by Qu et al. [111] on the potential of ionic liquids with ammonium cations as lubricants or additives, highlight the importance of reducing friction for energy savings. These studies provide useful information about the role of lubricants in energy efficiency. Furthermore, research by Kurdi et al. [112] on the practice of simulation and life cycle assessment in tribology, as well as study by Qu et al. [113] on the synergistic effects of ionic liquids and zinc-dialkyl-dithiophosphate (ZDDP) as lubricant additives, shows the potential for energy savings through advancements in lubrication technology. These researches provide significant contributions to understanding the impact of lubricants on energy efficiency and cost reduction. Finally, the references provide useful information about the possibilities for increased energy efficiency and savings through the development of low-temperature lubricants with lower friction and improved performance [114–118].

12.5.2.4 Extended service life

The development of lubricants with improved oxidation stability, wear protection, and self-replenishing additives is projected to contribute to longer service life by lowering the frequency of lubricant changes and minimizing downtime for maintenance. Somers et al. [119] conducted research on ionic liquid lubricants and their impact on friction and wear, while Zheng-lin and Li [120] investigated the role of CuO nanorods in promoting lubricity through good dispersion stability, demonstrating the potential of additives in improving lubricant performance. Furthermore, the work of Yu et al. [121] on the boundary lubrication of chemically grafted and cross-linked hyaluronic acid, as well as of Guegan et al. [122] on the synergies and antagonisms of friction modifier additives, emphasizes the importance of additives in improving wear protection and lowering friction. Finally, the references provide useful insights into the production of lubricants with improved qualities by using novel additives and base oils, resulting in longer service life and lower maintenance downtime.

12.5.2.5 Intelligent lubrication systems

The combination of smart technologies, condition monitoring systems, and advanced analytics will allow for the creation of intelligent lubrication systems that can adapt and optimize performance in real time based on data and operating conditions.

12.5.3 Practical implications for industries

Advances in low-temperature lubricants and associated technologies will have far-reaching practical ramifications for sectors that operate in extremely cold conditions. These advances may lead to:

1. **Increased equipment reliability:** Improved lubricant performance and condition monitoring systems will help to increase equipment reliability, lowering unexpected downtime and associated expenses.
2. **Lower maintenance costs:** Extended lubricant service life, predictive maintenance procedures, and optimal lubricant usage will all result in lower maintenance costs for sectors operating in very cold temperatures.
3. **Improved operational efficiency:** Energy-efficient lubricants, decreased friction, and optimized lubrication systems will result in enhanced productivity and lower energy usage. Low-temperature lubricants are used in drug-eluting devices [123] and various biomedical equipment to ensure smooth mechanical operation and to maintain stability during controlled drug release, even in environments with varying temperatures, without affecting the efficacy of the drug or contaminating the system.
4. **Improved environmental sustainability:** The development of biodegradable and ecologically friendly low-temperature lubricants would help companies meet severe environmental laws and contribute to sustainable practices, especially in sensitive ecosystems.
5. **Expansion into new areas:** Advances in harsh weather lubrication will allow enterprises to enter previously inaccessible or difficult areas, such as deep space exploration, Arctic resource exploitation, and high-altitude construction projects.

12.5.4 Recommendations for further research

To further advance the field of extreme weather lubrication and address the ongoing challenges, several suggestions for future research can be offered:

1. **Collaborative research efforts:** Fostering cooperation among academia, industry, and research institutions can speed up the development of innovative lubrication solutions and facilitate the transfer of information and technology.
2. **Multidisciplinary approach:** Combining expertise from chemistry, materials science, tribology, engineering, and environmental sciences can result in more comprehensive solutions for extreme weather lubrication. Vacuum technology [124] and food industries where foods are fried in a vacuum [125, 126] rely on low-temperature lubricants to ensure optimal performance and prevent contamination or outgassing in sensitive vacuum environments, which is crucial for applications in aerospace, semiconductor manufacturing, and scientific research.
3. **Advanced characterization and modeling:** Advanced characterization techniques and computational modeling can provide insights into the underlying principles driving lubricant behavior under extreme cold conditions, allowing for the rational development of high-performance formulations.

4. **Life cycle assessment and sustainability:** Conducting life cycle assessments and sustainability analyses for low-temperature lubricants can help guide the development of environmentally beneficial and long-term solutions while considering environmental, economic, and social factors.
5. **Pilot-scale and field testing:** Prior to commercial deployment, new low-temperature lubricants and technologies must undergo extensive pilot-scale and field testing to validate their performance in real-world situations and identify potential difficulties.
6. **Regulatory collaboration:** Working with regulatory agencies and industry groups can help guarantee that novel lubricant formulas and technologies fulfill applicable standards and regulations, allowing for wider adoption while minimizing environmental and safety risks. Addressing these recommendations and developing a collaborative and diverse approach will allow the field of harsh weather lubrication to continue to advance, creating creative solutions that enable reliable and sustainable operations in the harshest of cold environments.

12.6 APPLICATIONS IN SEVERAL FIELDS

Applications for low-temperature lubricants are numerous, ranging from industrial machinery to spacecraft in deep space to scientific expeditions in the Arctic regions. They solve difficulties such as low-temperature solidification and increased viscosity. They enhance performance and allow cold starts in automobiles. These lubricants save maintenance costs and downtime for industries, aircraft, and marine vessels operating in severe, cold conditions and ensure safety [127–135]. In Table 12.2, a thorough description of their uses and the special qualities that are needed are provided.

12.7 SUMMARY AND CONCLUSION

Extreme cold weather presents considerable problems for lubricants, necessitating specific formulations and advanced technology to maintain consistent performance and equipment protection. This chapter provides a comprehensive overview of the state of the art in lubricants designed for extreme cold weather applications, including lubricant types, properties, low-temperature effects, ideal lubricant characteristics, additive packages, formulation strategies, testing methods, regulatory compliance, and environmental considerations.

The chapter emphasized the necessity of customized additive packages, base oil selection, and the promise of future technologies such as nanotechnology, biodegradable lubricants, and smart lubricants in tackling the

Table 12.2 Applications and required properties of lubricants in different fields

Field	*Unique properties required*	*Applications/uses*	*Examples of lubricants*	*Key considerations*
Outer space	Low vapor pressure, radiation resistance, wide temperature range, compatibility with materials like titanium and aluminum alloys	Spacecraft mechanisms, robotic arms, bearings, gears, and other components	Perfluoropolyether (PFPE) lubricants, silahydrocarbon fluids, lead-based greases	Must withstand vacuum, extreme temperatures, and microgravity conditions
Automobiles	Low viscosity at cold temperatures, high viscosity index, excellent pumpability, antiwear additives, compatibility with engine materials	Engine oils, transmission fluids, differential fluids, brake fluids	Synthetic lubricants like polyalphaolefins (PAOs), esters, polyalkylene glycols (PAGs)	Important for cold starts, fuel efficiency, and component protection
Machineries	High load-carrying capacity, extreme pressure (EP) properties, corrosion inhibition, shear stability, compatibility with seals and metals	Construction equipment, mining machinery, oil and gas equipment, heavy-duty vehicles	Mineral oil-based or synthetic lubricants with specialized additive packages	Lubricants must withstand harsh operating conditions and prevent equipment failure
Aircraft	Low volatility, thermal and oxidation stability, antiwear properties, compatibility with aluminum and titanium alloys	Turbine engines, landing gear, hydraulic systems, flight control mechanisms	Synthetic esters, polyalphaolefins (PAOs), polyol esters	Safety-critical applications, lubricants must meet stringent performance requirements
Water transport	Seawater resistance, corrosion inhibition, biodegradability, low toxicity, antiwear properties	Propulsion systems, thrusters, deck machinery, cargo handling equipment	Biodegradable lubricants, synthetic esters, polyalkylene glycols (PAGs)	Lubricants must withstand exposure to seawater and marine environments
Scientific research	Wide temperature range, low volatility, compatibility with specialized materials, low outgassing, radiation resistance	Research equipment, instruments, and machinery used in extreme environments	Perfluoropolyether (PFPE) lubricants, silahydrocarbon fluids, synthetic hydrocarbons	Lubricants must ensure reliable data collection and minimize downtime in challenging conditions

issues of extreme weather lubrication. The chapter also explored the use of low-temperature lubricants in a variety of sectors, such as outer space, autos, machinery, airplanes, water transport, and scientific research trips.

As companies expand into difficult climate regions and pursue new frontiers, such as deep space exploration, the demand for sophisticated low-temperature lubricants will rise. Continuous R&D efforts, together with collaboration among academia, industry, and regulatory agencies, will be critical in generating innovation and resolving the issues connected with extreme weather lubrication.

The future looks promising for the development of more efficient, sustainable, and environmentally friendly lubricant systems that can resist the toughest cold temperatures. Advancements in nanotechnology, biodegradable lubricants, smart lubricant systems, and new formulation strategies are projected to push the limits of low-temperature lubrication performance, allowing for dependable and efficient operations in even the harshest settings.

By embracing these advancements and encouraging a collaborative and multidisciplinary approach, the field of extreme weather lubrication can continue to contribute to increased equipment reliability, lower maintenance costs, improved operational efficiency, and increased environmental sustainability in a variety of industries and applications.

REFERENCES

1. G. Huang, Q. Yu, Z. Ma, M. Cai, F. Zhou, and W. Liu, "Oil-Soluble Ionic Liquids as Antiwear and Extreme Pressure Additives in Poly-A-Olefin for Steel/Steel Contacts," *Friction*, 2017, doi: 10.1007/s40544-017-0180-8.
2. M. John and P. L. Menezes, "Self-Lubricating Materials for Extreme Condition Applications," *Materials*, 2021, doi: 10.3390/ma14195588.
3. X. Dai, B. B. Stogin, S. Yang, and T. S. Wong, "Slippery Wenzel State," *ACS Nano*, 2015, doi: 10.1021/acsnano.5b04151.
4. J. Petersson, K. Kuklane, and C. Gao, "Is There a Need to Integrate Human Thermal Models With Weather Forecasts to Predict Thermal Stress?," *Int J Environ Res Public Health*, 2019, doi: 10.3390/ijerph16224586.
5. B. Lin, M. Ding, T. Sui, Y. Cui, S. Yan, and X. Liu, "Excellent Water Lubrication Additives for Silicon Nitride to Achieve Superlubricity Under Extreme Conditions," *Langmuir*, 2019, doi: 10.1021/acs.langmuir.9b02337.
6. S. Adera, J. Alvarenga, A. V Shneidman, C. T. Zhang, A. Davitt, and J. Aizenberg, "Depletion of Lubricant From Nanostructured Oil-Infused Surfaces by Pendant Condensate Droplets," *ACS Nano*, 2020, doi: 10.1021/acsnano.9b10184.
7. M. Freschi, M. Di Virgilio, G. Zanardi, M. Mariani, N. Lecis, and G. Dotelli, "Employment of Micro- And Nano-Ws2 Structures to Enhance the Tribological Properties of Copper Matrix Composites," *Lubricants*, 2021, doi: 10.3390/lubricants9050053.

8. F. Yan, Z. Wang, Y. Du, S. Su, Y. Zheng, and L. Qinfeng, “Research on Rheological and Flow Behavior of Lubricating Grease in Extremely Cold Weather,” *Ind Lubr Tribol*, 2017, doi: 10.1108/ilt-01-2017-0011.
9. Y. Su, Y. Zhang, J. Song, and L. Hu, “Novel Approach to the Fabrication of an Alumina-MoS2 Self-Lubricating Composite via the in Situ Synthesis of Nanosized MoS_2,” *ACS Appl Mater Interfaces*, 2017, doi: 10.1021/acsami.7b09000.
10. Y. Lin, H. Zhang, S. Zhao, G. Hua, and W. Tian, “The Influence of Extreme Cold Ambient Temperature on Out of Hospital Cardiac Arrest: A Systemic Review and Meta-Analysis,” *Frigid Zone Medicine*, 2022, doi: 10.2478/fzm-2022-0025.
11. J.Li,H.Zhao,andL.Wang,“BioinspiredDepletion-ResistantLubricant-Infused Surfaces With Self-Replenishing Lubrication Through Capillary Filament,” *Adv Mater Interfaces*, 2021, doi: 10.1002/admi.202100561.
12. M. L. B. Palacio and B. Bhushan, “A Review of Ionic Liquids for Green Molecular Lubrication in Nanotechnology,” *Tribol Lett*, 2010, doi: 10.1007/s11249-010-9671-8.
13. B. A. Noble, A. Ovcharenko, and B. Raeymaekers, “Quantifying Lubricant Droplet Spreading on a Flat Substrate Using Molecular Dynamics,” *Appl Phys Lett*, 2014, doi: 10.1063/1.4898140.
14. D. Gasni, I. H. Mulyadi, J. Affi, and A. Y. Miswar, “Investigation of Wear Mechanism of Ball Bearings Lubricated by Bio-Lubricant,” *Int J Technol*, 2017, doi: 10.14716/ijtech.v8i7.688.
15. E. D. Renzis, V. Mariani, G. M. Bianchi, S. Falfari, and G. Cazzoli, “Application of a One-Dimensional Fuel-Oil Dilution Model Coupled With an Empirical Droplet-to-Film Formation Strategy for Predicting in-Cylinder Oil Effects in a Direct Injection Engine,” *J Phys Conf Ser*, 2022, doi: 10.1088/1742-6596/2385/1/012063.
16. S. Blanck, S. Loehlé, S. N. Steinmann, and C. Michel, “Adhesion of Lubricant on Aluminium Through Adsorption of Additive Head-Groups on Γ-Alumina: A DFT Study,” *Tribol Int*, 2020, doi: 10.1016/j.triboint.2019.106140.
17. R. J. Waltman, N. Kobayashi, T. Shimizu, A. Inoue, H. Deng, and Y. Wu, “The Role of Configuration Entropy in Boundary Lubricants Based on the N-Perfluoropropylene Oxide Main Chain Structure,” *Tribol Lett*, 2008, doi: 10.1007/s11249-008-9398-y.
18. W. Liu and X. Wang, “Bio-Lubricants Derived From Waste Cooking Oil With Improved Oxidation Stability and Low-Temperature Properties,” *J Oleo Sci*, 2015, doi: 10.5650/jos.ess14235.
19. B. G. P. Ravensteijn *et al.*, “Triple Function Lubricant Additives Based on Organic–Inorganic Hybrid Star Polymers: Friction Reduction, Wear Protection, and Viscosity Modification,” *ACS Appl Mater Interfaces*, 2018, doi: 10.1021/acsami.8b16849.
20. T. W. Park, C. S. Ho, S. Jeong, Y.-S. Choi, S. Park, and C. K. Song, “Different Characteristics of Cold Day and Cold Surge Frequency Over East Asia in a Global Warming Situation,” *J Geophys Res Atmos*, 2011, doi: 10.1029/2010jd015369.
21. M. J. Covitch and K. Trickett, “How Polymers Behave as Viscosity Index Improvers in Lubricating Oils,” *Adv Chem Engineer Sci*, 2015, doi: 10.4236/aces.2015.52015.

22. X. Paredes, A. S. Pensado, M. J. P. Comuñas, and J. Fernández, "How Pressure Affects the Dynamic Viscosities of Two Poly(propylene Glycol) Dimethyl Ether Lubricants," *J Chem Eng Data*, 2010, doi: 10.1021/je100285a.
23. M. Fan *et al.*, "High-Performance Lubricant Base Stocks From Biorenewable Gallic Acid: Systematic Study on Their Physicochemical and Tribological Properties," *Ind Eng Chem Res*, 2017, doi: 10.1021/acs.iecr.7b01452.
24. R. Shah, S. C. Tung, R. Chen, and R. A. Miller, "Grease Performance Requirements and Future Perspectives for Electric and Hybrid Vehicle Applications," *Lubricants*, 2021, doi: 10.3390/lubricants9040040.
25. B. Pohrer, I. Killmann, M. Zuercher, and E. Schlücker, "Practical Application of CO_2 as a Viscosity Index Improver for Lubricants," *Chem Eng Technol*, 2016, doi: 10.1002/ceat.201500281.
26. A. S. Sarpal, M. I. S. Sastry, V. Bansal, I. P. Singh, S. K. Mazumdar, and B. Basu, "Correlation of Structure and Properties of Groups I to III Base Oils," *Lubr Sci*, 2012, doi: 10.1002/ls.1172.
27. E. D. Bonnevie, M. L. Delco, D. Galesso, C. Secchieri, L. A. Fortier, and L. J. Bonassar, "Sub-Critical Impact Inhibits the Lubricating Mechanisms of Articular Cartilage," *J Biomech*, 2017, doi: 10.1016/j.jbiomech.2016.12.034.
28. W. Wang, S. Jiang, Y. Shen, S. Xia, and J. Xu, "Synthesis of High Viscosity Index Base Stock and Study on the Lubricating Properties," *Ind Lubr Tribol*, 2016, doi: 10.1108/ilt-02-2015-0020.
29. H. Li *et al.*, "State-of-the-Art of Extreme Pressure Lubrication Realized With the High Thermal Diffusivity of Liquid Metal," *ACS Appl Mater Interfaces*, 2017, doi: 10.1021/acsami.6b15825.
30. L. Peña-Parás *et al.*, "Extreme Pressure Properties of Nanolubricants for Metal-Forming Applications," *Ind Lubr Tribol*, 2016, doi: 10.1108/ilt-05-2015-0069.
31. M. A. Dandan, A. Yahaya, and S. Syahrullail, "The Effect of the Different Percentage of Pour Point Depressant (PPD) on the Tribological Properties of Palm Kernel Oil," *Tribol Ind*, 2019, doi: 10.24874/ti.2019.41.03.06.
32. R. S. Kamal, M. M. Shaban, G. S. S. Raju, and R. K. Farag, "High-Density Polyethylene Waste (HDPE)-Waste-Modified Lube Oil Nanocomposites as Pour Point Depressants," *ACS Omega*, 2021, doi: 10.1021/acsomega.1c04693.
33. C. F. W. Müller, F. L. Redondo, M. Dennehy, A. E. Ciolino, and W. R. Tuckart, "Bismuth (III) Sulfide as Additive: Towards Better Lubricity without Toxicity," *Ind Lubr Tribol*, 2018, doi: 10.1108/ilt-03-2017-0051.
34. S. Sharma, V. Mahto, and V. K. Sharma, "Effect of Flow Improvers on Rheological and Microscopic Properties of Indian Waxy Crude Oil," *Ind Eng Chem Res*, 2014, doi: 10.1021/ie403050y.
35. J. Zhao *et al.*, "Using Green, Economical, Efficient Two-Dimensional (2D) Talc Nanosheets as Lubricant Additives Under Harsh Conditions," *Nanomaterials*, 2022, doi: 10.3390/nano12101666.
36. W. Guan, H. Zhou, B. Wang, Y. Xia, and W. Liu, "Tribological Behavior of Aeronautical Steel Under Oil–air Lubrication Containing Extreme-Pressure and Anti-Wear Additives," *Adv Mech Eng*, 2016, doi: 10.1177/1687814016638299.
37. S. D. Leventini, B. S. Martin-Gutierrez, A. Kumar, A. S. Mittman, S. M. Kim, and A. Martini, "Tactile Perception of Vellum Quantified by Friction and Surface Roughness," *Tribol Lett*, vol. 70, no. 4, p. 127, 2022.

38. A. Kumar, S. Keerti, J. Jain, S. Sinha, S. Tekumalla, and M. Gupta, "Investigations of Wear Response of Pure Mg and Mg-0.4 Ce-Y2O3/ZnO Nanocomposites Using a Single and Repeated Scratch Tests," *Tribol Trans*, vol. 61, no. 5, pp. 951–959, 2018.
39. D. Bulut, T. Krups, G. Poll, and U. Giese, "Lubricant Compatibility of FKM Seals in Synthetic Oils," *Ind Lubr Tribol*, 2019, doi: 10.1108/ilt-02-2019-0065.
40. W. Zhao, J. Zhang, and G. Dong, "Friction and Wear Behavior of Different Seal Materials Under Water-Lubricated Conditions," *Friction*, 2020, doi: 10.1007/s40544-020-0364-5.
41. X. Meng, W. Zhao, M. Shen, and X. Peng, "Thermohydrodynamic Analysis on Herringbone-Grooved Mechanical Face Seals With a Quasi-3d Model," *Proc Inst Mech Eng Part J J Eng Tribol*, 2018, doi: 10.1177/1350650117752952.
42. P. Zhang, B. Gu, J. Zhou, and L. Wei, "Coupled Heat Transfer Characteristics in a Spiral Groove Mechanical Seal Lubricated by Magnetic Fluid," *Ind Lubr Tribol*, 2019, doi: 10.1108/ilt-06-2018-0225.
43. J. Zhang and H. A. Spikes, "On the Mechanism of ZDDP Antiwear Film Formation," *Tribol Lett*, 2016, doi: 10.1007/s11249-016-0706-7.
44. W. Wang *et al.*, "Tribological Performance of Organic Molybdenum in the Presence of Organic Friction Modifier," *PLoS One*, 2021, doi: 10.1371/journal.pone.0252203.
45. W. Dzięgielewski, J. Kowalczyk, A. Kulczycki, M. Madej, and D. Ozimina, "Tribochemical Interactions Between Carbon Nanotubes and ZDDP Antiwear Additive During Tribofilm Formation on Uncoated and DLC-Coated Steel," *Materials*, 2020, doi: 10.3390/ma13102409.
46. A. Martini, U. S. Ramasamy, and M. Len, "Review of Viscosity Modifier Lubricant Additives," *Tribol Lett*, 2018, doi: 10.1007/s11249-018-1007-0.
47. J. W. Robinson, Y. Zhou, J. Qu, J. T. Bays, and L. Cosimbescu, "Highly Branched Polyethylenes as Lubricant Viscosity and Friction Modifiers," *React Funct Polym*, 2016, doi: 10.1016/j.reactfunctpolym.2016.10.003.
48. N. Marx, L. Fernández, F. Barceló, and H. A. Spikes, "Shear Thinning and Hydrodynamic Friction of Viscosity Modifier-Containing Oils. Part II: Impact of Shear Thinning on Journal Bearing Friction," *Tribol Lett*, 2018, doi: 10.1007/s11249-018-1040-z.
49. H. Meheust, J.-F. Le Meins, E. Grau, and H. Cramail, "Bio-Based Polyricinoleate and Polyhydroxystearate: Properties and Evaluation as Viscosity Modifiers for Lubricants," *ACS Appl Polym Mater*, 2021, doi: 10.1021/acsapm.0c01153.
50. A. Tomala, B. Vengudusamy, M. R. Ripoll, A. N. Suarez, M. Remškar, and R. Rosentsveig, "Interaction Between Selected MoS2 Nanoparticles and ZDDP Tribofilms," *Tribol Lett*, 2015, doi: 10.1007/s11249-015-0552-z.
51. S. R. Shrivastava Prateek, A. Kumar, and others, "Investigation of Torsional Rod to Minimize Vibration in Automobile using ANSYS," 2020 1st International Conference on Innovative Research in Applied Science, Engineering and Technology (IRASET), 2020, pp. 1–6.
52. A. K. P. S. Saurabh Gupta Ruchika Saini, "Performance Analysis of Gudgeon Pin of Various Cross Sections by FEM," *Int J Recent Technol Eng*, vol. 8, no. 5, pp. 4569–4573, 2020.
53. A. Kumar, "Advancements in Emerging Superlubricity: A Review of the Atomistic Models, Simulation Techniques and Their Applications to

Explore the State of Ultra-Low Friction," *Mater Today Proc*, vol. 42, no. 4, 2021.884–892.
54. A. Kumar, A. Kumar, M. F. Luther King, Avinash Kumar, and Mohit Byadwal, "Laser Micromachining in Biomedical Industry," in Avinash Kumar, Ashwani Kumar, and Abhishek Kumar (Eds.), *Laser-based Technologies for Sustainable Manufacturing*, CRC Press, 2023, p. 38.
55. A. Muhammad, N. F. Mohd Yusof, and Z. M. Ripin, "The Tribological Performance Analysis of Palm Olein-Based Grease Lubricants Containing Copper Nanoparticle Additive," *Pro Inst Mech Eng Part J J Eng Tribol*, 2023, doi: 10.1177/13506501231190695.
56. K. Vyavhare, R. B. Timmons, A. Erdemir, B. L. Edwards, and P. B. Aswath, "Robust Interfacial Tribofilms by Borate- And Polymer-Coated ZnO Nanoparticles Leading to Improved Wear Protection Under a Boundary Lubrication Regime," *Langmuir*, 2021, doi: 10.1021/acs.langmuir.0c02985.
57. S. Asadauskas, R. Kreivaitis, G. Bikulčius, A. Grigucevičienė, and J. Padgurskas, "Tribological Effects of Cu, Fe and Zn Nano-Particles, Suspended in Mineral and Bio-Based Oils," *Lubr Sci*, 2015, doi: 10.1002/ls.1307.
58. F. Yan, Z. Wang, Y. Du, S. Su, Y. Zheng, and L. Qinfeng, "Research on Rheological and Flow Behavior of Lubricating Grease in Extremely Cold Weather," *Ind Lubr Tribol*, 2017, doi: 10.1108/ilt-01-2017-0011.
59. Y. Hou *et al.*, "The Hydrological Impact of Extreme Weather-Induced Forest Disturbances in a Tropical Experimental Watershed in South China," *Forests*, 2018, doi: 10.3390/f9120734.
60. W. Botzen *et al.*, "Economic Valuation of Climate Change–induced Mortality: Age Dependent Cold and Heat Mortality in the Netherlands," *Clim Change*, 2020, doi: 10.1007/s10584-020-02797-0.
61. Z. Jiang *et al.*, "Tribological Properties of Tungsten Disulfide Nanoparticles Surface-Capped by Oleylamine and Maleic Anhydride Dodecyl Ester as Additive in Diisooctylsebacate," *Ind Eng Chem Res*, 2017, doi: 10.1021/acs.iecr.6b04074.
62. M. I. Jordan and T. M. Mitchell, "Machine Learning: Trends, Perspectives, and Prospects," *Science (1979)*, 2015, doi: 10.1126/science.aaa8415.
63. F. Jiang *et al.*, "Artificial Intelligence in Healthcare: Past, Present and Future," *Stroke Vasc Neurol*, 2017, doi: 10.1136/svn-2017-000101.
64. Y. Subramanian *et al.*, "Artificial Intelligence Technique Based Performance Estimation of Solid Oxide Fuel Cells," *Mater Today Proc*, vol. 80, pp. 2573–2576, 2021.
65. A. Pathak, A. Kumar, A. Kumar, and A. Kumar, "Application of Laser Technology in the Mechanical and Machine Manufacturing Industry," in Avinash Kumar, Ashwani Kumar, and Abhishek Kumar (Eds.),*Laser-based Technologies for Sustainable Manufacturing*, CRC Press, 2023, pp. 107–155.
66. A. Kumar, A. Parihar, S. N. Basha, and U. Panda, "Clinically Available/Under Trial Drugs and Vaccines for Treatment of SARS-COV-2," in Arpana Parihar, Raju Khan, Ashok Kumar, Ajeet Kumar Kaushik and Hardik Gohel (Eds.), *Computational Approaches for Novel Therapeutic and Diagnostic Designing to Mitigate SARS-CoV2 Infection*, Elsevier, 2022, pp. 451–488.
67. A. Kumar, S. Datta, and D. Kalyanasundaram, "Liquid Slippage in Con Fined Flows: Effect of Periodic Micropatterns of Arbitrary Pitch and Amplitude,"

ASME 2016 5th Micro/Nanoscale Heat and Mass Transfer International Conference (MNHMT2016), Biopolis, Singapore, 2016.
68. K. Holmberg and A. Erdemir, "Influence of Tribology on Global Energy Consumption, Costs and Emissions," *Friction*, 2017, doi: 10.1007/s40544-017-0183-5.
69. S. T. Pham, V. Sencadas, W. Lei, D. Liu, S. Wan, and J. Hao, "Smart-Responsive Colloidal Capsules as an Emerging Tool to Design a Multifunctional Lubricant Additive," *ACS Appl Mater Interfaces*, 2021, doi: 10.1021/acsami.0c20759.
70. Y. Wen, J. Yuan, X. Ma, S. Wang, and Y. Liu, "Polymeric Nanocomposite Membranes for Water Treatment: A Review," *Environ Chem Lett*, 2019, doi: 10.1007/s10311-019-00895-9.
71. S. Fischer, A. Salcher, A. Kornowski, H. Weller, and S. Förster, "Completely Miscible Nanocomposites," *Angewandte Chemie*, 2011, doi: 10.1002/anie.201006746.
72. A. K. Mehra, R. Saini, and A. Kumar, "The Effect of Fibre Contents on Mechanical and Moisture Absorption Properties of Gourd Sponge/Coir Fibre Reinforced Epoxy Hybrid Composites," *Compos Commun*, vol. 25, p. 100732, 2021.
73. A. Kumar, A. Datta, A. Kumar, and A. Kumar, "Recent Advancements and Future Trends in Next-Generation Materials for Biomedical Applications," in Ashwani Kumar, Yatika Gori, Avinash Kumar, Chandan Swaroop Meena, and Nitesh Dutt. *Advanced Materials for Biomedical Applications*, CRC Press, 2022, pp. 1–19.
74. M.-A. Tehfe, R. Jamois, P. Cousin, S. Elkoun, and M. Robert, "In Situ Synthesis and Characterization of Silver/Polymer Nanocomposites by Thermal Cationic Polymerization Processes at Room Temperature: Initiating Systems Based on Organosilanes and Starch Nanocrystals," *Langmuir*, 2015, doi: 10.1021/la504518c.
75. K. Abé, D. Nagao, A. Watanabe, and M. Konno, "Fabrication of Highly Refractive Barium-titanate-incorporated Polyimide Nanocomposite Films with High Permittivity and Thermal Stability," *Polym Int*, 2012, doi: 10.1002/pi.4285.
76. A. Kumar, R. Kant, A. Gupta, and S. Bhattacharya, "Fabrication and Parameter Optimization Of Co2 Laser Machined Photo Mask for Photolithography Process," International Conference on Innovations in Design and Manufacturing, Jabalpur, India, 2012.
77. A. Parihar *et al.*, "3D Printing: Advancement in Biogenerative Engineering to Combat Shortage of Organs and Bioapplicable Materials," *Regen Eng Transl Med*, vol. 8, pp. 1–27, 2021.
78. A. Parihar, A. Kumar, U. Panda, R. Khan, D. S. Parihar, and R. Khan, "Cryopreservation: A Comprehensive Overview, Challenges, and Future Perspectives," *Adv Biol*, vol. 7, p. 2200285, 2023.
79. A. Kumar *et al.*, "Optimization of Laser Machining Process for the Preparation of Photomasks, and Its Application to Microsystems Fabrication," *Journal of Micro/Nanolithography, MEMS, and MOEMS*, vol. 12, no. 4, 2013, doi: 10.1117/1.JMM.12.4.041203.
80. R. K. Singh, A. Kumar, R. Kant, A. Gupta, E. Suresh, and S. Bhattacharya, "Design and Fabrication of 3-dimensional Helical Structures in

Polydimethylsiloxane for Flow Control Applications," *Microsys Technol*, vol. 20, no. 1, pp. 101–111, Jan. 2014, doi: 10.1007/s00542-013-1738-7.
81. S. Chakravarthy and G. Ananthasuresh, "Design of an Endoscopic Haptic Display System using an Integrated Ring-actuator," Accessed: Jan. 22, 2018. Online.. http://www.inacomm2013.ammindia.org/Papers/052-inacomm2013_submission_101.pdf
82. I. Etsion, "State of the Art in Laser Surface Texturing," *J Tribol*, 2005, doi: 10.1115/1.1828070.
83. Z. Na, Z. Li, M. Hao, Y. Liu, B. Ren, and Z. Wang, "Numerical Simulation and Experimental Investigation on Tribological Performance of SiC Surface With Squamous Groove Micro Texture," *Lubr Sci*, 2022, doi: 10.1002/ls.1608.
84. A. Kumar, A. Pathak, A. Kumar, and A. Kumar, "Physics of Laser--Matter Interaction in Laser-Based Manufacturing," in Avinash Kumar, Ashwani Kumar, and Abhishek Kumar, *Laser-based Technologies for Sustainable Manufacturing*, CRC Press, 2023, pp. 45–54.
85. A. Datta, A. Kumar, A. Kumar, A. Kumar, and V. P. Singh, "Advanced Materials in Biological Implants and Surgical Tools," in Ashwani Kumar, Yatika Gori, Avinash Kumar, Chandan Swaroop Meena, and Nitesh Dutt. *Advanced Materials for Biomedical Applications*, CRC Press, 2022, pp. 21–43.
86. A. Kumar, A. Kumar, and A. Kumar, *Laser-based Technologies for Sustainable Manufacturing*. CRC Press, 2023.
87. S. Wang *et al*., "Durable Nanofluids-Infused Hierarchical Surfaces With High Corrosion and Abrasion Resistance," *Adv Eng Mater*, 2023, doi: 10.1002/adem.202201292.
88. J. Taha-Tijerina, S. Shaji, S. S. Kanakkillam, M. I. Mendivil Palma, and K. Aviña, "Tribological and Thermal Transport of Ag-Vegetable Nanofluids Prepared by Laser Ablation," *Appl Sci*, 2020, doi: 10.3390/app10051779.
89. A. Kumar, A. Kumar, and A. Kumar, "Introduction to Optics and Laser-Based Manufacturing Technologies," in Avinash Kumar, Ashwani Kumar, and Abhishek Kumar. *Laser-based Technologies for Sustainable Manufacturing*, CRC Press, 2023, pp. 1–43.
90. S. Z. Erhan, B. K. Sharma, Z. Liu, and A. Adhvaryu, "Lubricant Base Stock Potential of Chemically Modified Vegetable Oils," *J Agric Food Chem*, 2008, doi: 10.1021/jf801463d.
91. R. Sharma, A. K. R. Somidi, and A. K. Dalai, "Preparation and Properties Evaluation of Biolubricants Derived From Canola Oil and Canola Biodiesel," *J Agric Food Chem*, 2015, doi: 10.1021/jf505825k.
92. M. T. Hernández-Sierra, J. E. Báez, L. D. Aguilera-Camacho, J. S. García-Miranda, and K. J. Moreno, "Friction and Wear Improvement by Using Curcumin as a Natural Additive in Green Lubricants," *Pro Inst Mech Eng Part J J Eng Tribol*, 2022, doi: 10.1177/13506501221121896.
93. T. Y. Woma, S. A. Lawal, A. S. Abdulrahman, M. A. Olutoye, and M. M. Ojapah, "Vegetable Oil Based Lubricants: Challenges and Prospects," *Tribology Online*, 2019, doi: 10.2474/trol.14.60.
94. A. Kumar, S. Datta, and D. Kalyanasundaram, "Reduction of Hydraulic Friction in Confined Flows by Laser Texturing: Experiments and Theoretical

Validation," ASME 2018 16th International Conference on Nanochannels, Microchannels, and Minichannels, ICNMM 2018, 2018. doi: 10.1115/icnmm2018-7740.
95. A. Kumar, S. Datta, and D. Kalyanasundaram, "Liquid Slippage in Confined Flows: Effect of Periodic Micropatterns of Arbitrary Pitch and Amplitude," *J Heat Transfer*, vol. 140, no. 1, 2018, doi: 10.1115/1.4037363.
96. A. Kumar, S. Datta, and D. Kalyanasundaram, "Permeability and Effective Slip in Confined Flows Transverse to Wall Slippage Patterns," *Phys Fluids*, vol. 28, no. 8, 2016, doi: 10.1063/1.4959184.
97. A. Kumar, S. Datta, and D. Kalyanasundaram, "Liquid Slippage in Confined Flows: Effect of Periodic Micropatterns of Arbitrary Pitch and Amplitude," ASME 2016 5th International Conference on Micro/Nanoscale Heat and Mass Transfer, MNHMT 2016, 2016. doi: 10.1115/MNHMT2016-6491.
98. S. Chakravarthy, K. Avinash, G. Ramu, and G. K. Ananthasuresh, "Design of an Endoscopic Haptic Display System Using an Integrated Ring-Actuator," 1st International and 16th National Conference on Machines and Mechanisms, iNaCoMM 2013, 2013.
99. V. Parmar, A. Kumar, G. V. Prakash, S. Datta, and D. Kalyanasundaram, "Investigation, Modelling and Validation of Material Separation Mechanism during Fiber Laser Machining of Medical Grade Titanium Alloy Ti6Al4V and Stainless Steel SS316L," *Mech Mater*, vol. 137, 2019, doi: 10.1016/j.mechmat.2019.103125.
100. V. Parmar *et al.*, "Oxidation Facilitated Antimicrobial Ability of Laser Micro-Textured Titanium Alloy against Gram-Positive Staphylococcus Aureus for Biomedical Applications," *J Laser Appl*, vol. 30, no. 3, 2018, doi: 10.2351/1.5039860.
101. M. Mousa, N. D. Evans, R. O. C. Oreffo, and J. I. Dawson, "Clay Nanoparticles for Regenerative Medicine and Biomaterial Design: A Review of Clay Bioactivity," *Biomaterials*, 2018, doi: 10.1016/j.biomaterials.2017.12.024.
102. M. Catauro, C. Scolaro, G. D. Poggetto, S. Pacifico, and A. Visco, "Wear Resistant Nanocomposites Based on Biomedical Grade UHMWPE Paraffin Oil and Carbon Nano-Filler: Preliminary Biocompatibility and Antibacterial Activity Investigation," *Polymers (Basel)*, 2020, doi: 10.3390/polym12040978.
103. G. Öncü and E. Durak, "Utilization of Waste Vegetable Oil Methyl Esters as Lubricant Oil," *Pro Inst Mech Eng Part J J Eng Tribol*, 2021, doi: 10.1177/1350650120987930.
104. D. Hou *et al.*, "Simultaneous Reduction and Polymerization of Graphene Oxide/Styrene Mixtures to Create Polymer Nanocomposites With Tunable Dielectric Constants," *ACS Appl Nano Mater*, 2019, doi: 10.1021/acsanm.9b01761.
105. A. Kumar, A. K. Shrama, T. V. K. Gupta, and J. K. Katiyar, "Influence of Hexagonal Boron Nitride Additive Nanocutting Fluid on the Machining of AA6061-T6 Alloy Using Minimum Quality Lubrication," *Proc Inst Mech Eng Part E J Eng Tribol*, p. 09544089221110980, 2022.
106. A. Kumar, A. Parihar, U. Panda, and D. S. Parihar, "Microfluidics-Based Point-of-Care Testing (POCT) Devices in Dealing with Waves of COVID-19 Pandemic: The Emerging Solution," *ACS Appl Bio Mater*, vol. 5, no. 5, pp. 2046–2068, 2022.

107. P. Vats, K. K. Gajrani, and A. Kumar, "Laser-Based Additive Manufacturing," in Avinash Kumar, Ashwani Kumar, and Abhishek Kumar. *Laser-based Technologies for Sustainable Manufacturing*, CRC Press, 2023, pp. 67–83.
108. A. Kumar, A. K. Sharma, and J. K. Katiyar, "State-of-the-Art in Sustainable Machining of Different Materials Using Nano Minimum Quality Lubrication (NMQL)," *Lubricants*, vol. 11, no. 2, p. 64, 2023.
109. S. Kumar, R. K. Verma, A. Kumar, and V. K. Patel, "Importance of Chemically Treated Natural Fibers in the Fabrication of Natural Fiber Reinforced Polymer Composites," in Vinay Kumar Patel, Rishi Kant, Pankaj Singh Chauhan, Shantanu Bhattacharya, *Trends in Fabrication of Polymers and Polymer Composites*, pp. 10–11, 2022.https://doi.org/10.1063/9780735423916
110. C. Greiner, J. Gagel, and P. Gumbsch, "Solids Under Extreme Shear: Friction-Mediated Subsurface Structural Transformations," *Adv Mater*, 2019, doi: 10.1002/adma.201806705.
111. J. Qu, J. J. Truhan, S. Dai, H. Luo, and P. J. Blau, "Ionic Liquids with Ammonium Cations as Lubricants or Additives," *Tribol Lett*, 2006, doi: 10.1007/s11249-006-9081-0.
112. A. Kurdi, N. Alhazmi, H. Alhazmi, and T. Tabbakh, "Practice of Simulation and Life Cycle Assessment in Tribology—A Review," *Materials*, 2020, doi: 10.3390/ma13163489.
113. J. Qu *et al.*, "Synergistic Effects Between Phosphonium-Alkylphosphate Ionic Liquids and Zinc Dialkyldithiophosphate (ZDDP) as Lubricant Additives," *Adv Mater*, 2015, doi: 10.1002/adma.201502037.
114. A. Kumar, S. Datta, and D. Kalyanasundaram, "Confinement Effects on Effective Slip of Patterned Surfaces," in American Society of Mechanical Engineers (ASME), *Fluids Engineering Division Summer Meeting*, 2017, p. V01CT23A004.
115. Y. Subramanian, R. Veena, S. A. Muhammed Ali, A. Kumar, et al., "Artificial Intelligence Technique Based Performance Estimation of Solid Oxide Fuel Cells," International Conference on Nanoelectronics, Nanophotonics, Nanomaterials, Nanobioscience \& Nanotechnology, Mangalam college of Engineering, Kottayam, Kerala, 2021.
116. A. Dhanasekaran, V. R, Y. Subramanian, and A. Kumar, "Deformation Analysis for BFO Thin Films with Different Thickness by Finite Element Method," ASEAN International Conference on Energy and Environment, Universiti Brunei Darussalam, Brunei, 2021.
117. A. Kumar, U. Panda, V. K. Patel, and R. Kant, "Laser-Assisted Fabrication of Polymers by Pushing Down the Limit of Resolution," in Vinay Kumar Patel, Rishi Kant, Pankaj Singh Chauhan, Shantanu Bhattacharya, *Trends in Fabrication of Polymers and Polymer Composites*, pp. 1–3, 2022. AIP Publishing LLC. https://doi.org/10.1063/9780735423916
118. M. Singh, A. Kumar, and A. R. Khan, "Capillary as a Liquid Diode," *Phys Rev Fluids*, vol. 5, no. 10, p. 102101, 2020.
119. A. Somers, P. C. Howlett, D. R. MacFarlane, and M. Forsyth, "A Review of Ionic Liquid Lubricants," *Lubricants*, 2013, doi: 10.3390/lubricants1010003.
120. T. Zheng-lin and S. Li, "A Review of Recent Developments of Friction Modifiers for Liquid Lubricants (2007–present)," *Curr Opin Solid State Mater Sci*, 2014, doi: 10.1016/j.cossms.2014.02.002.

121. J. Yu, X. Banquy, G. W. Greene, D. D. Lowrey, and J. N. Israelachvili, "The Boundary Lubrication of Chemically Grafted and Cross-Linked Hyaluronic Acid in Phosphate Buffered Saline and Lipid Solutions Measured by the Surface Forces Apparatus," *Langmuir*, 2011, doi: 10.1021/la203851w.
122. J. Guegan, M. Southby, and H. A. Spikes, "Friction Modifier Additives, Synergies and Antagonisms," *Tribol Lett*, 2019, doi: 10.1007/s11249-019-1198-z.
123. A. K. Gupta, et al., *Composites for Drug Eluting Devices: Emerging Biomedical Applications*, in *Application of Biotribology in Biomedical System*. Springer, 2024.
124. A. Choudhari, et al., "Design and Development of Vacuum Frying Machine for the Production of High-Quality Fried Products," Proceedings of International Conference on Intelligent Manufacturing and Automation: ICIMA, Springer 2022, doi.org/10.1007/978-981-19-7971-2_50.
125. A. Gupta, et al., "Design and Analysis of Vertical Vacuum Fryer," in *Proceedings of International Conference on Intelligent Manufacturing and Automation. Lecture Notes in Mechanical Engineering*, eds. H. Vasudevan, V. Kottur, and A. Raina, Springer, doi.org/10.1007/978-981-13-2490-1_13.
126. D. Dias, J. Bons, A. Kumar, M. H. Kabir, and H. Liang, "Forever Chemicals, Per-and Polyfluoroalkyl Substances (PFAS), in Lubrication," *Lubricants*, vol. 12, no. 4, p. 114, 2024.
127. A. Kumar, A. Kumar, A. Kumar, A. Choudhari, A. K. Gupta, and S. Faisal, "Introduction to Biotribology: A Science of Surface Interaction," in *Applications of Biotribology in Biomedical Systems*, eds. A. Kumar, A. Kumar, and A. Kumar, Springer Press, Springer Nature Group, ISBN 978-3-031-58329-2, 2024, https://link.springer.com/9783031583261.
128. N. Chowdhury, S. Faisalb, A. Kumar, A. Choudhari, A. K. Gupta, A. Kumar, and A. Kumar, "Characterization of Hydrogel Properties in the Advancement of Biotribology," in *Applications of Biotribology in Biomedical Systems*, eds. A. Kumar, A. Kumar, and A. Kumar, Springer Press, Springer Nature Group, ISBN 978-3-031-58329-2, 2024, https://link.springer.com/9783031583261.
129. A. Pal, R. Gour, A. Kumar, A. Kumar, A. Kumar, and N. Chowdhury, "Tribological Measurements of Human Skin," in *Applications of Biotribology in Biomedical Systems*, eds. A. Kumar, A. Kumar, and A. Kumar, Springer Press, Springer Nature Group, ISBN 978-3-031-58329-2, 2024, https://link.springer.com/9783031583261.
130. A. Kumar, P. Gajjar, K. S. Gudadur, A. Kumar, A. K. Gupta, A. Choudhari, N. Chowdhury, and A. Kumar, "Tribological Hurdles in Biomedical Manufacturing: A Comprehensive Examination," In *Applications of Biotribology in Biomedical Systems*, eds. A. Kumar, A. Kumar, and A. Kumar, Springer Press, Springer Nature Group, ISBN 978-3-031-58329-2, 2024, https://link.springer.com/9783031583261.
131. M. Heidari, S. Thangavel, P. Rahmanivahid, M. Khashehchi, A. Kumar, A. Kumar, and A. Kumar, "Navigating the Landscape: Cutting-edge Biomedical Manufacturing Techniques," in *Applications of Biotribology in Biomedical Systems*, eds. A. Kumar, A. Kumar, and A. Kumar, Springer Press, Springer Nature Group, ISBN 978-3-031-58329-2, 2024, https://link.springer.com/9783031583261.

132. R. Gour, N. Baliyan, A. Pal, A. Kumar, A. Kumar, and A. Kumar, "Animal Tribology," in *Applications of Biotribology in Biomedical Systems*, eds. A. Kumar, A. Kumar, and A. Kumar, Springer Press, Springer Nature Group, ISBN 978-3-031-58329-2, 2024, https://link.springer.com/9783031583261.
133. A. Choudhari, A. K. Gupta, A. Kumar, A. Kumar, A. Gupta, N. Chowdhury, and A. Kumar, "Wear and Friction Mechanism Study in Knee and Hip Rehabilitation: A Comprehensive Review," in *Applications of Biotribology in Biomedical Systems*, eds. A. Kumar, A. Kumar, and A. Kumar, Springer Press, Springer Nature Group, ISBN 978-3-031-58329-2, 2024, https://link.springer.com/9783031583261.
134. A. K. Gupta, A. Choudhari, A. Kumar, A. Kumar, A. Gupta, S. Faisal, and A. Kumar, "Composites for Drug Eluting Devices: Emerging Biomedical Applications," in *Applications of Biotribology in Biomedical Systems*, eds. A. Kumar, A. Kumar, and A. Kumar, Springer Press, Springer Nature Group, ISBN 978-3-031-58329-2, 2024, https://link.springer.com/9783031583261.
135. Darrius Dias, Jake Bons, Abhishek Kumar, M. Humaun Kabir, and Hong Liang, "Forever Chemicals, Per-and Polyfluoroalkyl Substances (PFAS), in Lubrication," *Lubricants*, vol. 12, no. 4, p. 114, 2024.

Chapter 13

Thermally stable lubricants

Neha Singh Raghuvanshi, Arpita Johri, Mudit Saxena, Ashish Kumar Gupta, Amit Choudhari, Abhishek Kumar, Avinash Kumar, and Ashwani Kumar

13.1 INTRODUCTION

Thermally stable lubricants, in essence, are specialized lubricating fluids or greases engineered to maintain their lubricating properties and structural integrity even under high temperatures. Unlike conventional lubricants that may degrade or evaporate when exposed to elevated temperatures, thermally stable lubricants endure such conditions without significant deterioration [1, 2]. These lubricants are designed to withstand thermal stress, ensuring adequate lubrication in environments where conventional options would fail [3]. They possess exceptional thermal stability, which allows them to function optimally at high temperatures without breaking down or losing effectiveness. Additionally, they exhibit resistance to oxidation, preventing the formation of harmful by-products that could compromise lubrication performance [4]. Furthermore, thermally stable lubricants have low volatility, minimizing the risk of evaporation, which is crucial for sustained lubrication in extreme temperatures [5].

Thermally stable lubricants' significance spans various industries, where maintaining equipment reliability is paramount. In extreme temperature environments such as those encountered in aerospace, automotive, and manufacturing industries, the role of thermally stable lubricants becomes even more critical [6–8]. They offer consistent performance and protection under harsh conditions, safeguarding machinery [9] and components from thermal degradation and ensuring uninterrupted operation. Thus, their application in such environments is essential for maintaining operational efficiency and preventing costly equipment failures [10, 11]. Formulated with premium base oils and additives, thermally stable lubricants exhibit notable attributes such as thermal stability, oxidation resistance, and evaporation prevention [12–15]. These properties ensure reliable performance in demanding environments by safeguarding against friction, wear, and corrosion, extending component lifespan and reducing maintenance requirements. They find applications in automotive engines, industrial machinery, aerospace, and high-speed bearings [16–20].

 DOI: 10.1201/9781003535447-13

Ongoing research focuses on enhancing thermally stable lubricants' thermal stability and overall performance. This ongoing enhancement underscores their significance in maintaining equipment reliability and efficiency across various industrial sectors. Despite enduring challenges, research endeavors strive to improve their efficacy, highlighting their pivotal role in sustaining critical components in demanding environments. Popular examples of thermally stable lubricants include synthetic lubricants such as polyalphaolefins (PAO), esters, and polyalkylene glycols (PAG) [21]. These lubricants are acclaimed for their remarkable thermal stability, making them ideal for demanding settings like industrial machinery and automotive engines. Perfluoropolyether (PFPE) lubricants, another category, exhibit outstanding thermal durability and chemical resilience, operating effectively in extreme temperature environments [22]. Polyphenyl ethers (PPE) are synthetic lubricants noted for their thermal endurance and resistance to oxidation, finding utility in high-temperature processes like metalworking and glass manufacturing [23–26]. Silicone-based lubricants, including silicone oils and greases, possess commendable thermal stability and resistance against oxidation at elevated temperatures, serving in applications ranging from automotive engines to food processing machinery [24, 27–29]. Liquid crystal polymer (LCP) lubricants represent another category of thermally stable polymers capable of functioning as solid lubricants in high-temperature scenarios [30, 31]. Offering low friction and wear resistance in elevated temperatures, they are prevalent in aerospace, automotive, and industrial applications. Overall, thermally stable lubricants represent a significant advancement in lubrication technology, ensuring consistent performance and safeguarding critical components in extreme conditions.

Thermally stable lubricants are indispensable across diverse industries due to their ability to withstand high temperatures and maintain lubrication effectiveness under demanding conditions. In the automotive sector, these lubricants play a crucial role in ensuring the smooth operation of engines and transmissions, contributing to improved fuel efficiency, extended engine life, and enhanced vehicle performance [32–34]. Similarly, in industrial machinery applications, thermally stable lubricants lubricate critical components such as gears and bearings, reducing friction and wear, and thereby optimizing productivity and equipment lifespan [35]. Furthermore, in aerospace and semiconductor manufacturing, where extreme temperatures are encountered, these lubricants ensure the safety, reliability, and integrity of aircraft operation and semiconductor fabrication processes [36]. Overall, the widespread use of thermally stable lubricants underscores their significance in sustaining industrial operations and technological advancements by minimizing downtime, reducing maintenance costs, and optimizing operational efficiency in challenging environments.

The objective of this chapter is to provide a comprehensive understanding of thermally stable lubricants, their characteristics, applications, and significance across various industries. This chapter aims to elucidate the definition and key properties of thermally stable lubricants and their importance in maintaining equipment reliability, reducing maintenance requirements, and extending component lifespan. Furthermore, the review explores the diverse applications of thermally stable lubricants in automotive, aerospace, industrial machinery, semiconductor manufacturing, and other extreme temperature environments. Additionally, this study aims to highlight ongoing research trends, challenges, and advancements in the field, with a focus on enhancing the thermal stability and overall performance of these lubricants.

13.2 FUNDAMENTALS OF THERMALLY STABLE LUBRICANTS

13.2.1 Mechanisms of lubrication

The fundamentals of thermally stable lubricants delve into the intricate mechanisms by which these specialized lubricants operate to minimize friction and wear between interacting surfaces under high-temperature conditions. At the core of lubrication science lies a nuanced understanding of the mechanisms that govern the behavior of lubricants at the molecular level, elucidating their role in enhancing the efficiency and longevity of mechanical systems operating in extreme thermal environments [20]. Figure 13.1 illustrates the lubrication mechanism facilitated by the diffusion process within a solid lubricant (SL) composed of soft metals, particularly silver (Ag), during high-temperature (HT) tribological operations. Kumar et al. [37] demonstrated that a higher silver content, typically around 15 wt%

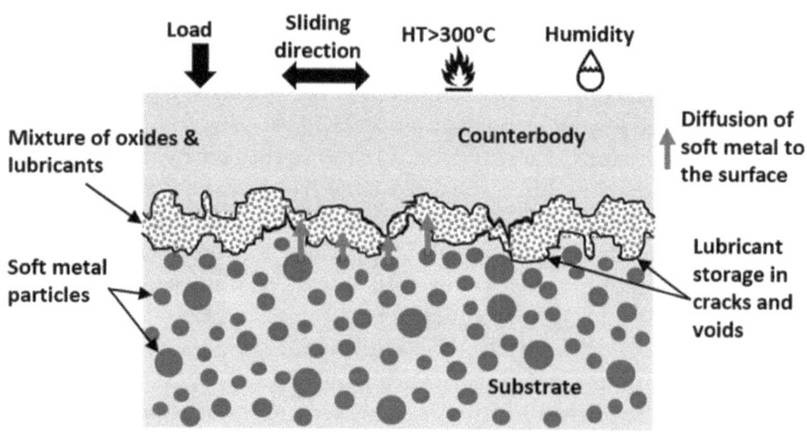

Figure 13.1 The diffusion process of solid lubrication derived from soft metals during high-temperature tribological operations [37].

or more within a host material, leads to a notable decrease in the coefficient of friction (CoF) and wear, particularly as temperature increases. This suggests that silver plays a crucial role in enhancing the lubricating properties of the composite material, especially at temperatures below 400°C. The mechanisms of lubrication encompass a spectrum of phenomena, each contributing to the overall effectiveness of thermally stable lubricants in mitigating friction and wear [38]. One of the primary mechanisms is hydrodynamic lubrication, wherein the lubricant forms a fluid film between moving surfaces, effectively separating them and reducing direct contact [39]. This fluid film acts as a cushion, absorbing the forces generated by surface interactions and minimizing wear on the contacting surfaces. In high-temperature environments, the viscosity of the lubricant becomes paramount, as it dictates the thickness and stability of the fluid film [40–45]. Thermally stable lubricants are engineered to maintain their viscosity and lubricating properties even at elevated temperatures, ensuring consistent performance and protection against wear under thermal stress.

Another critical mechanism is boundary lubrication, which occurs when the lubricant molecules adhere to the surfaces of interacting components, forming a protective boundary layer that reduces friction and wear [46]. In extreme temperature conditions, the integrity of this boundary layer becomes crucial, as thermal degradation or evaporation of the lubricant could compromise its effectiveness [47, 48]. Thermally stable lubricants exhibit exceptional resistance to thermal degradation and oxidation, preserving the integrity of the boundary layer and enhancing lubrication

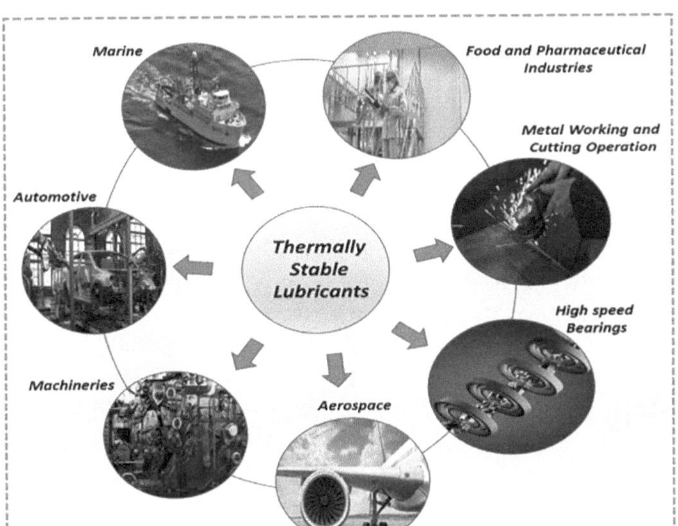

Figure 13.2 Illustration of the wide range of uses for thermally stable lubricants across different industries.

performance in high-temperature environments [49, 50]. The fundamentals of thermally stable lubricants revolve around understanding the complex interplay of mechanisms that govern lubrication under high-temperature conditions. From hydrodynamic lubrication to boundary lubrication and the role of additives, these mechanisms collectively contribute to the ability of thermally stable lubricants to minimize friction, reduce wear, and ensure the smooth operation of mechanical systems operating in challenging thermal environments. A deep understanding of these fundamentals is essential for developing and optimizing thermally stable lubricants tailored to specific industrial applications.

13.2.2 Factors affecting lubricant performance at high temperatures

Considering the factors affecting lubricant performance at high temperatures is crucial in designing high-temperature lubricants. Understanding these factors allows lubricant chemists to tailor the formulation to withstand the specific challenges posed by elevated temperatures, ensuring optimal performance and longevity of the lubricant under such conditions. For example, selecting synthetic base oils with superior thermal stability can mitigate the risk of thermal degradation and oxidation [51–56]. Adjusting viscosity levels appropriately helps maintain the lubricant film's thickness, preventing excessive friction or drag. Furthermore, incorporating additives that resist oxidation and shear ensures the lubricant can maintain its lubricating properties and effectively protect machinery components. By considering these factors during the design process, lubricant chemists can develop formulations that offer reliable performance and extended service life in high-temperature applications, ultimately contributing to improved equipment reliability, efficiency, and reduced maintenance costs [57].

Table 13.1 provides an overview of the key factors influencing the performance of lubricants at elevated temperatures. Firstly, the base oil type significantly impacts thermal stability, with synthetic oils generally offering superior performance to mineral oils [58, 59]. Viscosity plays a crucial role, as excessively low and high viscosities can lead to detrimental effects such as increased friction or drag. Oxidation, a chemical reaction with oxygen, can result in viscosity changes and deleting additives essential for lubricant performance [60–65]. Additive depletion further exacerbates this issue, compromising anti-wear and extreme pressure protection functionalities. Shear forces can thin the lubricant film, reducing its effectiveness, while volatility may lead to lubricant evaporation, necessitating more frequent top-ups. Contamination by water, dirt, or foreign substances can accelerate wear and viscosity changes, further impacting lubricant performance [1]. Overall, understanding and managing these factors are crucial for optimizing lubricant performance and ensuring the reliability of equipment operating in high-temperature environments.

Table 13.1 The summary of the factors that affect the performance of the lubricants at higher temperatures

Factor	*Description*	*Impact on Performance*
Base oil type	The chemical makeup of the lubricant's base fluid	Synthetic base oils generally offer better thermal stability compared to mineral oils
Viscosity	The thickness and resistance to the flow of the lubricant	Too low: thinning film, leading to increased friction and wear. Too high: increased drag, reducing efficiency and potentially causing overheating
Oxidation	The chemical reaction between the lubricant and oxygen at high temperatures	Increased viscosity, sludge formation, and depletion of anti-wear additives
Additive depletion	The breakdown or evaporation of performance-enhancing additives	Loss of specific functionalities like anti-wear, anti-rust, or extreme pressure protection
Shear	The mechanical force that breaks down the lubricant's molecular structure	Thinning of the lubricant film, reducing lubrication effectiveness
Volatility	The tendency of the lubricant to evaporate	Loss of lubricant volume, requiring more frequent top-ups
Contamination	Water, dirt, or other foreign substances entering the lubricant	Can accelerate wear, sludge formation, and viscosity changes

13.2.3 Key properties of thermally stable lubricants

Table 13.2 outlines the essential properties required for a lubricant to exhibit high stability in high-temperature environments. High oxidation stability is crucial to resist breakdown when exposed to oxygen at elevated temperatures, thereby reducing thickening, sludge formation, and varnish buildup, ultimately extending the lubricant's lifespan. A high viscosity index also ensures consistent viscosity across a wide temperature range, guaranteeing proper lubrication during startup and continuous operation [2]. Moreover, a high flash point and autoignition temperature are vital to resist ignition from heat sources, enhancing safety by reducing fire risk. Lubrication is crucial in biomedical applications, particularly in biotribology [66–68], where its significance is emphasized, especially under ambient conditions [69–76]. Shear stability is essential as it maintains the lubricant's molecular structure even under high shear forces, preventing thinning of the lubricant film and ensuring adequate lubrication. Finally, low volatility minimizes evaporation at high temperatures, reducing lubricant loss and the frequency of top-ups [77]. These properties collectively contribute to thermally stable lubricants' overall stability and performance in challenging high-temperature environments.

Table 13.2 The ideal properties required for a lubricant to have high stability at a higher temperature environment

Property	*Description*	*Benefit*
High oxidation stability	Resists breakdown due to oxygen exposure at high temperatures	Reduces thickening, sludge formation, and varnish buildup, extending lubricant life
High viscosity index	Maintains consistent viscosity across a wide temperature range	Ensures proper lubrication during startup (low temperatures) and continuous operation (high temperatures)
High flash point and autoignition temperature	Resists ignition from heat sources	Improves safety by reducing the risk of fire
Shear stability	Maintains its molecular structure even under high shear forces	Prevents thinning of the lubricant film, ensuring effective lubrication
Low volatility	Minimizes evaporation at high temperatures	Reduces lubricant loss and top-up frequency

13.3 TYPES OF THERMALLY STABLE LUBRICANTS

Table 13.3 presents a classification of thermally stable lubricants based on the primary components of the base oil, along with their respective advantages and disadvantages. Synthetic lubricants, including polyalphaolefins (PAO) [78, 79], esters, and polyalkylene glycols (PAG) [80–85], offer the highest thermal stability, excellent resistance to oxidation and shear, and a wide operating temperature range. However, they are more expensive and may not be compatible with all machinery. Perfluoropolyether (PFPE) lubricants [86] provide exceptional thermal and chemical stability but have a high cost and limited availability. Polyphenyl ethers (PPE) offer excellent high-temperature stability but can be costly, toxic, and incompatible with certain plastics [87]. Silicone-based lubricants exhibit outstanding thermal stability and resistance to oxidation but may have lower lubricity and higher costs for some applications. Liquid crystal polymer (LCP) lubricants offer unique properties, such as excellent thermal and shear stability and self-healing properties, but they are still under development and may be expensive [30, 88, 89]. Overall, understanding the advantages and disadvantages of each type of thermally stable lubricant is crucial for selecting the most suitable option for specific applications.

Table 13.4 provides a detailed further classification of thermally stable lubricants based on core factors such as base oil type, application, tribology, additive packages, specialty formulations, and environmental considerations. Firstly, lubricants are categorized by their base oil type, which includes mineral oil, synthetic oil (such as PAO, ester, PAG, silicone),

Table 13.3 Classification of the thermally stable lubricants based on the primary components of the base oil

Type	*Description*	*Advantages*	*Disadvantages*
A. Synthetic lubricants	Engineered fluids formulated for specific performance characteristics	Highest thermal stability, Excellent resistance to oxidation and shear, Wide operating temperature range, Reduced friction and wear	Higher cost compared to mineral oils, It may not be compatible with all machinery
(i) Polyalphaolefins (PAO)	Synthetic hydrocarbons with excellent high-temperature performance	Excellent thermal and oxidative stability, Low volatility, Good compatibility with many seal materials	It can be expensive, It may not have the best lubricity under extreme pressure
(ii) Esters	Derived from organic acids and alcohols	Excellent high and low-temperature performance, Biodegradable options available, Good lubricity and extreme pressure properties	Higher cost, Can be susceptible to hydrolysis (breakdown by water)
(iii) Polyalkylene glycols (PAG)	Synthetic lubricants based on polyethylene glycol	Excellent fire resistance, High thermal and shear stability, Non-toxic and biodegradable options available	It can be hygroscopic (attract moisture), It may not be compatible with some seal materials
B. Perfluoropolyether (PFPE) lubricants	Synthetic lubricants with perfluorinated polyether backbones	Exceptional thermal and chemical stability, Non-flammable, Wide operating temperature range	Extremely high cost, Limited availability, May not have the best lubricity
C. Polyphenyl ethers (PPE)	Synthetic lubricants with high aromatic content	Excellent high-temperature stability and oxidation resistance, Good radiation resistance	High cost, Can be toxic, May not be compatible with some plastics
D. Silicone-based lubricants	Synthetic lubricants based on silicone polymers	Excellent thermal stability and resistance to oxidation, Inert (chemically inactive), Wide operating temperature range	Lower lubricity compared to some other types, Can be expensive for some applications
E. Liquid crystal polymer (LCP) lubricants	Emerging class of lubricants with unique properties	Excellent thermal and shear stability, Low friction coefficients, Self-healing properties	Limited availability, Still under development, May be expensive

Table 13.4 Further classification of thermally stable lubricants based on core factors such as base oil, application, tribology, additives, and eco-friendliness

Sr No.	*Classification*	*Description*	*Examples*
1	Lubricants base oil type	Categorized by the primary component of the lubricant	Mineral oil (conventional), Synthetic oil (PAO, ester, PAG, silicone), Semi-synthetic oil, Bio-based oil, Group III base oil
2	Lubricants based on application	Designed for specific machinery and industries	Automotive (engine oil, transmission fluid, gear oil, brake fluid), Industrial (hydraulic fluid, compressor oil, turbine oil, metalworking fluids), Marine (engine oil, grease), aviation (engine oil)
3	Lubricants based on tribology	Classified by performance characteristics	Viscosity, Temperature stability, Oxidation resistance, Anti-wear properties, Load-carrying capacity, Water resistance, Environmental compatibility, Compatibility with materials
4	Lubricants based on additive packages [106]	Enhanced with specific chemical compounds	Anti-wear additives (zinc dialkyldithiophosphate (ZDDP), MoS_2), Antioxidants (phenols, amines, phosphites), Extreme pressure (ep) additives (sulfur, phosphorus, chlorine compounds), Viscosity index improvers (Polymethylacrylate (PMA), olefin copolymers, Polyisobutylene (PIB)), Detergents and dispersants (sulfonates, succinimides, alkyl phenates), Anti-foaming agents (silicones, PDMS, organic polymers), Rust and corrosion inhibitors (zinc, phosphorus compounds, organic acids), Friction modifiers (Organic Friction Modifiers (OFMs) ef, esters, fatty acids)
5	Specialty lubricants	Tailored formulations for unique applications	Food-grade lubricants (NSF H1, H2), High-temperature lubricants, Low-temperature lubricants, Specialty greases (lithium-based, calcium-based, etc.), Electrical conductive lubricants
6	Environmentally friendly lubricants	Designed to minimize environmental impact	Biodegradable base oils (vegetable oils, synthetic esters), Low-toxicity and biodegradable additives, Bio-based hydraulic fluid

semi-synthetic oil, bio-based oil, and Group III base oil [90–95]. Next, lubricants are classified based on their intended application, spanning the automotive, industrial, marine, and aviation sectors. Additionally, lubricants are categorized according to tribological performance characteristics like viscosity, temperature stability, oxidation resistance, anti-wear properties, load-carrying capacity, water resistance, environmental compatibility, and material compatibility [96]. Furthermore, lubricants can be differentiated based on their additive packages, which may include anti-wear additives, antioxidants, extreme pressure additives, viscosity index improvers, detergents, dispersants, anti-foaming agents, rust inhibitors, corrosion inhibitors, and friction modifiers [20]. Specialty lubricants are tailored formulations for unique applications, such as food-grade lubricants [97–99], high-temperature lubricants, low-temperature lubricants, specialty greases, and electrically conductive lubricants. Lastly, environmentally friendly lubricants are designed to minimize environmental impact by using biodegradable base oils, low-toxicity and biodegradable additives, and bio-based hydraulic fluids [100–105]. This comprehensive classification scheme aids in understanding the diverse range of thermally stable lubricants available and their suitability for specific applications and environmental considerations. Thermally stable lubricants can also be classified based on their composition, application, and performance characteristics. The following are common categories.

13.4 APPLICATIONS OF THERMALLY STABLE LUBRICANTS

The applications of thermally stable lubricants span a wide range of industries and machinery, where reliable lubrication is essential for optimal performance and longevity. From automotive engines and industrial machinery to marine vessels and aerospace systems, these specialized lubricants are critical in mitigating friction, reducing wear, and ensuring smooth operation even in extreme temperatures. By offering superior thermal stability, oxidation resistance, and load-carrying capacity, thermally stable lubricants significantly enhance equipment reliability, minimize downtime, and extend component lifespan across diverse industrial sectors. The following are some of the typical applications.

13.4.1 Automotive engines

Lubricants are indispensable for maintaining automotive engine performance, efficiency, and durability. They serve multiple functions, including lubrication, heat dissipation, corrosion prevention, and cleanliness, ensuring smooth engine operation. Lubricants reduce friction and minimize wear by forming a protective film between moving parts, thus safeguarding critical components like pistons, cylinders, and bearings. Moreover, they dissipate heat generated during engine operation, preventing overheating and preserving optimal engine temperatures. This effective heat dissipation prevents engine breakdowns and extends the engine lifespan [12]. Additionally, lubricants contain additives that shield engine components from corrosion and rust, prolonging their lifespan [33]. Furthermore, lubricants assist in maintaining the integrity of engine seals and gaskets, preventing oil leaks and ensuring engine efficiency. Detergents and dispersants in lubricants help keep engine components clean by preventing the buildup of sludge, deposits, and contaminants, resulting in improved fuel economy and reduced emissions. Anti-wear additives in lubricants offer extra protection to engine components under high-stress conditions, reducing metal-to-metal contact and extending component life. Lubricants are carefully formulated to be compatible with various engine materials, ensuring optimal performance without adverse effects on engine components [14, 107, 108].

13.4.2 Industrial machinery

Thermally stable lubricants are crucial in industrial applications, such as hydraulic systems, compressors, turbines, and gearboxes, where they facilitate smooth operation and minimize friction, thus reducing maintenance needs and prolonging equipment lifespan [17]. Operating in high temperatures due to friction and mechanical stress, industrial machinery requires lubricants engineered to withstand elevated temperatures without deteriorating or losing lubricating properties. These lubricants prevent harmful deposits and maintain optimal viscosity, ensuring machinery components operate smoothly [109]. Reducing friction between moving parts minimizes wear and extends the lifespan of critical components like bearings and gears. Thermally stable lubricants also reduce oxidation and thermal degradation, preserving machinery performance. Furthermore, they contain additives that inhibit oxidation by-products, such as sludge and varnish, which can compromise machinery reliability. By maintaining thermal stability, these lubricants provide consistent lubrication even in demanding conditions. They also protect against corrosion, extending machinery lifespan and reducing maintenance requirements. Additionally, thermally stable lubricants are formulated to be compatible with seals, O-rings, and other elastomeric materials commonly used in industrial machinery, preventing seal deterioration and ensuring effective sealing to prevent leaks and lubricant contamination [110–115]. Thermally stable lubricants extend maintenance intervals, increase productivity, and promote cost savings for industrial operations by delivering superior lubrication and protection. Some formulations are environmentally friendly, minimizing environmental impact and complying with regulations regarding toxicity and biodegradability.

13.4.3 Aerospace components

Aerospace applications often entail extreme temperatures and high-speed operations, necessitating thermally stable lubricants in aircraft engines, landing gear systems, and hydraulic actuators to uphold performance and reliability under challenging conditions. These lubricants play a pivotal role in aerospace components, where demanding operational settings and stringent performance criteria require outstanding lubrication and protection

[116]. Components like jet engines, landing gear, and control systems function in environments characterized by extreme temperatures and pressures, demanding lubricants engineered to endure without compromising their lubricating properties [117]. They ensure seamless operation and thwart friction-induced wear even under high temperatures. In aerospace, mitigating friction between moving parts is pivotal for maintaining efficiency and reliability. Thermally stable lubricants create a protective layer between components, lessening friction and wear, elongating critical parts' lifespan, and ensuring optimal flight performance. Aerospace components contend with harsh environments prone to corrosion and rust. Thermally stable lubricants incorporate corrosion inhibitors safeguarding metal surfaces from degradation triggered by moisture, salt, and environmental elements [118, 119]. This corrosion protection preserves component integrity and staves off premature failure. Aerospace components, fashioned from diverse materials like aluminum, titanium, and composites, require lubricants compatible with these materials to prevent degradation or compromise of structural integrity. Operating across diverse conditions—from high speeds and temperatures during flight to low temperatures at high altitudes—thermally stable lubricants furnish consistent lubrication, guaranteeing reliable performance and safety throughout the flight envelope [120]. Aerospace equipment mandates stringent maintenance to ensure safety and efficiency. Thermally stable lubricants curtail maintenance needs by furnishing enduring lubrication and protection, thus reducing the necessity for frequent inspections and lubricant replacements, ultimately minimizing downtime and maintenance expenses. Adhering to stringent environmental regulations, aerospace lubricants are formulated to be environmentally friendly, featuring biodegradable base oils and additives, posing a minimal environmental risk in case of spillage or leakage.

13.4.4 High-speed bearings

Within high-speed bearings, characterized by extreme operational conditions and precision performance demands, thermally stable lubricants play a pivotal role in furnishing exceptional lubrication and protection. These bearings generate substantial heat from friction and mechanical stress during operation, necessitating thermally stable lubricants to dissipate heat and maintain optimal operating temperatures within the bearing assembly, thus averting overheating and thermal lubricant degradation, which could precipitate premature bearing failure [121]. High-speed bearings achieve rotational speeds exceeding thousands of revolutions per minute (RPM) and encounter friction between rolling elements and raceways, causing wear and diminishing efficiency. Thermally stable lubricants establish a protective film between moving parts, mitigating friction and wear, ensuring smooth operation, and prolonging bearing lifespan. In high-speed bearing applications, lubricants distribute load uniformly across bearing surfaces

[122]. Thermally stable lubricants exhibit excellent load-carrying capacity, enabling bearings to support applied loads sans premature fatigue or failure. Bearing lubricants also combat oxidation and degradation due to exposure to oxygen and other reactive species. Laden with antioxidants and additives, thermally stable lubricants inhibit oxidation, forestalling harmful by-products that could impede bearing performance. Lubricant viscosity is critical for proper lubrication and load distribution in high-speed bearings. Thermally stable lubricants maintain viscosity over diverse temperatures and operating conditions, ensuring consistent lubrication performance and shielding bearings from excessive wear and friction [123]. High-speed bearings are susceptible to contamination by dust, dirt, and other particles that infiltrate the bearing assembly and cause damage. Thermally stable lubricants incorporate additives that ensnare and suspend contaminants, preventing their adherence to bearing surfaces and forestalling abrasion or pitting. Furnishing enduring lubrication and protection, thermally stable lubricants extend maintenance and lubricant replacement intervals for high-speed bearings, curbing downtime and maintenance costs while ensuring seamless operation and reliability [124].

13.4.5 Metalworking and cutting operations

In metalworking and cutting operations, thermally stable lubricants are crucial due to the intense heat, friction, and tool wear that can impact efficiency, quality, and tool longevity. During metalworking and cutting, temperatures rise significantly due to the friction between cutting tools and workpiece materials. Thermally stable lubricants play a key role in dissipating heat by forming a protective barrier between the tool and the workpiece [125]. This prevents overheating of the cutting tool and the workpiece material, thereby reducing thermal deformation and enhancing dimensional accuracy. Friction between the cutting tool and the workpiece material can lead to tool wear, surface imperfections, and compromised machining quality. Thermally stable lubricants mitigate friction by providing a lubricating film between the tool and the workpiece. This minimizes tool wear, extends tool life, and enhances machined components' surface finish and dimensional precision [126]. By reducing friction, heat, and wear during metalworking operations, thermally stable lubricants prolong the lifespan of cutting tools, enabling them to maintain sharp cutting edges for extended periods. This results in increased machining efficiency and reduced costs associated with tool replacement. Metalworking and cutting processes generate chips and swarf, which can interfere with machining operations and cause damage to tools [127, 128]. Thermally stable lubricants aid chip evacuation by minimizing friction and preventing chip buildup on cutting tools and workpiece surfaces. This ensures smooth and uninterrupted machining processes, boosting productivity and minimizing the risk of tool failure. Furthermore, thermally stable lubricants contribute to achieving superior surface finish quality by reducing friction and minimizing tool wear. They help produce smoother, more uniform surfaces with fewer tool marks, burrs, and defects, thereby enhancing the overall quality of machined components. Thermally stable lubricants are formulated to be compatible with a wide array of materials commonly encountered in metalworking applications, including ferrous and non-ferrous metals, alloys, and composites. This compatibility minimizes the risk of material degradation, discoloration, or corrosion, thereby preserving the integrity of machined components and workpiece surfaces. Many thermally stable lubricants are environmentally friendly, incorporating biodegradable base oils and additives that pose minimal environmental risk. They adhere to stringent environmental standards and certifications, ensuring compliance with biodegradability, toxicity, and sustainability criteria [127].

13.4.6 Food and pharmaceutical industries

Within the food and pharmaceutical sectors, thermally stable lubricants are paramount owing to stringent regulations, hygiene standards, and product safety prerequisites. Lubricants used in these industries must adhere to rigorous rules and standards to ensure product safety and hygiene. Specialized thermally stable lubricants formulated for the food and pharmaceutical sectors comply with food-grade and pharmaceutical-grade standards, guaranteeing their safety in environments where direct or incidental contact with food or pharmaceutical substances may occur. Processing equipment in food and pharmaceutical facilities, such as pumps, mixers, conveyors, and packaging machinery, necessitate lubrication to maintain smooth operation and prevent equipment malfunctions [129, 130]. Thermally stable lubricants minimize the risk of contamination by adhering to stringent purity standards and avoiding ingredients that could compromise the safety of food or pharmaceutical products. Equipment utilized in these sectors may encounter various temperatures, from ambient conditions to high-temperature sterilization procedures. Thermally stable lubricants ensure consistent lubricating performance and safeguard equipment components from thermal degradation, ensuring dependable operation across varying temperature ranges [17]. The food and pharmaceutical processing equipment may be subjected to corrosive substances, moisture, and harsh cleaning agents. Thermally stable lubricants incorporate corrosion inhibitors that shield metal surfaces from corrosion, rust, and deterioration, thereby preserving the integrity and durability of processing equipment and machinery. Continuous operation of processing equipment can lead to wear and abrasion of moving parts, necessitating adequate lubrication and wear protection provided by thermally stable lubricants. These lubricants reduce friction, minimize wear on critical components like bearings and gears, and enhance operational efficiency. Maintaining cleanliness and adherence to hygiene standards is paramount in food and pharmaceutical processing environments. Thermally stable lubricants are formulated to be non-toxic, odorless, and tasteless, ensuring they do not compromise the quality or safety of food or pharmaceutical products [131, 132]. They are designed to resist microbial growth and contamination, upholding a sanitary environment throughout the production process. Thermally stable lubricants exhibit compatibility with various materials commonly employed in food and pharmaceutical processing equipment, including stainless steel, plastic, and elastomers. They are engineered to minimize the risk of material degradation, swelling, or leaching, ensuring compatibility with equipment components and sealing systems.

13.4.7 Marine applications

Thermally stable lubricants play a pivotal role in marine settings, where equipment operates amidst challenging conditions characterized by exposure to saltwater, elevated temperatures, heavy loads, and adverse weather. Marine machinery, encompassing engines, gearboxes, winches, and bearings, contend with constant exposure to saltwater and moisture, leading to potential corrosion and rusting of metal surfaces [133]. Thermally stable lubricants feature corrosion inhibitors that create protective barriers, thwarting metal degradation and prolonging the operational life of marine equipment. Marine environments subject equipment to frequent encounters with water and humidity. Thermally stable lubricants are engineered to resist water washout, emulsification, and deterioration, ensuring continuous lubrication and safeguarding critical components even in wet conditions. Given the vast temperature fluctuations in marine settings, from freezing cold to scorching heat, thermally stable lubricants retain their viscosity and lubricating qualities across a wide temperature range, delivering dependable performance and safeguarding against extreme marine conditions. Marine machinery, such as propulsion systems, thrusters, and winches, routinely faces heavy loads and demanding torque requirements. Thermally stable lubricants exhibit exceptional load-bearing capacity and shear stability, mitigating friction, wear, and fatigue in heavily loaded components. Considering the sensitivity of marine ecosystems, lubricants must minimize environmental impact and adhere to regulatory standards [134]. Thermally stable lubricants are crafted to be biodegradable, non-toxic, and environmentally friendly, mitigating the risk of harm to marine life and ecosystems in the event of spills or leaks. In the maritime industry, equipment reliability is paramount to ensure safe and efficient vessel operations. Thermally stable lubricants provide enduring lubrication and protection, thereby reducing downtime, maintenance expenses, and the likelihood of equipment malfunctions at sea. Beyond general marine equipment, thermally stable lubricants find application in specialized marine tasks such as hydraulic systems, wire ropes, deck machinery, and stern tube bearings. Tailored formulations address each application's unique requirements and challenges, guaranteeing optimal performance and longevity of marine equipment [135].

13.5 PERFORMANCE CHARACTERISTICS

In the realm of lubrication, the performance characteristics of lubricants serve as the bedrock upon which the efficiency and reliability of machinery are built. These characteristics encompass a spectrum of vital properties, each pivotal in ensuring optimal lubricant performance across diverse operating conditions. Viscosity, a fundamental attribute, dictates the lubricant's ability to maintain a stable film thickness and effectively reduce friction between moving components. Temperature stability is paramount, as lubricants must withstand the rigors of fluctuating temperatures without compromising their lubricating properties. Oxidation resistance is another critical factor, as lubricants exposed to oxygen at high temperatures can degrade, leading to increased viscosity, sludge formation, and diminished performance [136]. Furthermore, a lubricant's load-carrying capacity prevents metal-to-metal contact and minimizes wear under heavy loads. These performance characteristics collectively determine the lubricant's ability to protect machinery components, minimize frictional losses, and ensure smooth and efficient operation. These make them indispensable considerations in lubricant selection and maintenance practices across various industrial sectors. Thermally stable lubricants exhibit different performance characteristics that make them well-suited for demanding applications in high-temperature environments. These critical attributes are given in Table 13.5.

Table 13.5 summarizes the performance characteristics essential for thermally stable lubricants, elucidating their significance in ensuring optimal machinery operation across diverse industrial applications. Thermal stability stands as a cornerstone attribute, crucial for preserving lubricant efficacy and structural integrity even amidst soaring temperatures. This characteristic safeguards against lubricant breakdown, ensuring consistent performance and averting detrimental consequences in high-temperature environments. Concurrently, oxidative resistance assumes paramount importance, as it combats the formation of deleterious by-products resulting from oxidation. By staving off oxidation-induced degradation, lubricants with superior oxidative resistance prolong their lifespan, diminish sludge formation, and curtail maintenance demands, bolstering equipment longevity and reliability [137]. Moreover, evaporation resistance emerges as a critical performance parameter, particularly in environments characterized by elevated temperatures. The ability to minimize evaporation ensures sustained lubrication, thereby preserving the integrity of lubricating films and mitigating the risk of component wear. Lubricity, another salient characteristic, serves to reduce friction and wear between moving components, thereby optimizing machinery performance and extending

Table 13.5 Performance characteristics of the thermally stable lubricants

Performance characteristic	*Description*	*Importance*
Thermal stability	Maintains lubricating efficacy and structural integrity at high temperatures	Ensures consistent performance and prevents lubricant breakdown in hot environments.
Oxidative resistance	Resists the formation of harmful by-products from oxidation	It extends lubricant life, reduces sludge formation, and minimizes maintenance needs.
Evaporation resistance	Minimizes evaporation at high temperatures	Maintains long-lasting lubrication and preserves the integrity of lubricating films.
Lubricity	Reduces friction and wear between moving components	Minimizes wear and tear, extending component lifespan.
Corrosion protection	Protects metal surfaces from corrosion and rust	Maintains equipment reliability and performance, especially in harsh environments.
Compatibility	Compatible with various materials used in industrial equipment	Prevents material degradation or swelling, ensuring compatibility with seals and other components.
Environmental considerations	May be formulated with biodegradability, non-toxicity, and eco-friendly properties	Suitable for environmentally sensitive applications and industries with strict regulations.

operational lifespan [59]. Additionally, corrosion protection assumes significance in safeguarding metal surfaces from corrosion and rust, thus upholding equipment reliability, particularly in harsh and corrosive environments. Ensuring compatibility with various materials utilized in industrial equipment is imperative to prevent material degradation or swelling, thereby safeguarding the functionality of seals and other crucial components [138]. Furthermore, environmental considerations, such as biodegradability, non-toxicity, and eco-friendliness, have garnered increasing prominence in lubricant formulations. Lubricants engineered with these properties are suitable for environmentally sensitive applications and industries subject to stringent regulatory standards, underscoring the imperative of sustainability and environmental stewardship in modern industrial practices [139].

13.6 CHALLENGES AND ONGOING RESEARCH

13.6.1 Challenges faced in developing thermally stable lubricants

Developing thermally stable lubricants is a complex endeavor fraught with challenges that necessitate meticulous attention to detail and innovative solutions. Despite their advanced formulations designed to withstand high temperatures and harsh operating conditions, these lubricants encounter several hurdles that must be addressed. One significant challenge lies in the higher cost associated with thermally stable lubricants compared to conventional alternatives. The specialized base oils and additives required for thermal stability contribute to elevated production costs, making these lubricants less accessible for some applications [140–142]. Moreover, limited availability from lubricant suppliers further restricts options for end-users, particularly in specialized industries or applications where specific performance requirements must be met. Compatibility issues represent another obstacle faced in the development of thermally stable lubricants. The unique compositions of these lubricants may not always be compatible with materials commonly found in industrial machinery components. Concerns such as swelling, degradation, or compromised sealing effectiveness may arise, necessitating careful consideration during lubricant selection and application [143]. Additionally, achieving thermal stability and other desired performance attributes often involves trade-offs between factors such as viscosity, shear strength, and lubricity, making it challenging to strike the right balance to meet specific application demands. Furthermore, environmental considerations pose significant challenges in the development and use of thermally stable lubricants. Some additives or base oils incorporated into these lubricants may pose environmental hazards if improperly handled or disposed [20]. Adhering to environmental regulations and adopting safe handling practices is imperative to mitigate potential risks and ensure environmental sustainability. Moreover, thermally stable lubricants are typically tailored for specific applications or operating conditions, requiring careful evaluation of variables such as temperature range, load capacity, and material compatibility to select the appropriate lubricant. Regular monitoring and upkeep are essential to ensure peak performance and equipment longevity, as neglecting maintenance can lead to premature wear and costly downtime [144].

Addressing these challenges demands collaboration among lubricant manufacturers, suppliers, and end-users to develop and implement lubrication strategies that are not only effective in meeting operational needs but also environmentally responsible and economically viable. By overcoming these hurdles through innovation and cooperation, the development and utilization of thermally stable lubricants can contribute to enhanced equipment reliability, reduced maintenance requirements, and improved sustainability in industrial operations.

13.6.2 Current research trends and advancements

Current research trends and advancements in the field of thermally stable lubricants are focused on improving their efficacy, durability, environmental sustainability, and cost-effectiveness. Scientists and engineers are engaged in ongoing research and development (R&D) efforts to explore new combinations of base oils and additives that can elevate the thermal stability, oxidation resistance, and lubricating properties of these lubricants. This process involves an in-depth analysis of the molecular structure of lubricant components and their behavior under high-temperature conditions, aiming to identify novel formulations that can withstand extreme environments.

One notable area of exploration in current research is the utilization of nanomaterials, such as nanoparticles and nanotubes, to enhance the lubricating properties and thermal resilience of lubricants [145–148]. Nanoparticles have shown promising results in improving wear resistance, reducing friction, and enhancing thermal conductivity, thereby enabling more efficient lubrication in challenging conditions [149]. Concurrently, efforts are being made to develop new additives or refine existing ones to further enhance the performance and lifespan of thermally stable lubricants. Additives play a critical role in preventing oxidation, corrosion, and wear, while also improving viscosity and shear stability. In response to growing environmental concerns, R&D initiatives are also focusing on the development of environmentally sustainable lubricants that minimize ecological impact without compromising effectiveness [150]. This involves exploring biodegradable base oils and additives derived from renewable sources, aligning with the broader industry shift toward sustainability and eco-friendliness. Additionally, tribology, the study of friction, wear, and lubrication, is playing a significant role in advancing thermally stable lubricant research. Researchers are delving into the fundamental mechanisms of lubrication, exploring innovative lubrication methods, surface treatments, and materials to enhance efficiency and reduce energy consumption. Rigorous testing and evaluation of lubricant formulations under simulated operational conditions are integral to the R&D process, with advanced testing methods such as bench tests, tribometers, and real-world field trials being employed to validate the effectiveness of new lubricant technologies [151]. Collaboration among lubricant manufacturers, research institutions, and end-users remains essential in driving innovation and addressing industry-specific challenges. Joint research initiatives, consortia, and partnerships facilitate knowledge exchange, technology transfer, and the introduction of novel lubricant solutions to the market, ensuring that thermally stable lubricants continue to evolve to meet the evolving needs of modern industries.

13.7 CONCLUSION

This chapter comprehensively explores the fundamentals, types, applications, performance characteristics, challenges, and current research trends of thermally stable lubricants. It delves into lubrication mechanisms, factors influencing lubricant performance at high temperatures, and crucial properties necessary for efficacy. Moreover, it discusses various thermally stable lubricants and their wide-ranging applications across industries such as automotive, industrial machinery, aerospace, metalworking, food, pharmaceuticals, marine, and high-speed bearings. Additionally, it addresses the critical performance characteristics required for thermally stable lubricants, the challenges encountered in their development, and ongoing research trends and advancements in the field. The insights from this chapter hold significant implications for industry practitioners and researchers. Understanding the properties and applications of thermally stable lubricants can aid industry professionals in optimizing machinery performance, reducing maintenance costs, and enhancing operational efficiency, particularly in high-temperature environments. Furthermore, identifying key performance characteristics and challenges can guide manufacturers in developing improved lubricant formulations tailored to specific industrial needs. For researchers, delineating current research trends and advancements offers valuable guidance for future investigations to address industry challenges and push the boundaries of thermally stable lubricant technology. Areas such as nanomaterial-enhanced lubricants, environmentally sustainable formulations, and innovative additive technologies present promising avenues for further exploration and development.

In conclusion, pursuing thermally stable lubricants is essential for ensuring industrial operations' reliability, efficiency, and sustainability across various sectors. While significant progress has been made in understanding their mechanisms, properties, and applications, challenges persist, necessitating continued research and collaboration among industry stakeholders and academia. Moving forward, prioritizing the development of environmentally sustainable lubricant solutions and fostering knowledge exchange and interdisciplinary cooperation can accelerate innovation and drive the adoption of advanced lubrication technologies in industrial settings. By collectively addressing industry challenges and embracing emerging research opportunities, we can pave the way for a more resilient, efficient, and sustainable future for lubrication technology.

REFERENCES:

1. Torbacke, M. and E. Kassfeldt, *Lubricants*. 2014. John Wiley & Sons.
2. Pirro, D.M., et al., *Lubrication Fundamentals, Revised and Expanded*. 2016. CRC Press.
3. Voevodin, A.A., et al., *Surface and Coatings Technology*, 2014. **257**: p. 247–265.
4. Boyde, S., *Green Chemistry*, 2002. **4**(4): p. 293–307.
5. Wan, S., et al., *Tribology International*, 2016. **102**: p. 620–635.
6. Wang, X., et al., *Tribology International*, 2015. **90**: p. 257–262.
7. dos Santos Politi, J.R., et al., *Journal of Thermal Analysis and Calorimetry*, 2013. **111**: p. 1437–1442.
8. Schlosberg, R.H., et al., *Tribology and Lubrication Technology*, 2001. **57**(2): p. 21.

9. Choudhari, A., et al. Design and manufacturing of compact and portable smart CNC machine. In: H. Vasudevan, V. Kottur, and A. Raina (eds.), *Proceedings of International Conference on Intelligent Manufacturing and Automation. Lecture Notes in Mechanical Engineering.* 2020. Springer. doi: 10.1007/978-981-15-4485-9_21.
10. Adetunla, A., et al. *The Development of Tribology in Lubrication Systems of Industrial Applications: Now and Future Impact.* 2023. EDP Sciences.
11. Mehra, A.K., R. Saini, and A. Kumar, *Composites Communications,* 2021. **25**: p. 100732.
12. Singh, Y., et al., *Process Safety and Environmental Protection*, 2017. **111**: p. 701–713.
13. Roy, D., et al., *Philosophical Magazine*, 2022. **102**(3): p. 189–209.
14. Gupta, A.K., B. Mallik, and D. Roy, *Materials Performance and Characterization*, 2020. **9**(1): p. 25–35.
15. Gupta, A.K., et al., *International Journal of Engineering Research and Applications*, 2016. **6**(4): p. 68–73.
16. Madanhire, I., et al., *Mitigating Environmental Impact of Petroleum Lubricants*, 2016: p. 59–72.
17. Kajdas, C., et al., *Chemistry and Technology of Lubricants*, 2010: p. 239–292.
18. Khonsari, M.M., and E.R. Booser, *Applied Tribology: Bearing Design and Lubrication.* 2017. John Wiley & Sons.
19. Choudhari, A., et al. Design and development of vacuum frying machine for the production of high-quality fried products. In: *Proceedings of International Conference on Intelligent Manufacturing and Automation: ICIMA 2022.* 2022. Springer. doi: 10.1007/978-981-19-7971-2_50.
20. Mang, T., and W. Dresel, *Lubricants and Lubrication.* 2007. John Wiley & Sons.
21. Biresaw, G., *Tribology Letters*, 2018. **66**: p. 1–16.
22. Sadeghi, F., E.J. Trope, and T.J. Schnell, *Tribology Transactions*, 1996. **39**(4): p. 849–854.
23. Gunsel, S., *Development and Evaluation of a High-Temperature Lubrication System.* 1986. The Pennsylvania State University.
24. Baumgarten, B., et al., *Forensic Chemistry*, 2018. **8**: p. 28–39.
25. Gupta, A., et al. Design and analysis of vertical vacuum fryer. In: H. Vasudevan, V. Kottur, and A. Raina (eds.), *Proceedings of International Conference on Intelligent Manufacturing and Automation. Lecture Notes in Mechanical Engineering.* Springer. doi.org/10.1007/978-981-13-2490-1_13.
26. Pawlak, Z., *Tribochemistry of Lubricating Oils.* 2003. Elsevier.
27. Ngaile, G. and F. Botz, *Journal of Tribology*, 2008. **130**(2). https://doi.org/10.1115/1.2805432
28. Chakraborty, S., et al., *Materials Letters*, 2020. **274**: p. 128004.
29. Roy, D., et al., *Materials Letters*, 2020. **271**: p. 127780.
30. Xu, F., Y. Xin, and T. Li, *Polymer*, 2018. **140**: p. 269–277.
31. Kumar, A., *Materials Today: Proceedings*, 2021. **42**: p. 884–892.
32. Tung, S.C. and M.L. McMillan, *Tribology International*, 2004. **37**(7): p. 517–536.
33. Caines, A.J., R.F. Haycock, and J.E. Hillier, *Automotive Lubricants Reference Book*. Vol. 354. 2004. John Wiley & Sons.

34. Roy, D., et al., *Journal of Alloys and Compounds*, 2021. **854**: p. 157174.
35. Srivastava, S.P., *Developments in Lubricant Technology*. 2014. John Wiley & Sons.
36. Liu, H., et al., *Mechanism and Machine Theory*, 2020. **145**: p. 103701.
37. Kumar, R., et al., *Materials*, 2022. **15**(5): p. 1695.
38. Wang, Y., et al., *Tribology International*, 2024. **191**: p. 109148.
39. Gropper, D., L. Wang, and T.J. Harvey, *Tribology International*, 2016. **94**: p. 509–529.
40. Lugt, P.M. and G.E. Morales-Espejel, *Tribology Transactions*, 2011. **54**(3): p. 470–496.
41. Datta, A., A. Kumar, A. Kumar, A. Kumar, and V.P. Singh, Advanced materials in biological implants and surgical tools. In: V. Sheel Rajput and J. Bhinder (eds), *Advanced Materials for Biomedical Applications*. 2022. CRC Press, p. 21–43.
42. Kumar, A., A. Datta, A. Kumar, and A. Kumar, Recent advancements and future trends in next-generation materials for biomedical applications. In: V. Sheel Rajput and J. Bhinder (eds), *Advanced Materials for Biomedical Applications*. 2022. CRC Press, p. 1–19.
43. Kumar, A., S. Datta, and D. Kalyanasundaram, Liquid slippage in confined flows: Effect of periodic micropatterns of arbitrary pitch and amplitude. ASME 2016 5th International Conference on Micro/Nanoscale Heat and Mass Transfer, MNHMT 2016, 2016.
44. Kumar, A., D. Datta, and D. Kalyanasundaram, *Physics of Fluids*, 2016. **28**, Article ID: 082002. https://doi.org/10.1063/1.4959184.
45. Kumar, A., S. Datta, and D. Kalyanasundaram, Reduction of hydraulic friction in confined flows by laser texturing: Experiments and theoretical validation. ASME 2018 16th International Conference on Nanochannels, Microchannels, and Minichannels, ICNMM 2018, 2018.
46. Hsu, S.M. and R.S. Gates, *Tribology International*, 2005. **38**(3): p. 305–312.
47. Spikes, H., Boundary lubrication and boundary lubricating films. In: G.E. Totten (ed), *Handbook of Lubrication and Tribology*. 2012. CRC Press, Vol. 2.
48. Talkar, S., A. Choudhari, and P. Rayar, Building envelope optimization and cost-effective approach in HVAC to support smart manufacturing. In: H. Vasudevan, V. Kottur, and A. Raina (eds.), *Proceedings of International Conference on Intelligent Manufacturing and Automation. Lecture Notes in Mechanical Engineering*. 2020. Springer. doi: 10.1007/978-981-15-4485-9_31.
49. Murru, C., et al., *ACS Sustainable Chemistry and Engineering*, 2021. **9**(4): p. 1459–1476.
50. Dias, D., et al., *Lubricants*, 2024. **12**(4): p. 114.
51. Kumar, A., S. Datta, and D. Kalyanasundaram, *Journal of Heat Transfer*, 2018. **140**: p. 012403.
52. Kumar, A., S. Datta, and D. Kalyanasundaram, Liquid slippage in con fined flows: Effect of periodic micropatterns of arbitrary pitch and amplitude. ASME 2016 5th Micro/Nanoscale Heat and Mass Transfer International Conference (MNHMT2016). 2016. Biopolis, Singapore.
53. Kumar, A., A.K.S. Gangwar, A. Kumar, C.S. Meena, V.P. Singh, N., Dutt, A. Prasad, and Y. Gori, Biomedical study of femur bone fracture and healing. In: V. Sheel Rajput and J. Bhinder (eds), *Advanced Materials for Biomedical Applications*. 2022. CRC Press, p. 235–250.

54. Gupta, A.K., A. Choudhari, A. Kumar, A. Kumar, A. Gupta, S. Faisal, and A. Kumar, Composites for drug eluting devices: Emerging biomedical applications. In: A. Kumar, A. Kumar, and A. Kumar (eds), *Applications of Biotribology in Biomedical Systems*. 2024. Springer Press, Springer Nature Group.
55. Choudhari, A., A.K. Gupta, A. Kumar, A. Kumar, A. Gupta, N. Chowdhury, and A. Kumar, Wear and friction mechanism study in knee and hip rehabilitation: A comprehensive review. In: A. Kumar, A. Kumar, and A. Kumar (eds), *Applications of Biotribology in Biomedical Systems*. 2024. Springer Press, Springer Nature Group.
56. Heredia-Cancino, J.A., et al., *Tribology International*, 2018. **124**: p. 230–237.
57. Zeng, Q., *Green Materials*, 2021. **10**(2): p. 51–58.
58. Bamberger, E.N., E.V. Zaretsky, and W.J. Anderson, *Journal of Lubrication Technology*, 1970. **92**(1): p. 23–31.
59. Suryanarayana, C., et al., *Progress in Materials Science*, 2013. **58**(4): p. 383–502.
60. Su, R., et al., *Lubricants*, 2024. **12**(4): p. 116.
61. Kumar, A., et al., *Journal of Micro/Nanolithography, MEMS, MOEMS*, 2013: p. 12.
62. Kumar, A., et al., *Tribology Transactions*, 2018. **61**: p. 951–959.
63. Kumar, A. and U. Panda, Microfluidics-based devices and their role on point-of-care testing. In: R. Khan, A. Parihar, and S.K. Sanghi (eds), *Biosensor Based Advanced Cancer Diagnostics*. 2022. Elsevier, p. 197–224.
64. Kumar, A., U. Panda, V.K. Patel, and R. Kant, *Trends in Fabrication of Polymers and Polymer Composites*. 2022c. AIP Publishing LLC, p. 1–3.
65. Kumar, A., A. Parihar, S.N. Basha, and U. Panda, Clinically available/under trial drugs and vaccines for treatment of SARS-COV-2. In: A. Parihar, R. Khan, A. Kumar, A. Kumar Kaushik, and H. Gohel (eds), *Computational Approaches for Novel Therapeutic and Diagnostic Designing to Mitigate SARS-CoV2 Infection*. 2022. Elsevier, p. 451–488.
66. Kumar, A., et al., Biotribology. In: A. Kumar, A. Kumar, and A. Kumar (eds), *Application of Biotribology in Biomedical System*. 2024. Springer.
67. Gupta, A.K., et al., Composites for drug eluting devices: Emerging biomedical applications. In: A. Kumar, A. Kumar, and A. Kumar (eds), *Application of Biotribology in Biomedical System*. 2024. Springer.
68. Avinash et al., Tribological hurdles in biomedical manufacturing: a comprehensive examination. In: A. Kumar, A. Kumar, and A. Kumar (eds), *Application of Biotribology in Biomedical System*. 2024. Springer.
69. Choudhari, A., et al., Wear and friction mechanism study in knee and hip rehabilitation: A comprehensive review. In: A. Kumar, A. Kumar, and A. Kumar (eds), *Application of Biotribology in Biomedical System*. 2024. Springer.
70. Chowdhury, N., et al., Characteristics of hydrogel properties in the advancement of biotribology. In: A. Kumar, A. Kumar, and A. Kumar (eds), *Application of Biotribology in Biomedical System*. 2024. Springer.
71. Kumar, A., A. Parihar, U. Panda, and D.S. Parihar, *ACS Applied Bio Materials*, 2022. **5**: p. 2046–2068.
72. Kumar, A, A. Pathak, A. Kumar, and A. Kumar, Physics of laser—matter interaction in laser-based manufacturing. In: A. Kumar (ed), *Laser-based Technologies for Sustainable Manufacturing*. 2023. CRC Press, p. 45–54.

73. Kumar, A, A.K. Sharma, and J.K. Katiyar, *Lubricants*, 2023. **11**: p. 64.
74. Heidari, M., S. Thangavel, P. Rahmanivahid, M. Khashehchi, A. Kumar, A. Kumar, and A. Kumar, Navigating the landscape: Cutting-edge biomedical manufacturing techniques. In: A. Kumar, A. Kumar, and A. Kumar (eds), *Applications of Biotribology in Biomedical Systems*. 2024. Springer Press, Springer Nature Group.
75. Gour, R., N. Baliyan, A. Pal, A. Kumar, A. Kumar, and A. Kumar, Animal tribology. In: A. Kumar, A. Kumar, and A. Kumar (eds), *Applications of Biotribology in Biomedical Systems*. 2024. Springer Press, Springer Nature Group.
76. Kumar, A., et al., Introduction to biotribology: A science of surface interaction. In: A. Kumar, A. Kumar, and A. Kumar (eds), *Application of Biotribology in Biomedical System*. 2024. Springer.
77. Lansdown, A.R., *Lubrication: A Practical Guide to Lubricant Selection*. 2013. Elsevier.
78. Rudnick, L.R., Polyalphaolefins. In: L.R. Rudnick (ed), *Synthetics, Mineral Oils, and Bio-Based Lubricants*. 2020. CRC Press, p. 1–36.
79. Hope, K., *Polyalphaolefin Lubricant Applications*. 2017.
80. Liu, W., et al., *Langmuir*, 2019. **35**(36): p. 11784–11790.
81. Kumar, A., A.K. Shrama, T.V.K. Gupta, and J.K. Katiyar, Influence of hexagonal boron nitride additive nanocutting fluid on the machining of AA6061-T6 alloy using minimum quality lubrication. In: *Proceedings of the Institution of Mechanical Engineers Part E—Journal of Process Mechanical Engineer*, 2022: p. 09544089221110980.
82. Kumar, S., et al., *Trends in Fabrication of Polymers and Polymer Composites*, 2022a, p. 10–11.
83. Leventini, S.D., et al., *Tribology Letters*, 2022. **70**: p. 127.
84. Mehra, A.K., R. Saini, and A. Kumar, *Composites Communications*, 2021. **25**: p. 100732.
85. Parihar, A., A. Kumar, U. Panda, R. Khan, D.S. Parihar, and R. Khan, *Advanced Biology*, 2023. p. 2200285.
86. Kono, R.N., et al., *IEEE Transactions on Magnetics*, 2001. **37**(4): p. 1827–1829.
87. Hamid, S. and S.A. Burian, Polyphenyl ether lubricants. In: L.R. Rudnick (ed), *Synthetics, Mineral Oils, and Bio-Based Lubricants*. 2020, CRC Press, p. 181–188.
88. Choudhari, A., et al. Designing and prototyping for conservation and effective utilization of waste heat from air conditioner. In: *IOP Conference Series: Materials Science and Engineering*. IOP Publishing, Vol. 1104, No. 1, p. 012007. doi: 10.1088/1757-899X/1104/1/012007.
89. Chung, T.-S., G.W. Calundann, and A.J. East, Liquid crystal polymers and their applications. In: N. Cheremisinoff (ed), *Handbook of Polymer Science and Technology*. 1989. CRC Press, p. 625–676.
90. Soleimani, M., et al., *Antioxidants Classification and Applications in Lubricants*. 2018. BoD-Books on Demand Norderstedt.
91. Parihar, A., et al., *Regenerative Engineering and Translational Medicine*, 2021. p. 1–27.
92. Parmar, V., et al., *Journal of Laser Applications*, 2018. **30**: p. 031001.
93. Parmar, V., A. Kumar, G.V. Prakash, S. Datta, and D. Kalyanasundaram, *Mechanics of Materials*, 2019. p. 137.

94. Pal, A., R. Gour, A. Kumar, A. Kumar, A. Kumar, and N. Chowdhury, Tribological measurements of human skin. In: A. Kumar, A. Kumar, and A. Kumar (eds), *Applications of Biotribology in Biomedical Systems.* 2024. Springer Press, Springer Nature Group, 2024.
95. Kumar, A., P. Gajjar, K.S. Gudadur, A. Kumar, A.K. Gupta, A. Choudhari, N. Chowdhury, and A. Kumar, Tribological hurdles in biomedical manufacturing: A comprehensive examination. In: A. Kumar, A. Kumar, and A. Kumar (eds), *Applications of Biotribology in Biomedical Systems.* 2024. Springer Press, Springer Nature Group.
96. Teh, J.L., et al., *Journal of Molecular Liquids*, 2022. **366**: p. 120274.
97. Sniderman, D., *Tribology and Lubrication Technology*, 2016. **72**(1): p. 26.
98. Krol, S.C., et al., Food-grade lubricants. In: L.R. Rudnick (ed), *Synthetics, Mineral Oils, and Bio-Based Lubricants.* 2020. CRC Press, p. 849–864.
99. Chakraborty, S., et al., *Materials Letters*, 2019. **257**: p. 126739.
100. Hörner, D., *Journal of Synthetic Lubrication*, 2002. **18**(4): p. 327–347.
101. Pathak, A., A. Kumar, A. Kumar, and A. Kumar, Application of laser technology in the mechanical and machine manufacturing industry. In: A. Kumar (ed), *Laser-based Technologies for Sustainable Manufacturing.* 2023. CRC Press, p. 107–155.
102. Saurabh Gupta Ruchika Saini, A.K.P.S., Performance analysis of gudgeon pin of various cross sections by FEM. *The International Journal of Recent Technology and Engineering*, 2020. **8**: p. 4569–4573.
103. Shrivastava Prateek, S.R. and A. Kumar A, others. Investigation of torsional rod to minimize vibration in automobile using ANSYS, 2020 1st International Conference on Innovative Research in Applied Science, Engineering and Technology (IRASET), 2020, p. 1–6.
104. Singh, M., A. Kumar, and A.R. Khan, *Physical Review Fluids*, 2020. **5**: p. 102101.
105. Singh, R.K., et al., *Microsystem Technologies*, 2014b. **20**: p. 101-111.
106. Puhan, D., Lubricant and lubricant additives. In: A. Patnaik, T. Singh, and Vikas Kukshal (eds), *Tribology in Materials and Manufacturing-Wear, Friction and Lubrication.* 2020. IntechOpen.
107. Tung, S.C. and V.W. Wong, *Engine Lubricants Overview and Development Trends.* 2017.
108. Atkinson, D., et al., *Chemistry and Technology of Lubricants*, 2010: p. 293–324.
109. Alimova, Z.X., et al. *Influence of the Antioxidant Properties of Lubricants on the Wear of Agricultural Machinery Parts.* 2021. IOP Publishing.
110. Rahman, M.H., S. Shahriar, and P.L. Menezes, *Lubricants*, 2023. **11**(7): p. 289.
111. Subramanian, Y., et al., *Journal of Electronic Materials*, 2023: p. 1–11.
112. Vats, P., K.K. Gajrani, and A. Kumar, Laser-based additive manufacturing. In: A. Kumar (ed), *Laser-based Technologies for Sustainable Manufacturing.* 2023. CRC Press, p. 67–83.
113. Dias, D., et al., *Lubricants*, 2024. **12**(4): p. 114.
114. Kumar, A., A. Kumar, A. Kumar, A. Choudhari, A.K. Gupta, and S. Faisal, Introduction to biotribology: A science of surface interaction. In: A. Kumar, A. Kumar, and A. Kumar (eds), *Applications of Biotribology in Biomedical Systems.* 2024. Springer Press, Springer Nature Group.

115. Chowdhury, N., S. Faisalb, A. Kumar, A. Choudhari, A.K. Gupta, A. Kumar, and A. Kumar, Characterization of hydrogel properties in the advancement of biotribology. In: A. Kumar, A. Kumar, and A. Kumar (eds), *Applications of Biotribology in Biomedical Systems*. 2024. Springer Press, Springer Nature Group.
116. Lansdown, A.R. and S. Lee, *Chemistry and Technology of Lubricants*. 2010, Springer, p. 345–373.
117. Xin, C.U.I., et al., *Chinese Journal of Aeronautics*, 2022. **35**(11): p. 85–112.
118. Berkey, K.L., et al., *Air Force Aero Propulsion Laboratory Efforts under the RTD Plan for Improved Aircraft Turbine Engine Lubricants*. 1967. Defense Technical Information Center.
119. Kumar, A., et al., Laser micromachining in biomedical industry. In: A. Kumar (ed), *Laser-based Technologies for Sustainable Manufacturing*. 2023. CRC Press, p. 169–206.
120. Kałdoński, T. and P.P. Wojdyna, *Journal of KONES*, 2011. **18**: p. 163–184.
121. Qiang, H., et al., Advances in Mechanical Engineering, 2015. 7(12): p. 1687814015618640.
122. Zeidan, F.Y. and D.J. Paquette. *Application of High Speed and High Performance Fluid Film Bearings in Rotating Machinery*. 1994. Texas A&M University. Turbomachinery Laboratories.
123. Jiang, S. and H. Mao, *Journal of Tribology*, 2011. **133**: p. 655–664.
124. Zeng, Q., et al., *Industrial Lubrication and Tribology*, 2016. **68**(3): p. 325–335.
125. Kumar, S., et al., Lubricant performance characteristics in metalworking manufacturing. In: J.K. Katiyar (ed), *Tribology in Sustainable Manufacturing*. 2024. CRC Press, p. 43–82.
126. Astakhov, V.P. and S. Joksch, *Metalworking Fluids (MWFs) for Cutting and Grinding: Fundamentals and Recent Advances*. 2012. Elsevier.
127. Sultana, M.N., et al., *American Journal of Mechanics and Applications*, 2019. 7(4): p. 71–87.
128. Roy, D., et al., *Journal of Materials Engineering and Performance*, 2020. **29**(12): p. 7952–7963.
129. Li, J. and Y. Wu, *Lubricants*, 2014. **2**(1): p. 21–43.
130. Mandal, S., et al., *Crystals*, 2021. **11**(12): p. 1493.
131. Andrew, J.M., *Tribology and Lubrication Technology*, 2022. **78**(6): p. 62–70.
132. Liu, Y., et al., *Food Chemistry*, 2022. **370**: p. 131026.
133. Carter, B.H. and D. Green, *Chemistry and Technology of Lubricants*, 2010: p. 389–409.
134. Ware, C.S., et al., *ACS Applied Materials and Interfaces*, 2018. **10**(4): p. 4173–4182.
135. Yang, Z., et al., *Progress in Organic Coatings*, 2023. **177**: p. 107458.
136. Litwin, W., *Tribology Transactions*, 2019. **62**(5): p. 839–849.
137. Sliney, H.E., *Tribology International*, 1982. **15**(5): p. 303–315.
138. Allam, I.M., *Journal of Materials Science*, 1991. **26**: p. 3977–3984.
139. Daniel, D., et al., *Applied Physics Letters*, 2013. **102**(23): p. 231603.
140. Horita, C.N., et al., *Trends in Food Science and Technology*, 2018. **72**: p. 45–61.
141. Datta, A., et al., *Advanced Materials in Biological Implants and Surgical Tools*. 2022. CRC Press Boca Raton, FL.

142. Kumar, A., et al., Recent advancements and future trends in next-generation materials for biomedical applications. In: V. Sheel Rajput and J. Bhinder (eds), *Advanced Materials for Biomedical Applications*. 2022. CRC Press, p. 1–19.
143. Martini, A., U.S. Ramasamy, and M. Len, *Tribology Letters*, 2018. **66**: p. 1–14.
144. Sharma, B.K. and A.J. Stipanovic, *Thermochimica Acta*, 2003. **402**(1–2): p. 1–18.
145. Opia, A.C., et al., *Jurnal Tribologi*, 2020. **27**: p. 116–131.
146. Mandal, S., et al., *Materials and Design*, 2022. **220**: p. 110882.
147. Mandal, S., et al., *Applied Surface Science*, 2023. **623**: p. 157011.
148. Kumar, A., et al., *Tribology Transactions*, 2018. **61**(5): p. 951–959.
149. Rahman, M.M., et al., *Lubricants*, 2022. **10**(11): p. 281.
150. Amiril, S.A.S., et al., *Journal of Cleaner Production*, 2017. **168**: p. 1571–1589.
151. Zhu, S., et al., *Tribology International*, 2019. **133**: p. 206–223.

Chapter 14

Extreme pressure lubricants

Upendra Maurya, Vinay Saini, and Nikhil Thoppil

14.1 INTRODUCTION

From the last few decades, global awareness regarding the imperative for better energy-efficient and eco-friendly transportation systems has increased multifolds [1, 2]. This heightened awareness is primarily driven by factors such as limited crude oil reserves, escalating petroleum prices, and increasingly stringent environmental regulations aimed at mitigating greenhouse emissions. Consequently, researchers have been actively investigating novel approaches to enhance the environmental sustainability and fuel economy of upcoming mechanical and transportation systems [3]. As part of these efforts, there is ongoing exploration of alternative methods to run modern transportation systems using low-carbon energy sources, which include natural gas, biofuels, electricity, and other emerging technologies. Simultaneously, advancements in lubrication technologies and materials are being investigated to minimize detrimental energy losses attributed to friction in the sliding components of mechanical worlds.

For instance, in a recent extensive examination concentrating on commercial cars, Holmberg et al. found that approximately 33.33% of fuel energy is utilized to mitigate friction in these vehicles [4, 5]. This study proposed that implementing more sophisticated friction control technologies could lead to an 18% reduction in parasitic fuel losses within the next decade due to friction in engines. This reduction may translate to 117,000 million liters of global fuel savings annually. Moreover, this study projected 61% over the next two decades, and it is anticipated that there will be a reduction resulting in annual fuel savings of 3,85,000 million liters. These results correspond to global monetary savings of 174,000 million euros in the next 5–10 years and 576,000 million euros in the subsequent 15–25 years. The improvements in fuel efficiency for passenger cars would also lead to a decrease in CO_2 emissions by 290 million and 960 million tons per year, respectively. This estimation of global saving potential is consistent with detailed energy calculations conducted for passenger cars in Japan by Nakamura [6]. These anticipated savings are expected to play a substantial role in bolstering global initiatives to mitigate the greenhouse effect and

DOI: 10.1201/9781003535447-14

address global warming. Tribological impacts, encompassing factors like friction and wear, are typically detrimental in contemporary machinery due to their adverse effects on energy efficiency and operational longevity. Lubricants serve a pivotal function in mitigating friction and wear by forming a protective barrier between rubbing surfaces.

14.2 BOUNDARY LUBRICATION

Boundary lubrication is a crucial aspect of lubrication science, playing a pivotal role in minimizing wear and friction between rubbing surfaces in various mechanical systems. This regime of lubrication is particularly relevant in conditions where extreme pressure and high loads are encountered, and traditional lubrication mechanisms, such as hydrodynamic lubrication, may be insufficient. In boundary lubrication, a thin layer of lubricant forms on the surface of interacting solid bodies, providing a protective film that reduces metal-to-metal contact and reduces friction through different physicochemical interactions (adsorption/wetting, and chemical interreactions on the solid–liquid interface of exposed asperities. However, the entire load is supported by direct asperity contact between rubbing surfaces. This lubrication regime is prevalent in scenarios where the operating conditions surpass the capabilities of hydrodynamic lubrication, which relies on a continuous fluid film. High loads, low speeds, and intermittent contact between surfaces are common conditions where boundary lubrication becomes crucial (Figure 14.1).

The lubricant used in boundary lubrication is often composed of additives and base oils specifically formulated to enhance the protective film at the surface. Additives like anti-wear agents, friction modifiers, and extreme pressure additives play a crucial role in improving the lubricant's ability to withstand severe operating conditions. These additives form chemical bonds with the metallic surfaces, creating a protective layer that minimizes friction and prevents wear. Spikes explored the mechanisms and factors influencing boundary lubrication and emphasized the importance of

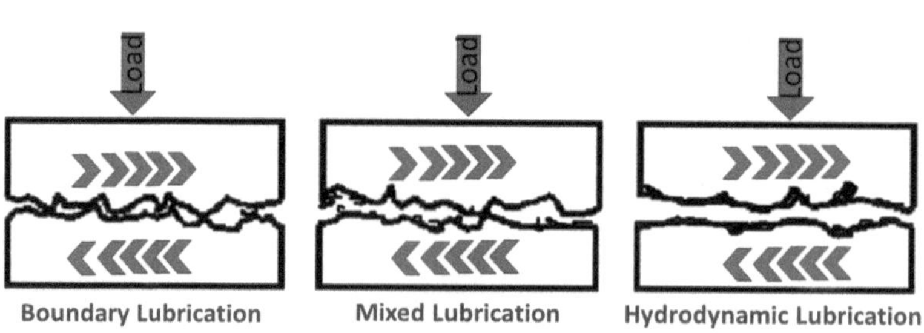

Figure 14.1 Lubrication regime between sliding surfaces.

additive chemistry in forming robust boundary lubrication films and highlighted the challenges associated with achieving effective lubrication under extreme conditions [7]. Boundary lubrication remains a critical aspect of lubrication science due to complex interactions at the interface, particularly in situations where traditional lubrication mechanisms prove inadequate. Through the collaborative efforts of researchers and tribology experts, a deeper understanding of the mechanisms and factors influencing boundary lubrication has been achieved. In the past, fatty acids and aliphatic alcohols (such as oleic acid) containing surface-active substances are generally considered by formulators to enhance the lubrication properties of lubes through adsorbed films at interfaces. However, the various compounds comprising elements like sulfur, chlorine, and phosphorus form potential sacrificial layers of sulfide chloride of low strength over exposed metallic surfaces through chemical bonding at elevated temperatures. Ongoing research and development in this field continue to contribute to the design of more effective lubricants, ultimately enhancing the efficiency and longevity of machinery in various industrial applications.

14.3 EXTREME PRESSURE ADDITIVES (EPAS)

EPAs serve to mitigate the pronounced impact of boundary lubrication by averting detrimental metal-to-metal interactions in lubrication under elevated pressure and temperature conditions. These conditions commonly arise in specific gear components of automotive vehicles and diverse industrial machinery, where heightened pressure can rupture the lubricant film. In instances where the pressure between two moving surfaces reaches levels that expel most of the lubricating oil film, i.e., boundary lubricant, alternative lubrication methods become essential. Lubricant compositions solely comprising high-molecular-weight hydrocarbon base stock are prone to failure, resulting in damage to the components. This challenge can be reduced by incorporating additives to enhance elevated-pressure performance. EPAs, typically comprised of derivatives such as sulfur, phosphorus, chlorine, or molybdenum, engage with the interacting surfaces to generate a surface layer possessing lower shear strength compared to the metal. This protective coating behaves as a solid lubricant upon direct contact with the rubbing surfaces, thereby diminishing friction, wear, and surface scoring.

At elevated local temperatures caused by metal-to-metal contact, an EPA undergoes a chemical reaction with the metallic surface, creating a protective film that possesses the necessary ductility to hinder the welding of opposing asperities [7–9]. This film effectively prevents scuffing or scoring, which can be detrimental to sliding surfaces experiencing high loads. Inorganic films are formed using chemically reactive groups, including phosphorus, sulfur, and occasionally chlorine. EPAs typically adhere to the metallic surface through chemical or physical attraction, reacting with the surface at the elevated local temperatures generated during contact of microscopically tiny rough parts (asperities) under a boundary lubrication regime. These additives create a eutectic with a common formula of $FeS_xP_yO_z$, characterized by a low melting point that is weaker than the underlying metallic substrate [10]. These surfaces undergo deformation upon contact, preventing the welding of metallic surfaces at the points of contact. In the past, EPAs were crafted using active sulfur, lead soaps, and later chlorinated groups. However, by the mid-1950s, lead soaps were discouraged from additive formulations and replaced with phosphorus and zinc. Recently formulated gear oil additives now incorporate thermally stable sulfur–phosphorus components that are noncorrosive to materials like nonferrous metals such as bronze.

EP agents find specific utility in applications prone to shock or high loading, such as gears, turbines, and roller and ball bearings. For effective performance, EP lubricants must exhibit qualities such as good lubricity, high film strength, effective cooling properties, robust load-bearing capacity, and compatibility with common types of base stocks. Bio-based products typically include natural wax esters, such as jojoba oil (although currently in short supply), synthetic wax esters (formed by esterifying an unsaturated fatty acid and an unsaturated fatty alcohol), sulfurized vegetable oils, or telomerized triglyceride vegetable oils [11]. These lubricants find applications in processes like the continuous casting of steel. In the realm of vegetable-based oils, sulfurized fatty acids and sulfurized jojoba oil (specifically oleic, C18:1, and erucic, C22:1) have demonstrated outstanding performance as EPAs [11]. Botanical additive packages offer the flexibility to formulate "load-carrying" capabilities tailored for specific applications. This optimization ensures performance on par with or superior to EP agents while also addressing economic considerations without environmental pollution concerns. It is noteworthy that the use of chlorinated paraffins is currently restricted in North America.

14.4 CLASSIFICATION OF POPULAR EPAS

Based on the additive element types, EPAs are broadly categorized as follows.

- Chlorine-containing EPA
- Sulfur-containing EPA
- Phosphorus-containing EPA
    1. Zinc dialkyl dithiophosphate
    2. Phosphonium ionic liquids
- Nanoparticles containing EPA

1. PTFE
2. TALC
3. Inorganics fluerene

14.4.1 Chlorine-containing EPA

The functionality of AW (anti-wear) and EP (extreme pressure) additives containing chlorine involves initial adsorption onto the friction pair surface. If subjected to elevated temperature and pressure situations, the chlorine component undergoes a reaction with the metallic surface of the rubbing pair, resulting in the creation of a boundary-lubricating protective film characterized by a layered structure. This film exhibits commendable performance in terms of EP and AW properties. The following chemical reaction scheme (Equation 14.1 or Equations 14.2 and 14.3) gives tribochemical reactions explaining the mechanism of film formation.

$$RCl_x + Fe \xrightarrow{\Delta} RCl_2 + FeCl_2 \tag{14.1}$$

$$RCl_x \xrightarrow{\Delta} RCL_{x-2} + 2HCL \tag{14.2}$$

$$2HCL + Fe \rightarrow FeCl_2 \tag{14.3}$$

The activation temperature for EP (extreme pressure) additives containing chlorine is around 200°C. The chlorine film, associated with these additives, possesses a comparatively low melting point, breaking down within the temperature range of 300–420 °C. Upon contact with water, a hydrolysis reaction occurs, leading to the generation of hydrochloric acid and hydroxide. This results in a loss of lubricating effectiveness and gives rise to chemical wear and corrosion. Therefore, the optimal efficacy of chlorine-containing AW and EP additives is realized in anhydrous conditions and at temperatures below 350°C [12]. In comparison to alternative EP (extreme pressure) additives, chlorinated paraffin exhibits consistent reactivity with various metals and boasts a broader spectrum of applications. Additionally, it offers a favorable balance between cost and performance [13]. Nonetheless, it undergoes a transformation into highly toxic dioxins when subjected to pressurizing and heating, leading to its classification as hazardous waste detrimental to both human health and the environment. Consequently, its exploration and application are subject to restrictions [14].

The EP (extreme pressure) performance of additives containing chlorine is less potent compared to those containing sulfur, and the environmental risks associated with sulfur are comparatively lower than those linked to chlorine [15]. Currently, sulfur-containing AW and EP additives are frequently employed as substitutes for their chlorine-containing counterparts. Nevertheless, the mechanism, as well as the chemical and physical properties of sulfur-containing AW and EP additives, still lack clarity.

14.4.2 Sulfur-containing EPA

In the historical context, EP additives were initially produced with active sulfur, lead soaps, and subsequently with chlorinated groups. However, around the 1950s, lead soaps were phased out from lubricant formulations and replaced by phosphorus and zinc additives. Contemporary gear oil additives now incorporate thermally stable sulfur–phosphorus components that exhibit noncorrosive properties, particularly toward nonferrous metals such as bronze. The mechanism of sulfur-containing EP (extreme pressure) additives involves a sequential process. Initially, the EP additives undergo polar adsorption onto the metallic surfaces, creating a physical adsorption film [16]. This film serves to isolate two sets of friction pairs and contributes to fluid lubrication. With an escalating load, the temperature within the operational zone rises, leading to the rupture and failure of the original physical adsorption film. Subsequently, the S–S bond undergoes initial breakage and reacts with the metallic surface, resulting in the creation of an iron mercaptan chemical adsorption film that exhibits AW effects. As the load further increases, and the temperature rises beyond 80°C, the C–S bond starts to rupture. During this phase, an iron sulfide film is generated, effectively suppressing the initiation of sintering [17]. Sulfides typically initiate reactions at approximately 200°C and operate up to the temperature range of 400°C. Their melting point surpasses that of ferric chloride. Consequently, sulfur-containing AW and EP additives exhibit a high resistance to sintering loads and demonstrate superior EP performance. At a concentration of 2%, these additives can enhance the EP performance by 8.7 times of distilled water [18]. However, the ferrous sulfide film lacks the layered structure seen in $FeCl_2$ and Fe_2PO_4, making it resistant to shear but characterized by brittleness and strong intergranular corrosion. So the pursuit of increased load-bearing capacity often comes at the expense of wear performance, resulting in subpar AW performance [19].

Sulfurized isobutylene stands out as the predominant sulfur-containing EP (extreme pressure) additive in widespread use. Categorized under sulfurized olefins, it is derived through the vulcanization of isobutylene as its primary raw material. This additive boasts notable features such as excellent oil solubility, elevated sulfur content, commendable thermal stability, and outstanding EP performance [20]. The conventional preparation process for sulfurized isobutylene involves two distinct stages: sulfurization dechlorination and sulfur chlorination of isobutene. This method is intricate and results in the generation of significant industrial "three wastes" (waste gas, waste liquid, and waste residue). Moreover, the reaction releases a substantial amount of corrosive gases such as H_2S and HCl, leading to severe environmental pollution, and both raw materials and by-products exhibit high corrosiveness to equipment. Currently, many researchers are directing their attention to utilizing isobutylene and sulfur as base materials to form sulfurized isobutylene through an elevated-pressure one-step method. In comparison to the conventional industrial processing approach for sulfurized isobutylene, the high-pressure one-step method is simpler and yields superior product performance.

In additives sharing similar structures, a higher sulfur content tends to facilitate the formation of a boundary-lubricating film, enhancing the EP property. However, an excess of sulfur content can lead to issues such as embrittlement, corrosion, and an accelerated wear rate on the metallic surface, consequently diminishing the EP property. While sulfur-containing AW and EP additives exhibit excellent EP performance, their AW property is generally inferior to that of phosphorus-containing counterparts. It is noteworthy, though, that there is an insufficient body of systematic research on phosphorus-containing AW and EP additives in existing literature, thereby limiting their guidance for theoretical exploration and industrial applications.

14.4.3 Phosphorus-containing EPA

14.4.3.1 Zinc dialkyl dithiophosphate (ZDDP)

Zinc dialkyl dithiophosphate (ZDDP) stands out as the most effective antiwear and EP additive employed in lubricants. With a history spanning over five decades since its introduction to the industry, significant and dedicated efforts have been invested in characterizing this material through a range of experimental techniques under diverse environmental conditions. Within lubricant formulations (shown in Figure 14.2), zinc additives commonly manifest as an organic compound called zinc dialkyl dithiophosphate (ZDDP), encompassing zinc and phosphorus. This particular additive, being inherently polar, exhibits a natural magnetic affinity to other polar compounds and materials such as iron and certain metals found in machinery components. This innate attraction facilitates the migration of the additive

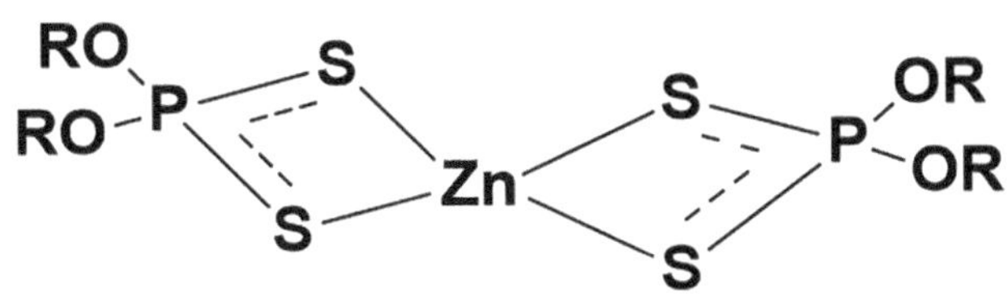

Figure 14.2 Molecular structure of ZDDP.

from the oil to metallic surfaces, creating a thin yet durable layer that envelops and adheres to these surfaces. A generic reaction is given as [21]:

$$Zn\,[(RO)_2\,PS_2]_2 + O_2\,(ROOH) \longrightarrow Zn_{n+2}\,[P_nX_{3n+1}]_2 + \text{sulfur compounds}$$

In the above chemical reaction, the symbol X predominantly represents oxygen (X = O) atoms. The resulting polyphosphate anion would exhibit the formula $O(PO_3)_n^{-n(n+2)}$. It is important to note that in this chain, sulfur (X = S) atoms have the capability to substitute for oxygen atoms [21]. Consequently, a strong poly(thio)phosphate tribofilm is formed and deposited on the rubbing surface. The majority of the organic sulfur components remain miscible in the lubricant, resulting in a film that is notably depleted in sulfur compared to the ZDDP molecule itself. When the substrate is composed of steel, the polyphosphate and native iron oxide films typically react, enhancing the adhesion of the film on the rubbing surface. However, the precise mechanism governing the formation of the film on the surface remains poorly understood. Some literature suggests that dispersed zinc phosphate is initially generated in the lubricant oil and subsequently forms a film [22]. Nevertheless, ZDDP tribofilms can be induced by friction at temperatures well below the threshold of thermal decomposition. This antiwear film can even be generated at ambient temperatures. Under such conditions, the material supplying the contacting zone is chemisorbed ZDDP, present as a very thin film, resulting in a significantly extended induction period for tribofilm formation.

In a study employing a pin-on-disk tribometer with lower speeds and, consequently, more modest flash temperatures, Georges et al. proposed that the magnitude of the flash temperature alone was insufficient to explain the formation of tribofilms [23]. Westerfield and Agnew employed infrared spectroscopy to investigate ZDDP solution reactions under very high shear and pressure. Their findings indicated that the decomposition rate of ZDDP increased with pressure up to 0.7 GPa but was then slowed at elevated pressures [24]. The thermal disintegration of ZDDP was identified as autocatalytic, and research revealed that the substitution of the zinc cation by metallic ions, such as Cu^{2+} and Fe^{3+}, facilitated the production of metallic dithiophosphates that disintegrate more rapidly than ZDDP and at lower temperatures [23]. Cumulatively, these observations infer that the presence of certain metal cations at nascent surfaces may accelerate ZDDP disintegration.

Most of the research conducted has focused on studying ZDDP behavior on iron surfaces. However, it has been demonstrated that ZDDPs also generate both tribofilms and thermal films on a wide array of materials, such as silicon, ceramics, and DLC coatings. Several hypotheses have been proposed to elucidate the driving forces behind the formation of ZDDP tribofilms, particularly their preferential formation on rubbing surfaces. These factors may include the rise in pressure, flash temperature, surface catalysis, and triboemission [25]. When rubbing surfaces are slides relatively, heat is produced, leading to a localized and transient temperature increase known as "flash temperature". There is a suggestion that this surge in temperature could initiate the formation of ZDDP tribofilms. The extent of this temperature increase is dictated by the moving heat source theory and is influenced by factors such as the friction-generated heat rate (a product of friction and sliding speed), the relative speed of the surfaces in contact, the thermal characteristics of the solid surfaces, and the dimensions of the contact area. While significant flash temperature increases can occur at high sliding speeds, they are generally minimal at speeds below 0.1 m/s. Notably, ZDDPs have been observed to generate tribofilms even at very low sliding speeds where the rise in flash temperature is negligible, suggesting that it may not be the primary factor driving ZDDP film formation. However, it is important to note that flash temperature calculations are predicated on the assumption of a Boltzmann energy distribution. Some propose that the intense energy dissipation at asperity junctions could lead to non-Boltzmann energy distributions, challenging this assumption.

There is also a suggestion that the exceptionally high pressures found in non-conforming contacts, such as those encountered in gears, rolling bearings, and the ball-on-flat geometry frequently employed in rubbing experiments, might facilitate the chemical reaction of ZDDP [26]. Based on quantum chemical simulations, Mosey et al. proposed that under extremely high pressures, cross-linking within a phosphate network may occur, leading to the formation of a highly interconnected network with enhanced mechanical properties [26]. Nevertheless, the pressures needed far surpass the yield pressures of most materials, making them improbable in rubbing contacts. Additionally, high-pressure infrared studies have failed to provide evidence of structural alterations in ZDDP up to 21 GPa [27].

It is a well-established fact that when solid surfaces rub against each other, plastic deformation and fracture processes at contacting asperities create localized charged regions. This phenomenon leads to the emission of energetic particles, including photons, electrons, ions, and even X-rays. In the presence of ZDDP-containing oils, there is a suggestion that this phenomenon, known as "triboemission", triggers the tribochemical reaction of ZDDP to form tribofilms. While triboemission is easily detected in dry rubbing contacts, it has thus far proven challenging to employ in lubricated systems, where emitted particles are assumed to be immediately captured by the lubricant. This difficulty complicates efforts to confirm or refute the potential impact of triboemission on tribofilm formation. While ZDDP offers various advantages, it is essential to consider its drawbacks. Notably, research has revealed its impact on the performance of a vehicle's catalytic converter, influencing the quality of the car's exhaust emissions [28, 29]. Consequently, recent lubricant performance requirements from the American Petroleum Institute (API) advocate for lower concentrations of this additive in lubricant formulations. The latest API quality standards for gasoline engine oils reflect reduced ZDDP concentrations, prompting the use of alternative compounds to compensate.

14.4.3.2 Phosphonium ionic liquid

In the field of tribology, phosphonium ionic liquids (ILs) have emerged as promising candidates for advanced lubrication and customized lubricating additives since their exploration for lubrication purposes began in 2001 [1, 30–32]. ILs exhibit exceptional physicochemical properties, including low melting points, low flammability, high thermal stability, and negligible vapor pressure, all of which align with the requirements of high-performance lubricants [33]. A key attribute of ILs is their tunable properties achievable by varying the cations and anions, resulting in numerous families suitable for diverse tribological systems. The effectiveness of ILs in lubrication can be ascribed to their intrinsic polarity, enabling the formation of stable ordered layers on metal surfaces in the liquid state to prevent contact. Additionally, ILs may undergo tribochemical reactions and the phosphous component of ILs can react with substrate materials to generate tribofilms, providing further protection against wear (Figure 14.3).

Ionic liquids (ILs) offer high tunability through the manipulation of cation and anion structures, or a combination of both, to meet specific engineering and manufacturing needs. Consequently, comprehending the correlation between ILs' chemical structures and their physicochemical properties, as well as their tribological characteristics, becomes imperative

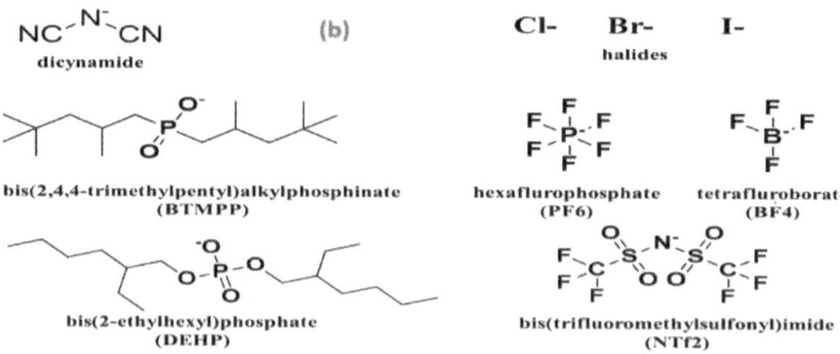

Figure 14.3 Most common (a) phosphonium cation and (b) several anion groups for lubricant.

for the molecular design of more efficient ILs. Their physicochemical properties can be readily altered by mixing different cations and anions or adjusting their alkyl chain lengths. Table 14.1 gives important phosphonium ILs explored as lubricant additives and their important findings.

14.4.3.3 Tricresyl phosphate ester (TCP)

Tricresyl phosphate ester (the general formula shown in Figure 14.4) is the earliest and most popular ester group used as a lubricant additive. Under mild conditions, tricresyl phosphate (TCP) initially adsorbs onto the rubbing surfaces of the friction pair. In the presence of water, TCP undergoes hydrolysis to form phosphoric acid, which subsequently reacts with metal to produce iron phosphate. With increasing temperature and load, TCP reacts with oxides under boundary regime conditions to form polyphosphate or phosphate compounds, impacting both EP (extreme pressure) and AW (anti-wear) performance. Additionally, phosphate ester AW and EP additives encompass phosphite ester. However, the anti-wear effectiveness of phosphate ester is inferior to that of phosphite ester. Barros-Bouchet et al. conducted a comparative study on the lubricating properties of trimethyl phosphite and trimethyl phosphate, revealing that trimethyl phosphite generates more active phosphorus than trimethyl phosphate, resulting in superior friction reduction performance [44].

14.5 NANO-OILS/NANOLUBRICANTS

Because of their extremely small size, dispersibility, large surface area, insolubility, durability, nonvolatility, and nonreactivity with other additives, researchers swiftly recognized the potential of nanoparticles (NPs) in lubricants. This realization led to the emergence of an exciting new area known as nanolubricants. The term "Nanofluids" was first introduced by Choi et al., referring to colloidal nanoparticles dispersed in a base fluid [45]. Nanoparticles offer advantages due to their nanometer size and higher aspect ratio. They have the potential to interact with the asperities of rubbing surfaces, thereby reducing friction and wear through mechanisms such as ball bearing, mending, polishing, and film formation. Another significant benefit of nanoparticles compared to sulfur and phosphorus-based commercial anti-wear additives is that nanoparticles do not rely on tribochemical reactions, which are sensitive to system conditions [4]. As per characteristic chemical compositions, Gulzar et al. have categorized these nanoparticles into seven major classes [46].

1. Metals (Cu, Ti, Fe, Al, Pb, etc.)
2. Metal oxide or sulfides (CuO, Fe_2O_3, TiO_2, ZnO, etc.)
3. Carbon based (graphene, graphite, etc.)

Table 14.1 Important phosphonium ionic liquids explored as lubricant additives and their observations

Sr No.	*Ionic liquids*	*Base oil*	*Comments*	*References*
1	[P_{66614}][DEHP]	PAO, 5W-30	• IL was fully miscible up to 5%. • IL showed a 60% reduction in friction. • IL blend decreased wear by three times compared to the base oil. • IL was stable up to 347°C.	[34]
2	([P6,6,6,14][BTMPP])	Mineral Oil	• IL improved EP properties by 34%. • ILs showed excellent synergy with oxide nanoparticles.	[8]
3	[P_{66614}][DEHP], [P_{26614}][DEHP]	PAO, 10W, 10W30	• Both the ILs were fully miscible up to 5% blend and stable up to around 350°C. • Both ILs decreased friction significantly and wear twice less than the base oil.	[35]
4	([P_{4444}]-[DEHP] , [P_{66614}][DEHP], [P_{66614}][i-C_7H_{15}COO], [P_{66614}][n-C_7H_{15}COO], [P_{66614}][i-C_9H_{19}COO], [P_{66614}][n-C_9H_{19}COO], [P_{66614}][n$C_{17}H_{35}$COO], [P_{66614}][RSO_3]	PAO	• For eight ILs temperature stability was up to 250, 310, 347, 348, 308, 330, 327, 308, 431°C, respectively. • Sulfonate IL was the most oxidation-stable and carboxylate was the least stable. • For eight ILs, solubility was found to be between 1% and 10%. • It was found that ILs with shorter alkyl chain lengths were having less solubility. • All the ILs showed excellent anti-wear properties. • The anti-wear ranking observed was phosphonium phosphate> phosphonium carboxylate > phosphonium sulfonate.	[36]
5	([P44414][DEHP]) ([P6,6,6,14][BTMPP]) ([P6,6,6,14][DBP])	PAO-6	• EP properties improved by 75%. • Hybrid nanolubricant outperformed commercial ZDDP. • Sacrificial regenerative tribofilm formed on nascent tribpair.	[29]
6	**5 aprotic ILs** i.e ([N_{2222}][DEHP]), ([N_{4444}][DEHP]) ([N_{6666}][DEHP]) ([N_{8888}-][DEHP]) ([N_{8881}][DEHP]) **4 protic ILs** i.e. ([$N_{222\text{-}H}$][DEHP]) ([N_{444H}][DEHP]) ([N_{666H}][DEHP]) ([N_{888H}][DEHP])	Engine oil without AW additives	• [N_{8888}][DEHP], [N_{8881}][DEHP], and [$N_{888\text{-}H}$][DEHP] were blended with base oil at concentrations of 2.04, 1.78, and 1.74 wt%, respectively. • ILs were also compared with ZDDP. • [N_{888H}][DEHP] and [N_{8881}][DEHP] both showed superior wear protection properties than ZDDP. • Protic [N_{888H}][DEHP] generated the least amount of wear.	[37]

(*Continued*)

Table 14.1 (Continued)

Sr No.	*Ionic liquids*	*Base oil*	*Comments*	*References*
7	Five ILs different cations but same anions.([P_{8888}][DEHP]) [P_{66614}][DEHP], ([P_{44414}][DEHP]) ([P_{4448}][DEHP]), ([P_{4444}][DEHP])	GTL 4 base oil	• It was observed that each alkyl chain should be of at least six carbon atoms long for oil solubility. • [P_{8888}][DEHP] and [P_{66614}][DEHP] enhanced wear protection of the base oil.	[38]
8	[$P_{4,4,4,2}$][$C_2C_2PO_4$], [$P_{6,6,6,14}$][$(C_2F_5)_3PF_3$]	Three easter oils and one vegetable oil	• Both ILs show improved antifriction and anti-wear properties at1% ILs blended in the base oil. • Due to very low melting points, both lubricants can be used at very low temperatures. • Pure [$P_{4,4,4,2}$][$C_2C_2PO_4$] shows low coefficient of friction but inferior anti-wear properties and bad temperature stability than [$P_{6,6,6,14}$] [$(C_2F_5)_3PF_3$].	[39]
9	N_DEHP, N_DBDTP, P_TMPP, P_TFSI, P_DMP, P_DEDTP compared with ZDDP	Blend of SN 150 and bright stock 90	• N_DEHP showed the lowest coefficient of friction. • N_DEHP and P_TFSI outperformed ZDDP, but N_DBDTP, P_TMPP, P_DMP, and P_DEDTP performed inferior to ZDDP for wear reduction.	[40]
10	[P_{66614}][$(iC_8)_2PO_2$], [P_{66614}][BEHP] compared with ZDDP	YUBASE	• Both the ILs enhanced the antifriction properties of the base oil. • It was observed that [P_{66614}][BEHP] blended with base oil performed better than [P_{66614}][$(iC_8)_2PO_2$]. • [P_{66614}][BEHP] showed superior wear reduction than [P_{66614}][$(iC_8)_2PO_2$]. • The ZDDP showed less friction than both ionic liquids.	[41]
11	P[$_{6,6,6,14}$]DPP, P[$_{6,6,6,14}$]BEHP, P[$_{1,4,4,4}$]DPP, C_3mpyrNTf_2 , P[$_{6,6,6,14}$]NTf_2, P[$_{6,6,6,14}$] $(iC_8)_2PO_2$	VO, TMP, PE MO, and PAO	• All the blended IL showed reduced friction and wear properties over the base oil. • The research shows that the majority of the properties of the blended lubricant depend on the unique performances of the base oils, e.g., if base oil shows high friction and wear, then IL blend in the base oil did not improve the base oil performances.	[42]
12	Two oil-miscible quaternary ammonium phosphite ILs (POPA and PTPA) compared with ZDDP	PAO 10	• Both the ILs outperformed the ZDDP in friction and wear reduction. • Both the ILs have better extreme pressure properties than ZDDP.	[43]

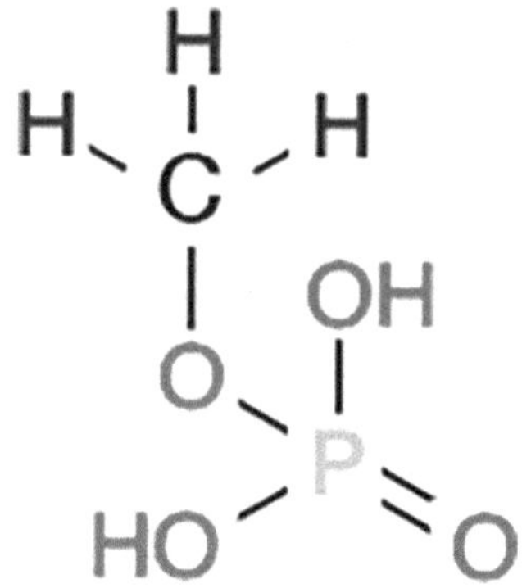

Figure 14.4 Phosphate ester (general formula).

4. Chalcogenides (MoS_2, WS_2, etc.)
5. Nanocomposites (Al_2O_3/SiO_2, SiO_2/ZrO_2, etc.)
6. Ceramics (hBN, Al_2O_3, SiO_2, etc.)
7. Others (PTFE, etc.)

14.5.1 Mechanism of nanoparticles in lubricants

Well-dispersed nanoparticles are drawn toward rubbing metallic surfaces due to two crucial factors. Firstly, undulation occurs due to the elevated temperature at the rubbing junctions, causing nanoparticles to move toward the rubbing surface. Secondly, a stronger electric field is generated due to the production of exo-electrons on the rubbing surface, which may result in the migration of nanoparticles [47]. Based on surface characterization reported in the literature, there are four known mechanisms for enhanced wear protection provided by nanoparticles [48, 49]. Two of these mechanisms could be accredited to the direct action of NPs (film formation, and ball bearing) while others are due to the secondary surface enhancement (polishing and mending effects).

(a) **Mending (self-repairing) effect:** This occurs if nanoparticles occupy the scars and grooves of the counter surfaces, compensating for the loss of material (NPs fill the groves and scars), which may result in reduced wear [48].
(b) **Polishing effect:** The average roughness of rubbing surfaces is minimized due to the abrasive action of the nanoparticles. Owing to lower surface roughness, this polishing or "artificial smooth" effect results in improved tribological performance [48].
(c) **Ball bearing or rolling effect:** This is the result of a direct action of nanoparticles, which occurs when quasi-spherical or spherical

nanoparticles roll between counter surfaces, converting sliding friction into a combination of rolling and sliding friction [48].

(d) **Protective film:** Film formation results in a coherent, smooth, and sacrificial thin film on top of the rubbing surface attributed to tribosintering/chemical reaction between substrate and nano-additives under tribostress. Due to very low shear resistance, this film may reduce wear and friction significantly. For effective protection of substrate, the tribofilm formation rate should be higher than the progressive material removal rate due to wear.

However, in various studies, the classification of active mechanisms is still a subject of discussion [7, 48, 49] (Figure 14.5).

Furthermore, the plausible interactions of different NPs as extreme pressure additves are depicted from synergistic studies (Figure 14.6).

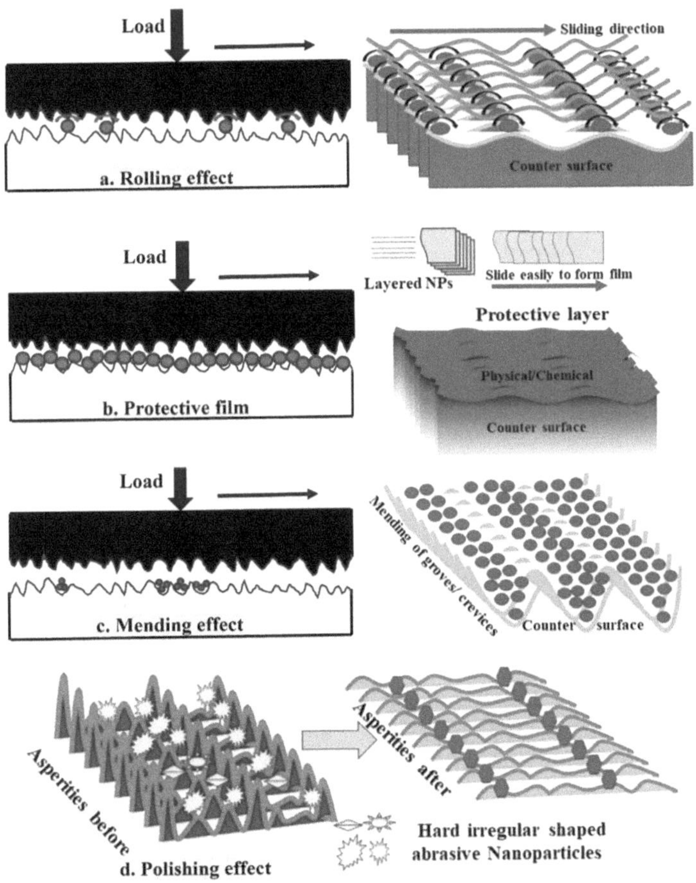

Figure 14.5 Proposed mechanism of NPs in general (adapted from Ref. [50]).

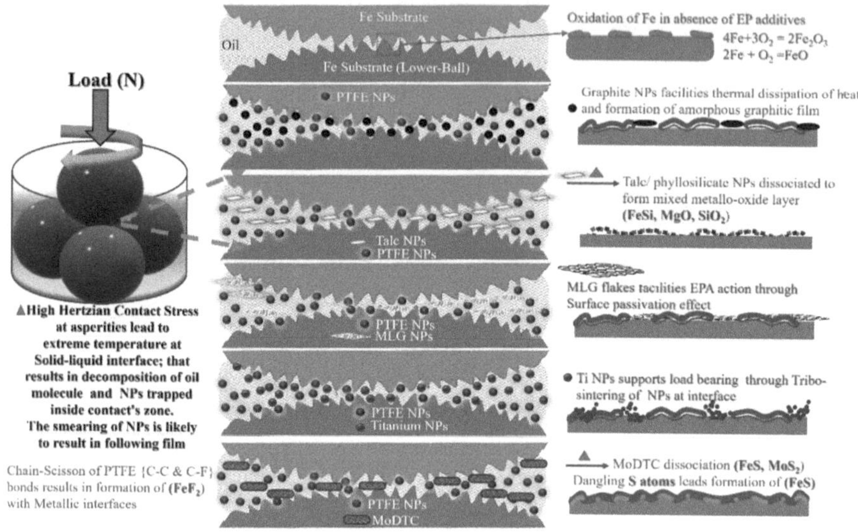

Figure 14.6 EPA action of Different NPs and their film forming ability (adapted from Ref. [51]).

REFERENCES

1. Maurya, U., Vasu, V.: Ionic liquids as a potential sustainable green lubricant for machining in the era of Industry 4.0. In: *Advances in Sustainable Machining and Manufacturing Processes*. pp. 135–155. CRC Press, Boca Raton (2022).
2. Rizvi, S.Q.A.: *A Comprehensive Review of Lubricant Chemistry, Technology, Selection, and Design*. ASTM International, West Conshohocken (2009).
3. Upendra, M., Ajay, K., Vinay, S., Gananath, T., Gulam Asharaf, H.: Unravelling the effect of chain and branch content on viscosity of polyisobutylene-mineral oil blends by modelling and its tribological properties. *Tribology in Industry*. 02, 45–56 (2024). https://doi.org/10.61552/JME.2024.03.005
4. Holmberg, K., Erdemir, A.: Influence of tribology on global energy consumption, costs and emissions. *Friction*. 5(3), 263–284 (2017). https://doi.org/10.1007/s40544-017-0183-5
5. Holmberg, K., Andersson, P., Erdemir, A.: Global energy consumption due to friction in passenger cars. *Tribology International*. 47, 221–234 (2012). https://doi.org/10.1016/j.triboint.2011.11.022
6. Nakamura, T.: Improvement of fuel efficiency of passenger cars by taking advantage of tribology. *Tribology Online*. 12, 76–81 (2017). https://doi.org/10.2474/trol.12.76
7. Spikes, H.: Friction modifier additives. *Tribology Letter*. 60 (2015). https://doi.org/10.1007/s11249-015-0589-z
8. Upendra, M., Vasu, V.: Synergistic effect between phosphonium-based ionic liquid and three oxide nanoparticles as hybrid lubricant additives. *Journal of Tribololgy*. (2020). https://doi.org/10.1115/1.4045769

9. Spikes, H.: The history and mechanisms of ZDDP. *Tribology Letter.* 17, 469–489 (2004). https://doi.org/10.1023/B:TRIL.0000044495.26882.b5
10. Spikes, H.A.: Additive-additive and additive-surface interactions in lubrication. *Lubrication Science.* 2, 3–23 (1989). https://doi.org/10.1002/ls.3010020102
11. Bart, J.C.J., Gucciardi, E., Cavallaro, S.: Formulating lubricating oils. In: *Biolubricants.* pp. 351–395. Elsevier (2013).Woodhead Publishing, Cambridge, Cambridgeshire, United Kingdom.
12. Li, H., Zhang, Y., Li, C., Zhou, Z., Nie, X., Chen, Y., Cao, H., Liu, B., Zhang, N., Said, Z., Debnath, S., Jamil, M., Ali, H.M., Sharma, S.: Extreme pressure and antiwear additives for lubricant: academic insights and perspectives. *International Journal of Advanced Manufacturing Technology.* 120, 1–27 (2022).
13. VoPham, T., Bertrand, K.A., Jones, R.R., Deziel, N.C., DuPré, N.C., James, P., Liu, Y., Vieira, V.M., Tamimi, R.M., Hart, J.E., Ward, M.H., Laden, F.: Dioxin exposure and breast cancer risk in a prospective cohort study. *Environmental Research.* 186, 109516 (2020). https://doi.org/10.1016/J.ENVRES.2020.109516
14. Glüge, J., Schinkel, L., Hungerbühler, K., Cariou, R., Bogdal, C.: Environmental risks of medium-chain chlorinated paraffins (MCCPs): A review. *Environmental Science & Technology.* 52, 6743–6760 (2018). https://doi.org/10.1021/acs.est.7b06459
15. Gao, F., Kotvis, P. V., Tysoe, W.T.: The surface and tribological chemistry of chlorine- and sulfur-containing lubricant additives. *Tribology International.* 37, 87–92 (2004). https://doi.org/10.1016/S0301-679X(03)00040-9
16. Parenago, O.P., Kuz'mina, G.N., Zaimovskaya, T.A.: Sulfur-containing molybdenum compounds as high-performance lubricant additives (Review). *Petroleum Chemistry.* 57, 631–642 (2017). https://doi.org/10.1134/S0965544117080102
17. Alves, S.M., Schroeter, R.B., Bossardi, J.C. dos S., Andrade, C.L.F. de: Influence of EP additive on tool wear in drilling of compacted graphite iron. *Journal of the Brazilian Society of Mechanical Sciences and Engineering.* 33, 197–202 (2011). https://doi.org/10.1590/S1678-58782011000200011
18. Ding, H., Yang, X., Xu, L., Li, M., Li, S., Zhang, S., Xia, J.: Analysis and comparison of tribological performance of fatty acid-based lubricant additives with phosphorus and sulfur. *Journal of Bioresources and Bioproducts.* 5, 134–142 (2020). https://doi.org/10.1016/J.JOBAB.2020.04.007
19. Lozinskii, A. V., Kuchin, D.P., Efimov, V.N., Pakhomova, M.I.: Solving the problems of high-temperature sulfur corrosion in primary oil refineries. *Chemistry and Technology of Fuels and Oils.* 57, 613–618 (2021). https://doi.org/10.1007/s10553-021-01283-2
20. Xu, X., Hu, J., Yang, S., Xie, F., Guo, L.: Extreme pressure synergistic mechanism of bismuth naphthenate and sulfurized isobutene additives. *Surface Review and Letters.* 24, 1750071 (2017). https://doi.org/10.1142/S0218625X17500718
21. Willermet, P.A., Dailey, D.P., Carter, R.O., Schmitz, P.J., Zhu, W., Bell, J.C., Park, D.: The composition of lubricant-derived surface layers formed in a lubricated cam/tappet contact II. Effects of adding overbased detergent and

dispersant to a simple ZDTP solution. *Tribology International*. 28, 163–175 (1995). https://doi.org/10.1016/0301-679X(95)98964-F
22. Martin, J.M.: Antiwear mechanisms of zinc dithiophosphate: A chemical hardness approach. *Tribology Letter*. 6, 1–8 (1999). https://doi.org/10.1023/A:1019191019134
23. Fujita, H., Spikes, H.A.: The formation of zinc dithiophosphate antiwear films. *Proceedings of the Institution of Mechanical Engineers, Part J: Journal of Engineering Tribology*. 218, 265–278 (2004). https://doi.org/10.1243/1350650041762677
24. Westerfield, C., Agnew, S.: IR study of the chemistry of boundary lubrication with high temperature and high pressure shear. *Wear*. 181–183, 805–809 (1995). https://doi.org/10.1016/0043-1648(95)90199-X
25. Zhang, J., Spikes, H.: On the mechanism of ZDDP antiwear film formation. *Tribology Letter*. 63, 1–15 (2016). https://doi.org/10.1007/s11249-016-0706-7
26. Mosey, N.J., Müser, M.H., Woo, T.K.: Molecular mechanisms for the functionality of lubricant additives. *Science* (1979). 307, 1612–1615 (2005). https://doi.org/10.1126/science.1107895
27. Tse, J.S., Song, Y., Liu, Z.: Effects of temperature and pressure on ZDDP. *Tribology Letter*. 28, 45–49 (2007). https://doi.org/10.1007/s11249-007-9246-5
28. Maurya, U., Vasu, V.: Boehmite nanoparticles for potential enhancement of tribological performance of lubricants. *Wear*. 498–499, 204311 (2022). https://doi.org/10.1016/J.WEAR.2022.204311
29. Maurya, U., Vasu, V., Kashinath, D.: Ionic liquid-nanoparticle-based hybrid-nanolubricant additives for potential enhancement of tribological properties of lubricants and their comparative study with ZDDP. *Tribology Letter*. 70, 11 (2022). https://doi.org/10.1007/s11249-021-01551-6
30. Zhou, Y., Qu, J.: Ionic liquids as lubricant additives: A review. *ACS Applied Materials & Interface*. 9, 3209–3222 (2017). https://doi.org/10.1021/acsami.6b12489
31. Maurya, U., Vasu, V., Kashinath, D.: Three-way compatibility study among nanoparticles, ionic liquid, and dispersant for potential in lubricant formulation. *Materials Today: Proceedings* (2022). https://doi.org/10.1016/J.MATPR.2022.03.329
32. Jyothish, P., Maurya, U., Vasu, V.: Advancements in metalworking fluid for sustainable manufacturing. In: *Sustainable Materials and Manufacturing Technologies*. pp. 58–81. CRC Press, London (2023).
33. M, U., Vasu, V.: Ionic liquids and its potential in lubricants: A review. *SSRN Electronic Journal*. (2018). https://doi.org/10.2139/ssrn.3320367
34. Qu, J., Bansal, D.G., Yu, B., Howe, J.Y., Luo, H., Dai, S., Li, H., Blau, P.J., Bunting, B.G., Mordukhovich, G., Smolenski, D.J.: Antiwear performance and mechanism of an oil-miscible ionic liquid as a lubricant additive. *ACS Applied Materials & Interface*. 4, 997–1002 (2012). https://doi.org/10.1021/am201646k
35. Yu, B., Bansal, D.G., Qu, J., Sun, X., Luo, H., Dai, S., Blau, P.J., Bunting, B.G., Mordukhovich, G., Smolenski, D.J.: Oil-miscible and non-corrosive phosphonium-based ionic liquids as candidate lubricant additives. *Wear*. 289, 58–64 (2012). https://doi.org/10.1016/j.wear.2012.04.015

36. Zhou, Y., Dyck, J., Graham, T.W., Luo, H., Leonard, D.N., Qu, J.: Ionic liquids composed of phosphonium cations and organophosphate, carboxylate, and sulfonate anions as lubricant antiwear additives. *Langmuir.* 30, 13301–13311 (2014). https://doi.org/10.1021/la5032366
37. Barnhill, W.C., Luo, H., Meyer, H.M., Ma, C., Chi, M., Papke, B.L., Qu, J.: Tertiary and quaternary ammonium-phosphate ionic liquids as lubricant additives. *Tribology Letter.* 63, (2016). https://doi.org/10.1007/s11249-016-0707-6
38. Barnhill, W.C., Qu, J., Luo, H., Meyer, H.M., Ma, C., Chi, M., Papke, B.L.: Phosphonium-organophosphate ionic liquids as lubricant additives: Effects of cation structure on physicochemical and tribological characteristics. *ACS Applied Materials & Interface.* 6, 22585–22593 (2014). https://doi.org/10.1021/am506702u
39. Otero, I., López, E.R., Reichelt, M., Villanueva, M., Salgado, J., Fernández, J.: Ionic liquids based on phosphonium cations As neat lubricants or lubricant additives for a steel/steel contact. *ACS Applied Materials & Interface.* 6, 13115–13128 (2014). https://doi.org/10.1021/am502980m
40. Sharma, V., Doerr, N., Aswath, P.B.: Chemical-mechanical properties of tribofilms and their relationship to ionic liquid chemistry. *RSC Advances.* 6, 22341–22356 (2016). https://doi.org/10.1039/c6ra01915c
41. González, R., Bartolomé, M., Blanco, D., Viesca, J.L., Fernández-González, A., Battez, A.H.: Effectiveness of phosphonium cation-based ionic liquids as lubricant additive. *Tribology International.* 98, 82–93 (2016). https://doi.org/10.1016/j.triboint.2016.02.016
42. Somers, A.E., Khemchandani, B., Howlett, P.C., Sun, J., Macfarlane, D.R., Forsyth, M.: Ionic liquids as antiwear additives in base oils: Influence of structure on miscibility and antiwear performance for steel on aluminum. *ACS Applied Materials & Interface.* 5, 11544–11553 (2013). https://doi.org/10.1021/am4037614
43. Fu, X., Sun, L., Zhou, X., Li, Z., Ren, T.: Tribological study of oil-miscible quaternary ammonium phosphites ionic liquids as lubricant additives in PAO. *Tribology Letter.* 60, 1–12 (2015). https://doi.org/10.1007/s11249-015-0596-0
44. De Barros-Bouchet, M.I., Righi, M.C., Philippon, D., Mambingo-Doumbe, S., Le-Mogne, T., Martin, J.M., Bouffet, A.: Tribochemistry of phosphorus additives: Experiments and first-principles calculations. *RSC Advances.* 5, 49270–49279 (2015). https://doi.org/10.1039/C5RA00721F
45. Chao, S.U.S., Eastman. J.A.: Enhancing thermal conductivity of fluids with Nanoparticles. ASME International Mechanical Engineering Congress & Exposition, November 12–17, 1995, San Francisco, CA. https://ecotert.com/pdf/196525_From_unt-edu.pdf
46. Gulzar, M., Masjuki, H.H., Kalam, M.A., Varman, M., Zulkifli, N.W.M., Mufti, R.A., Zahid, R.: Tribological performance of nanoparticles as lubricating oil additives. *Journal of Nanoparticle Research.* 18, 1–25 (2016). https://doi.org/10.1007/s11051-016-3537-4
47. Kong, L., Sun, J., Bao, Y.: Preparation, characterization and tribological mechanism of nanofluids. *RSC Advances.* 7, 12599–12609 (2017). https://doi.org/10.1039/c6ra28243a

48. Uflyand, I.E., Zhinzhilo, V.A., Burlakova, V.E.: Metal-containing nanomaterials as lubricant additives: State-of-the-art and future development. *Friction.* 7, 93–116 (2019). https://doi.org/10.1007/s40544-019-0261-y
49. Dai, W., Kheireddin, B., Gao, H., Liang, H.: Roles of nanoparticles in oil lubrication. *Tribology International.* 102, 88–98 (2016). https://doi.org/10.1016/j.triboint.2016.05.020
50. Saini, V., Bijwe, J., Seth, S., Ramakumar, S.S.V.: Interfacial interaction of PTFE sub-micron particles in oil with steel surfaces as excellent extreme-pressure additive. *Journal of Molecular Liquids.* 115238 (2020). https://doi.org/10.1016/j.molliq.2020.115238
51. Saini, V., Bijwe, J., Seth, S., Ramakumar, S.: Augmenting the lubrication performance of nano-oils through synergistic co-functioning of nanoparticles. *Tribology International.* 182, 108332 (2023). https://doi.org/10.1016/j.triboint.2023.108332

Chapter 15

Friction and wear study to evaluate performance of polymethacrylate in lubricated contacts

Yatika Gori, Chandra Kishore, Pravin P. Patil, and M. R. Tyagi

15.1 INTRODUCTION

Viscosity index (VI) is a key metric used to measure a fluid's resistance to flow at different temperatures. There is an inherent inverse relationship: as temperature increases, viscosity decreases. To address this challenge and ensure consistent lubrication performance across a wide temperature range, VI improvers are employed. These additives, typically soluble polymer chains, are incorporated into various lubricants like engine oils, gear oils, hydraulic fluids, steering fluids, and transmission oils. VI improvers essentially work by thickening the oil at higher temperatures, maintaining a critical film thickness that separates moving parts and minimizes friction and wear.

Beyond VI improvers, lubricants often contain friction modifier (FM) additives. These organic or polymeric compounds further enhance lubrication performance, particularly in thin film conditions. FMs work by physically or chemically interacting with the contacting surfaces, reducing friction and wear. Exploring combinations of different VI and FM additives can be beneficial for optimizing lubrication performance. To evaluate the effectiveness of lubricants, tribological testing plays a crucial role. Two common test methods are employed:

- **Four-ball tester:** This method represents a non-conformal contact scenario. In this test, a steel ball is rotated against three stationary steel balls under controlled conditions of temperature, speed, and load. The extent of wear scar formation on the stationary balls provides a direct measure of the lubricant's ability to prevent wear. A smaller wear scar signifies superior wear protection offered by the lubricant.
- **Journal-bearing test:** This setup simulates a conformal contact condition, ideal for evaluating lubricants used in bearings and other components with sliding contact.

DOI: 10.1201/9781003535447-15

A lubricant's viscosity (resistance to flow) changes with temperature. The viscosity index (VI) indicates this change, with a higher VI signifying a smaller viscosity decrease at high temperatures [1–3]. To improve base oil's viscosity across temperatures, viscosity index improvers (VIIs) or viscosity modifiers (VMs) are added [4–10]. At high pressure and shearing forces (shear rates), VII solutions can become thinner (shear thinning) [11]. This phenomenon contributes to the formation of elastohydrodynamic (EHD) films, which are crucial for lubrication [12–15]. Notably, research by Johnston, G. J. [15] focused on improving optical interferometry techniques to measure these thin films in rolling contact situations. VIIs can also be used to create thicker boundary layers at lower operating speeds [16–19]. The formation of these boundary films on metal surfaces helps reduce friction and wear during contact [20–24].

Accurate measurement of viscosity is crucial for various applications, from oil production to everyday use. Errors in viscosity measurement can lead to problems during oil processing. To understand viscosity better, researchers study wear and friction [25–30]. In recent years, combined experimental and numerical methods have been used to analyze viscosity and boundary-lubricating films [31–34]. Yash et al. [35] investigated the viscosity of oils using various methods and concluded that the SARA method provides the most accurate results, with an average absolute percent deviation (AAPD) below 5%.

Researchers have been actively studying the behavior and structure of oils to improve their performance. Regueira et al. [36] and Ioanna et al. [37] employed modified models for oil analysis, achieving accurate results aligned with experimental data. Yuanping et al. [38] demonstrated that adding lighter oils can effectively reduce the viscosity of heavy crude oil under specific conditions. For heavy crude oil viscosity studies, researchers successfully implemented the PR FT model [39, 40]. The authors propose future research on the chemical properties of VII additives and their mathematical modeling for further optimization [41].

In a separate study, Kharka et al. [42] investigated green lubricants, comparing synthetic ester and fatty alcohol-based lubricants. They found that fatty alcohol-based lubricants resulted in less wear and smoother surfaces during gear hobbing, while synthetic ester lubricants excelled in cooling and heat dissipation through convection.

Tomar et al. [43] focused on the mathematical modeling of lubricant behavior in journal bearings. They analyzed the effects of different recess geometries in hybrid journal bearings using FEA to solve the modified Reynolds equation. Their findings revealed a direct correlation between recess geometry and journal-bearing performance.

Cyriac et al. [44] studied polymeric friction modifier additives (PFMs) and compared them to organic friction modifiers (OFMs) for boundary lubrication on steel surfaces. PFMs demonstrated superior friction reduction compared to OFMs, but both exhibited drawbacks in terms of anti-wear properties. These studies highlight the need for a deeper understanding of wear and friction behavior of VI improvers (VIIs) to achieve energy savings through friction reduction.

15.2 EXPERIMENTAL SETUP

Tribological investigations were conducted to assess the influence of polymeric methyl methacrylate (PMA) on the lubrication performance of a specific base oil (N150). Eight lubricant formulations were prepared by blending N150 with varying PMA concentrations (2%, 4%, 6%, and 8%). Non-conformal contact conditions were simulated using a four-ball testing apparatus to achieve both hydrodynamic and mixed lubrication regimes. This experimental design aimed to evaluate the effectiveness of PMA in enhancing the tribological properties, such as friction and wear, of the base oil.

15.2.1 Four-ball geometry test

The schematic diagram in Figure 15.1 illustrates a four-ball tester. Under lubricated conditions, the balls and test oil establish a point of contact. The system allows for the input of operational parameters (load, speed, and temperature) at desired settings. In return, it provides real-time monitoring and recording of output parameters (force, displacement, and contact area). During the experiment, a load of 50 N was applied to an 8-mm-diameter pin, with its rotational speed progressively increased from 25 rpm to 620 rpm. This corresponds to a sliding speed range of 0.004 m/s to 0.26 m/s. To ensure consistency across all test conditions, the oil bath temperature was

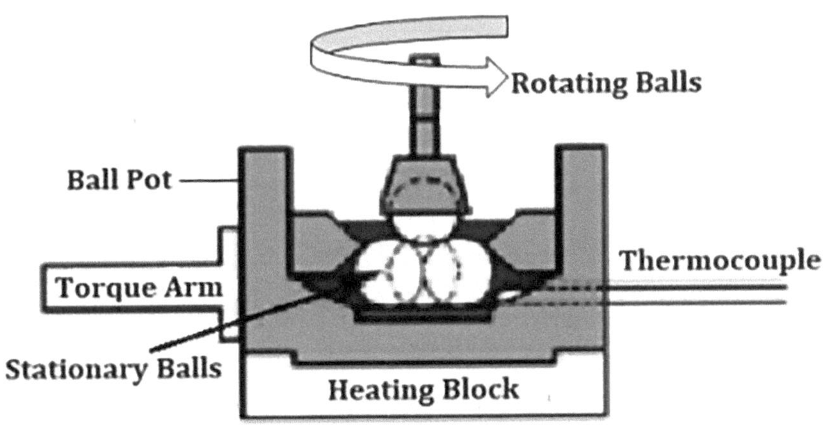

Figure 15.1 Schematic diagram of four-ball tester.

maintained at a constant 30°C. The experiments were repeated multiple times to confirm the repeatability of the results.

15.3 TEST LUBRICANTS

This study investigates the effect of polymethylmethacrylate (PMMA) on the viscosity of paraffinic mineral base oil. The properties of the base oil (N150) are presented in Table 15.1. PMMA was added to the base oil at concentrations of 2, 4, 6, and 8 wt% to improve viscosity. The viscosity coefficients of the base oil were obtained from the *CRC Handbook of Lubrication*. The notations used for the oil samples are listed in Table 15.1.

15.4 EXPERIMENTAL RESULTS AND DISCUSSION

Tribological investigations of polymethylacrylate (PMA) as a viscosity index (VI) improver have been conducted to assess its effectiveness in reducing wear and friction in lubricated contacts. The findings, presented in Figures 15.2–15.5, demonstrate the significant role of PMA VII as an additive in enhancing lubricant performance. The study involved incorporating PMA at various weight concentrations into an N150 base oil to create solutions with optimized viscosity properties.

Figure 15.2 illustrates the influence of different polymethacrylate (PMA) concentrations in N150 base oil on the coefficient of friction (CoF). Experimental results showed that the base oil exhibited the highest CoF (0.17), whereas the addition of 4% PMA concentration resulted in the lowest CoF (0.08) across the entire range of tested speeds. This observed decrease in CoF (from 0.17 to 0.08) with increasing PMA concentration (2%, 4%, 6%, and 8%) at a constant sliding speed can be attributed to the enhanced boundary lubrication effect provided by PMA. Additionally, increasing the sliding speed (from 0 m/s to 0.028 m/s) led to a reduction in CoF (0.17–0.08). This indicates the development of a hydrodynamic lubrication component, although remaining relatively weak under the chosen

Table 15.1 Viscometric property of oil samples

Sample code	*Description*	*Viscosity in cst*		*Viscosity index*	*α [GPa⁻¹]*
		40°C	*100°C*		
N1	N150	33.3	5.9	100	28.4 E−9
P2	98% N150 + 2% PMA	36.08	5.99	110	
P4	96% N150 + 4% PMA	44.66	7.44	132	
P6	94% N150 + 6% PMA	60.99	10.15	154	
P8	92% N150 + 8% PMA	76.06	12.69	167	

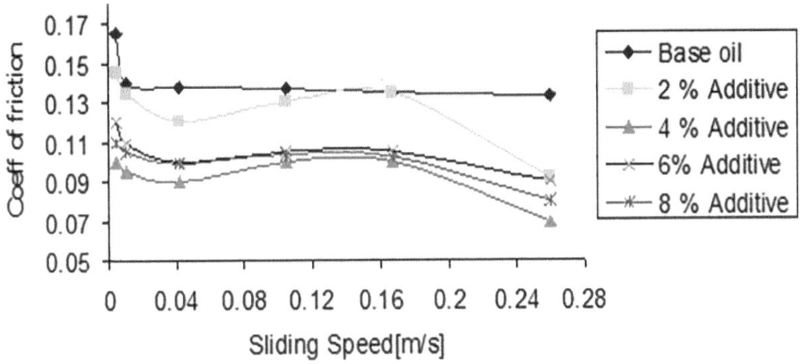

Figure 15.2 Friction behavior four-ball geometry—PMA.

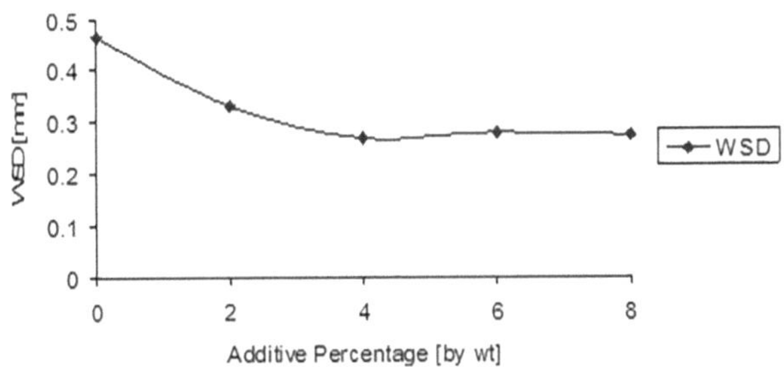

Figure 15.3 Wear scar diameter—PMA.

experimental conditions. As shown in Figure 15.2, the minimum CoF value of 0.08 is achieved with the 4% PMA additive (N04, representing 96% N150 base oil and 4% PMA).

Initial wear is high due to the base oil's limited lubrication properties, as evidenced by the large wear scar (0.5 mm) on the contacting surfaces (Figure 15.3). This initial wear, however, reduces contact severity, allowing for slightly better lubricant retention. Conversely, the presence of PMA molecules in varying concentrations leads to significantly lower wear rates and smaller wear scars (around 0.3 mm) (Figure 15.3). Notably, this wear reduction is inversely proportional to the concentration of the PMA additive.

The study observed a lower percentage of metal contact for the base oil, likely due to increased wear (Figure 15.4). Conversely, the 4% PIB blend exhibited lower wear, resulting in a higher severity of contact (more metal-to-metal interaction) under the same load conditions. This explains the high metal contact (up to 90%) observed for this blend. Interestingly, the 4% PIB blend still displayed a low coefficient of friction (CoF), possibly due to the presence of PMA molecules that minimized boundary film shear within the contact area.

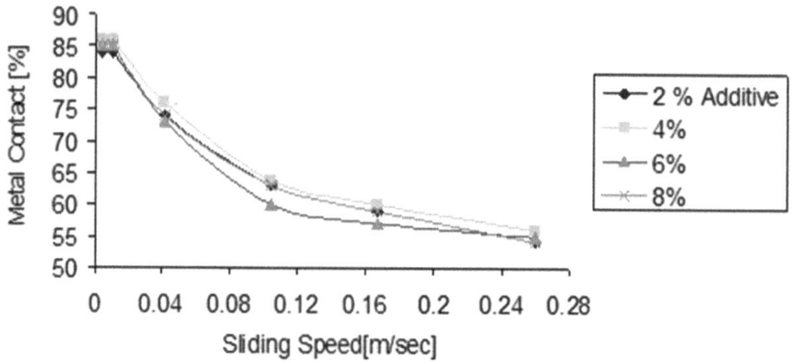

Figure 15.4 Metal contact behavior—PMA.

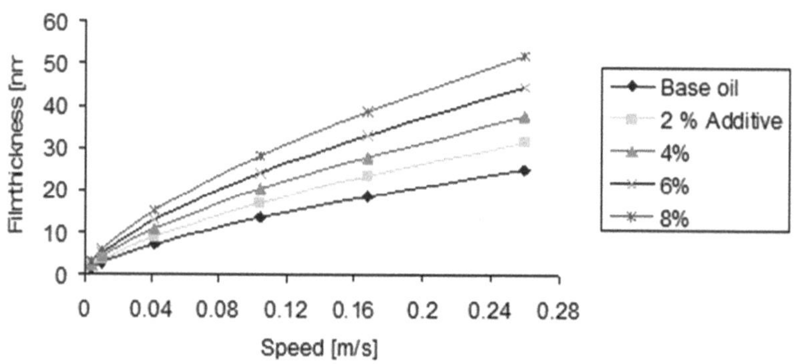

Figure 15.5 Theoretical film thickness four-ball—PMA.

Figure 15.4 further reveals that the percentage of metal contact decreases (from 90% to 50%) as sliding speed increases (0–0.28 m/s). This trend suggests improved lubrication with increasing speed. However, the differences in metal contact percentage at various speeds and additive concentrations are not substantial. Notably, at the highest sliding speed, the 8% additive blend exhibits the lowest metal contact (55%), while the 4% PMA blend shows the highest (60%). Wear reduction in this study is quantified by measuring the wear scar diameter (WSD). This parameter is directly related to the extent of material loss due to wear.

An experiment investigating the effect of PMA concentration on film thickness in N150 base oil is depicted in Figure 15.5. The results reveal a decreasing film thickness with increasing PMA concentration at the highest sliding speed (0.26 m/s). While the base oil exhibits a film thickness of 50 nm, a 6% PMA concentration reduces it to 45 nm. Further research, potentially through four-ball and pin-on-disk tribometry tests, is warranted to explore the optimal PMA concentration for minimizing friction under mixed and boundary lubrication conditions. This optimal concentration likely results from a combination of the boundary lubrication properties of the additives and their shear behavior, which requires further independent investigation.

Future research directions could involve developing new lubricant formulations with superior tribological properties and exploring advanced modeling techniques that incorporate real-time sensor data for more accurate performance predictions [45–47].

15.5 CONCLUSION

This study examined the effectiveness of VI improver as a friction modifier in lubricated contacts under light load conditions. The focus was on how VI improver concentration in a mineral base oil for light-duty applications impacts friction and wear behavior. Both conformal and counter-formal contact conditions were investigated. The results indicate that VI improver addition improves the anti-wear and friction properties of low-viscosity mineral base oils.

The following conclusions have been derived from the present study:

1. A four-ball test was conducted to investigate the tribological behavior of a base oil with the addition of a friction modifier, PMA VII.
2. The results revealed a dependence of the coefficient of friction (CoF) on lubrication regimes. In the boundary lubrication regime, CoF decreased with increasing speed for all contact conditions. This aligns with classical behavior where lubricants provide better separation at higher speeds. However, under hydrodynamic lubrication, CoF increased with speed. This suggests that the base oil and PMA VII combination might have a specific viscosity range optimal for hydrodynamic lubrication.
3. Furthermore, the presence of PMA VII at 4% concentration demonstrated a positive impact, reducing friction and wear across all contact conditions. This implies that PMA VII effectively modifies the tribological properties of the base oil.

REFERENCES

1. T. W. Selby (1958), The Non-Newtonian Characteristics of Lubricating Oils. *ASLE Transactions*, 1(1), pp. 68–81.https://doi.org/10.1080/05698195808972315
2. P. A. Asseff (1971), Some performance characteristics of hydrorefined lubricating oils. *Pet. Hydrocarbons*, Vol. 5, No. 4, pp. 168–170.
3. M. Smeeth et al (1996), "Performance of viscosity index improvers in lubricated contacts," *Langmuir*, Vol. 12, No. 19, pp. 4594–4598.

4. P. P Patil and A. Kumar (2017), "Dynamic structural and thermal characteristics analysis of oil lubricated multi speed transmission gearbox: Variation of load, rotational speed and convection heat transfer," *ISTM*, Vol. 41, pp. 281–291. https://doi.org/10.1007/s40997-016-0063-z.
5. A. Kumar (2018), "Low to high speed transient structural and thermal temperature measurement of oil-lubricated multi-speed heavy vehicle transmission gearbox system based on FEA," pp. 1–21. https://doi.org/10.4018/978-1-5225-3722-9.ch001.
6. S. Bhoi, A. Prasad, A. Kumar, R. B. Sarkar, B. Mahto, C. S. Meena, and C. Pandey (2022), "Experimental study to evaluate the wear performance of UHMWPE and XLPE material for orthopedics application," *Bioengineering*, Vol. 9, p. 676. https://doi.org/10.3390/bioengineering9110676.
7. A. K. Paul, A. Prasad, and A. Kumar (2022), "Review on artificial neural network and its application in the field of engineering," *Journal of Mechanical Engineering*, Vol. 1, No. 1, pp. 53–61. https://doi.org/10.56697/JMEP.2022.1107.
8. A. G. Pandey, A. Kumar, P. P. Patil, and V. Kumar (2022), "FEA based contact analysis of railway track and wheel," *AIP Conference Proceedings*, Vol. 2481, pp. 050010-1–050010-8. https://doi.org/10.1063/5.0103726.
9. A. Shukla and A. Kumar (2021), "Design and thermal characteristic analysis of gearbox system based on finite element analysis (FEA)," *AIP Conference Proceedings*, pp. 2317, 030002-1–030002-5. https://doi.org/10.1063/5.0036235.
10. A. Kumar, Y. Gori, and P. P. Patil (2020), "Finite element analysis based thermo-mechanical performance study of heavy vehicle medium duty transmission gearbox," https://doi.org/10.4018/978-1-7998-4939-1.
11. S. Rana and A. Kumar (2018), "FEA based design and thermal contact conductance analysis of steel and Al rough surfaces," *IJAER*, Vol. 13, No. 16, pp. 12715–12724.
12. A. Kumar, H. Jaiswal and P. P. Patil (2017), "Parametric optimization of vibrating heavy vehicle medium duty transmission gearbox housing using response surface method," *International Journal of Vehicle Structures and Systems*, Vol. 9, No. 3, pp. 149–153. https://doi.org/10.4273/ijvss.9.3.04.
13. A. Kumar and P. P. Patil (2016), "FEA simulation based performance study of multi-speed transmission gearbox," *IJMMME*, Vol. 6, No. 1, pp. 57–67. https://doi.org/10.4018/IJMMME.2016010103.
14. A. Kumar, R. Jain, and P. P. Patil (2016), "Dynamic analysis of heavy vehicle medium duty drive shaft using conventional and composite material," *IOP Conference Series: Materials Science and Engineering*, Vol. 149, pp. 1–11. https://doi.org/10.1088/1757-899X/149/1/012156.
15. G. J. Johnston, R. Wayte, and H. A. Spikes (1991), "The measurement and study of very thin lubricant films in concentrated contact," *Tribology Transactions*, Vol. 34, pp. 187–194.
16. A. Dyson and R. Wilson (1965–1966), *Film thickness in elastohydrodynamic lubrication by silicone fluids*, The Proceedings of the Institution of Mechanical Engineers, London, Vol. 180, No. 3, pp. 97–112.
17. J. D. Novak and W. O. Winer (1969), "The effects of pressure on the non-Newtonian behavior of polymer blended petroleum oils," *Journal of Lubrication Technology*, Vol. 91, pp. 459–463.

18. J. M. Palacios and M. L. Bajon (1983), "Effective viscosity of oils containing VI improver and its relation to wear," *Tribology International*, Vol. 16, pp. 97–31.
19. J. Briant, J. Denis, and D. Parc (1989), *Rheological properties of lubricants*, Institut Francais du Petrole, Paris.
20. A. Kumar and P. P. Patil (2015), "FEA based dynamic vibration response analysis of loose mounted heavy vehicle transmission gearbox casing," *IJAMMC*, Vol. 5, No. 1, pp. 136–141. http://dx.doi.org/10.11127/ijammc.2015.03.02.
21. A. Kumar and P. P. Patil (2014), "Dynamic vibration analysis of heavy vehicle truck transmission gearbox housing using FEA," *JESTR*, Vol. 7, No. 4, pp. 66–72.
22. A. Kumar, H. Jaiswal, R. Jain, and P. P. Patil (2014), "Free vibration and material mechanical properties influence based frequency and mode shape analysis of transmission gearbox casing," *Procedia Engineering*, Vol. 97, pp. 1097–1106. https://doi.org/10.1016/j.proeng.2014.12.388.
23. A. Kumar, Y. Gori, C. S. Meena, V. P. Singh, and V. K. Sharma (2022), "Design and analysis of heavy vehicle medium duty transmission gearbox system," pp. 67–86. https://doi.org/10.1201/9781003360001-4.
24. A. Kumar, S. Rana, Y. Gori, and N. K. Sharma (2021), "Thermal contact conductance prediction using FEM based computational techniques," pp. 183–220. https://doi.org/10.1201/9781003202233-13.
25. A. Kumar and P. P Patil (2017), "Free vibration and connecting bolt constraint based FEA analysis of heavy vehicle medium duty transmission gearbox housing made from AISI 4130 alloy," pp. 133–146. http://dx.doi.org/10.1016/B978-0-12-810998-4.00007-7.
26. A. Kumar, A. Dwivedi, H. Jaiswal, and P. P. Patil (2015), "Material-based vibration characteristic analysis of heavy vehicle transmission gearbox casing using finite element analysis," Vol. 308, pp. 527–533. https://doi.org/10.1007/978-81-322-2012-1_56.
27. F. Ahmad, V. Tomer, A. Kumar, and P. P. Patil (2016), "FEA simulation based thermo-mechanical analysis of tractor exhaust manifold," pp. 173–181. https://doi.org/10.1007/978-81-322-2740-3_18.
28. A. Kumar, Y. Gori, N. Dutt, Y. K. Singla, and A. Maurya (eds) (2021), "Advanced computational methods in mechanical and materials engineering," https://doi.org/10.1201/9781003202233.
29. K. K. Gajrani, A. Prasad, and A. Kumar (eds) (2022), *Advances in sustainable machining and manufacturing processes*. Publisher Taylor & Francis (CRC Press). https://doi.org/10.1201/9781003284574.
30. S. Singh, S. K. Behura, A. Kumar, and K. Verma (eds) (2022), "Nanomanufacturing and nanomaterials design: Principles and applications," https://doi.org/10.1201/9781003220602.
31. V. K. Sharma, A. Kumar, M. Gupta, V. Kumar, D. K. Sharma, and S. Sharma (eds) (2022), "Additive manufacturing in industry 4.0: Methods, techniques, modeling, and nano aspects," https://doi.org/10.1201/9781003360001.
32. A. Kumar, V. P. Singh, C. S. Meena, and N. Dutt (eds) (2023), "Thermal energy systems: Design, computational techniques and applications," https://doi.org/10.1201/9781003395768.
33. R. Kumar, A. Prasad, and A. Kumar (eds) (2023), "Sustainable smart manufacturing processes in industry 4.0," https://doi.org/10.1201/9781003436072.

34. V. P. Singh, A. Kumar, C. S. Meena, and G. Dwivedi (eds) (2024), *Energy efficient vehicles: Technologies and challenges* (1st ed.). CRC Press. https://doi.org/10.1201/9781003464556.
35. Y. Khemka (2021), "Accurate prediction of the viscosity of light crude oils using one-parameter friction theory: Effect of crude oil characterization methods and property correlations," *Fuel*, Vol. 283, pp. 118926.
36. V. B. Regueira (2018), "Improvement of the expanded fluid viscosity model for crude oils: Effects of the plus-fraction characterization method and density," *Energy Fuels*, Vol. 32, pp. 1624–33.
37. Ioanna-Panagiota Kanaveli, Maria Atzemi, and Evripidis Lois (2017), "Predicting the viscosity of diesel/biodiesel blends," *Fuel*, Vol. 199, pp. 248–263.
38. G. Yuanping and L. Kewen (2012), "New models for calculating the viscosity of mixed oil," *Fuel*, Vol. 95, pp. 431–437.
39. I. L. Mohammed, J. Abutaqiya, F. M. Zhang, and M. Vargas (2019), "Viscosity modeling of reservoir fluids using the friction theory with PC-SAFT crude oil characterization," *Fuel*, Vol. 235, pp. 113–129.
40. S. E. Quinones-Cisneros, S. I. Andersen, and J. Creek (2005), "Density and viscosity modeling and characterization of heavy oils," *Energy Fuels*, Vol. 19, pp. 1314–1318.
41. P. Patil, Y., Gori, A., Kumar and M. R. Tyagi (2021), "Experimental analysis of tribological properties of polyisobutylene thickened oil in lubricated contacts," *Tribology International*, Vol. 159, No. 106983, pp. 1–7. https://doi.org/10.1016/j.triboint.2021.106983.
42. V. Kharka, N. K. Jain, and K. Gupta (2022), "Performance comparison of green lubricants in gear hobbing with minimum quantity lubrication," *Tribology International*, Vol. 173, pp. 107582. https://doi.org/10.1016/j.triboint.2022.107582.
43. A.K. Tomar and S. C. Sharma (2022), "Non-Newtonian lubrication of hybrid multi-recess spherical journal bearings with different geometric shapes of recess," *Tribology International*, Vol. 171, No. 2022, pp. 107579. https://doi.org/10.1016/j.triboint.2022.107579.
44. F. Cyriac, T. X. Yi, S.K. Poornachary, and P.S. Chow (2022), "Boundary lubrication performance of polymeric and organic friction modifiers in the presence of an anti-wear additive," *Tribology International*, Vol. 165, pp. 107256. https://doi.org/10.1016/j.triboint.2021.107256.
45. V.P. Singh, A. Kumar, C.S. Meena, and G. Dwivedi (2024), "Introduction to energy-efficient vehicles for sustainable transportion," in *Energy efficient vehicles*. CRC Press, Boca Raton, pp. 1–18.
46. R.K. Upadhyay, V.P. Singh, and A. Kumar (2024), *An introduction to sustainable transportion: Policy, planning and implementation* in Energy Efficient Vehicles, Varun Pratap Singh, Ashwani Kumar, Chandan Swaroop Meena and Gaurav Dwivedi (Eds.), CRC Press, Boca Raton FL USA , pp. 74–95. https://doi.org/10.1201/9781003464556-5
47. V.P. Singh and A. Kumar (2024), "Techno-economic and future aspects of the HEV-EV-FCV," in *Energy efficient vehicles*. CRC Press, Boca Raton, pp. 1–20.

Chapter 16

Performance estimation of non-circular bearing in thin film lubrication

Sandeep Soni

16.1 INTRODUCTION

Hydrodynamic bearing design has drawn interest from the analytical and applied engineering communities for many years. Robust hydrodynamic bearing technology is required for wider applications, like machine tools, aeroplanes, nuclear reactors, and textile machinery spindles, in order to resist a variety of operating circumstances.

According to a microscopic perspective, lubricant film development is an essential component of a journal bearing's successful operation. The development of theoretical relationships to reveal real-world engineering challenges has been made possible by the advancements in thin film lubrication. Numerous researchers have examined the elliptical bearings' performance properties [1–4]. The static and dynamic characteristics of elliptical bearings have been theoretically and experimentally explored by Kumar et al. [1], Hashimoto et al. [2], Nair et al. [3], and Chauhan et al. [4] using the finite element and finite difference approaches. They have computed the fluid-film coefficients, oil flow rate, coefficient of friction, and load-carrying capability.

It has been reported in the thin film lubrication reports [5–7] that the narrow range decreases the distance between the two surface walls, changing the liquid lubricant's viscosity. The viscosity changes brought on by distance differ for each type of lubricant. A rheological hypothesis for evaluating lubrication qualities in the boundary lubrication domain was proposed by Tichy [8]. Considerations from continuum mechanics served as the foundation for the model. The friction and load estimations for hydrodynamic contact have been presented by the author.

A surface layer model for thin-film lubrication was created by Tichy [9]. In the suggested model, the author has suggested the surface contact

DOI: 10.1201/9781003535447-16

layer viscosity, surface layer thickness, and conventional viscosity parameters. Wen [10] discussed on the unique occurrences connected to EHL and micro-EHL circumstances for the lubricating layer. They have studied the significance of thin film lubrication research, both analytically and experimentally.

In order to assess the film thickness, Luo et al. [11] investigated the mechanism of the lubricant film's shift from elastohydrodynamic lubrication (EHL) to thin film lubrication (TFL). The experimental examination of the changing of film thickness in the TFL and EHL regimes has been provided by them.

The Reynolds equation for thin film lubrication was obtained by Qingwen et al. [12]. They described how the lubricant film's behavior is influenced by the adsorption and viscosity of bearing surfaces.

The lubricant velocity profiles that were produced on the contacting surfaces were explained by Qingwen et al. [13]. They explained how, as the shear rate increased, speed effects became increasingly important in thin film flow. An analogous viscosity model was developed by Qu et al. [14] using the notion of adsorption. They provided the findings of their calculations under the thin-film range after applying the model to alter the Reynolds lubrication equation. The thin-film lubrication's characteristic adsorbent thickness was calculated and given by Qingwen et al. [15]. They displayed a graphical representation of a bearing's load capacity and frictional characteristics when it operates in the thin layer regime.

Zhang [16] provided an overview of the development in thin-film lubrication (TFL) research as it pertained to the concepts, computational strategies, and measurement methodologies used. The theory of thin-film lubrication (TFL), which fills the gap between the boundary lubrication (BL) and elastohydrodynamic lubrication (EHL) regimes, was thoroughly reviewed by Liran and Jianbin [17]. Ono [18] has exhibited the effects of surface viscosity layers in thin film lubrication simulations. They took into account the surface layers with viscosity on both sides of the bearing surfaces. For the micro-tapered bearing, they contrasted the one-side and two-side surface layer models.

A key factor in the hydrodynamic lubrication evaluation, particularly for the journal bearing design, is the lubricant flow viscosity. Further viscosity changes are practically necessary for large speed and large load situations. The properties of the surface adsorbent and the fluid velocity have a significant influence on the viscosity distribution in the fluid flow throughout the non-circular bearing.

Therefore, it is imperative to examine the elliptical hydrodynamic bearing's steady-state properties in the thin-film fluid regime. The performance of a short elliptical hydrodynamic bearing in thin-film lubrication is described in this book chapter. The theoretical calculation of the unique performance characteristics at steady state conditions takes into account the repercussion of adsorbent thickness (h_{aL}) in the short elliptical bearing.

16.2 GOVERNING LUBRICATION EQUATIONS

16.2.1 Modeling of Reynolds equation

The incompressible flow and Newtonian lubricant in the clearance zone of an elliptical bearing are governed by the Reynolds equation, which can be expressed in the following dimensional form [19]:

$$\frac{\partial}{\partial \hat{x}}\left(\hat{h}^3 \frac{\partial \hat{p}}{\partial \hat{x}}\right)+\frac{\partial}{\partial \hat{z}}\left(\hat{h}^3 \frac{\partial \hat{p}}{\partial \hat{z}}\right)=6\hat{U}\hat{\mu}\left(\frac{d\hat{h}}{d\hat{x}}\right) \tag{16.1}$$

Equation (16.1) can be solved analytically or numerically for the short elliptical hydrodynamic bearing approximation, which disregards the first component of LHS and takes into account an appropriately small axial direction.

$$\frac{\partial}{\partial \hat{x}}\left(\hat{h}^3 \frac{\partial \hat{p}}{\partial \hat{x}}\right)\approx 0 \tag{16.2}$$

For the elliptical bearing, the reduced version of the Reynolds lubrication equation is identified as follows [19]:

$$\frac{\partial}{\partial \hat{z}}\left(\hat{h}^3 \frac{\partial \hat{p}}{\partial \hat{z}}\right)=6\hat{U}\hat{\mu}\left(\frac{d\hat{h}}{d\hat{x}}\right) \tag{16.3}$$

In fact, according to Equation (16.3), the pressure is unaffected by lubricant flow conditions in the rotational direction. First, by appropriately substituting in Equation (16.3), the governing lubrication equation has been transformed into a non-dimensional form. This allowed for the enumeration of pressure variations in the clearance field of the short elliptical bearing.

Using Gumbel's boundary technique, the short elliptical bearing (L/D = 0.50) has been figured out. Atmospheric pressure is preserved at the elliptical bearing surface ends. Equation (16.4) has been applied as boundary conditions to determine the circumferential pressure distribution in the clearance zone of the short elliptical bearing.

$$\begin{gathered}\hat{p}=0;\ \ \hat{z}=0 \ \ to \ \ (\pm L/2)\\ \hat{p}=0;\ \ \theta=0,\ 2\pi\end{gathered} \tag{16.4}$$

For the elliptical bearing, the lubricant layer thickness (Figure 16.1) is enumerated by [19]:

$$\hat{h}=\left[\hat{C}_V+\hat{e}*cos\theta+\left(\hat{C}_H-\hat{C}_V\right)*sin^2\left(\theta-\phi\right)\right] \tag{16.5}$$

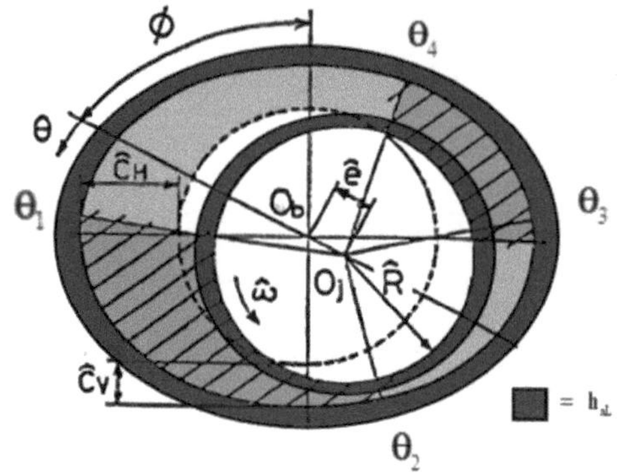

Figure 16.1 Illustration of elliptical bearing with adsorbed layer thickness h_{aL} .

The dimensionless lubricant layer thickness for the elliptical bearing is obtained through the following equation:

$$H = \left[1 + \varepsilon * \cos\theta + \varepsilon_p * \sin^2(\theta - \phi)\right] \tag{16.5a}$$

The shape and coordinates of a short elliptical bearing with an adsorbent layer thickness (h_{aL}) are depicted in Figure 16.1 for the proposed thin-film lubrication analysis.

16.2.2 Modeling of thin film lubrication

The aftermath of adsorbent layer thickness (h_{aL}) of thin-film lubrication regime to dynamic viscosity $(\hat{\mu})$is presented by assuming equal adsorption qualities at the journal and lobed outer bearing surface in the equivalent viscosity revisal model for the performance evaluation of non-circular bearing. For elliptical bearing, Equation (16.6) establishes a relationship betwixt lubricant film thickness $(\hat{h})$ and adsorbent layer thickness (h_{aL}) [14, 20]:

$$\hat{\mu} = \left(\frac{\hat{h}}{\hat{h} - 2 * \hat{h}_{aL}}\right)\hat{\mu}_0 \tag{16.6}$$

It is possible to attain a fluctuation in the dynamic viscosity $(\hat{\mu})$ with lubricant film variation $(\hat{h})$ and adsorbed thickness (h_{aL}).

The impact of the thin-film regime in the short elliptical hydrodynamic bearing has been determined using the simultaneous solution of Equations (16.3), (16.5), and (16.6). The following formulation yields the updated Reynolds equation that incorporates the adsorbent thickness (h_{aL}) and the short elliptical hydrodynamic bearing approximation:

$$\frac{\partial}{\partial\hat{z}}\left(\hat{h}^2\left\{\hat{h}-2\hat{h}_{aL}\right\}\frac{\partial\hat{p}}{\partial\hat{z}}\right)=\frac{6\hat{U}\hat{\mu}_0}{\hat{R}}\left(\frac{d\hat{h}}{d\theta}\right) \tag{16.7}$$

The short elliptical hydrodynamic bearing's lubrication behavior in the thin-film lubrication regime has been accurately simulated by Equation (16.7), which has been solved using a finite difference approach to produce the elliptical bearing's steady-state characteristics.

16.3 SOLUTION APPROACH

Using a finite difference approach for the utilized static load, the lubricant flow field solution of the short elliptical bearing assembly under the thin-film regime with Newtonian lubricant is derived. In addition to a specific boundary condition (Equation 16.4), lubricant layer thickness (Equation 16.5), and viscosity model (Equation 16.6), the Reynolds lubrication equation (Equation 16.3) has been enumerated simultaneously for the short elliptical bearing.

Equation (16.7) has transformed into a dimensionless form that can be solved numerically by adding the subsequent dimensionless parameters:

$$\left.\begin{array}{l}\theta=\left(\hat{x}/\hat{R}\right);\quad H=\left(\hat{h}/\hat{C}_V\right);\quad \varepsilon=\left(\hat{e}/\hat{C}_V\right);\\ \mu=\left(\hat{\mu}/\hat{\mu}_0\right);\quad \lambda=\left(\hat{D}/\hat{L}\right);\quad z=\left\{\hat{z}/(L/2)\right\};\\ h_{aL}=\left(\hat{h}_{aL}/\hat{C}_V\right);\quad p=\left\{\hat{p}/\left(6\hat{\mu}\hat{U}\hat{R}/\hat{C}^2\right)\right\}\\ \varepsilon_p=\left\{\left(\hat{C}_H-\hat{C}_V\right)/\hat{C}_V\right\}\end{array}\right\} \tag{16.8}$$

The following dimensionless Reynolds' lubrication equation is employed in the proposed thin-film lubrication analysis of short elliptical hydrodynamic bearing.

$$\left\{(\lambda)^2\left(H^3-2*h_{aL}*H\right)\right\}\left(\frac{\partial^2 p}{\partial z^2}\right)=6\left(\frac{dH}{d\theta}\right) \tag{16.9}$$

The circumferential pressure distribution in the short elliptical bearing is attained by using the central difference approach to the Reynolds lubrication equation and approximating the unwrapped short elliptical bearing film as systematic rectangular grid spacing. The iterative numerical elucidation of Equation (16.9) has been carried out through the FDM. A computer technique has been developed to resemble the short elliptical bearing's performance parameters, both with and without taking the thickness of the adsorbed layer (h_{aL}) into account during the solution stage.

16.4 PERFORMANCE PARAMETERS

In order to enumerate distinct performance parameters such as load-carrying capacity, attitude angle, frictional characteristics, and lubricant flow rate, the finite difference solution of Equation (16.9) pertaining to the adsorbed thickness (h_{aL}) furnished circumferential pressure fluctuation in the bearing lubricant film.

16.4.1 Load-carrying capacity and attitude angle

For the short elliptical hydrodynamic bearing, the components of the lubricant-film reactions ($\hat{W}_x$ and $\hat{W}_y$) in the bearing center's direction and normal to it can be enumerated, appropriately by:

$$\left.\begin{aligned} \hat{W}_x &= -2\int_0^{\pi}\int_0^{\hat{L}/2}\left(\hat{p}\cos\theta \hat{R} d\theta d\hat{z}\right) \\ \hat{W}_y &= 2\int_0^{\pi}\int_0^{\hat{L}/2}\left(\hat{p}\sin\theta d\theta d\hat{z}\right) \end{aligned}\right\} \tag{16.10}$$

In order to figure out $\hat{W}$ and ϕ under thin-film lubrication of the proposed elliptical bearing, the components of lubricant-film reactions in the direction of $\hat{W}_x$ and $\hat{W}_y$ are altered, taking into account the consequence of adsorbed thickness (h_{aL}). These components can be written as follows:

$$\left.\begin{aligned} \hat{W} &= \left(\hat{W}_x^{\,2} + \hat{W}_y^{\,2}\right)^{0.5} \\ \phi &= \left[\tan^{-1}\left(\hat{W}_y / \hat{W}_x\right)\right] \end{aligned}\right\} \tag{16.11}$$

16.4.2 Frictional characteristics

The frictional force surrounding the bearing surface is provided by the integration of shear stresses. In the short elliptical hydrodynamic bearing, the shear stress is computed using

$$\hat{\tau} = \hat{\mu}\left(\frac{\hat{U}}{\hat{h}}\right)$$

By integrating the ensuing relationship over the journal surface area, the frictional force (F_f) is obtained:

$$\hat{F}_f = \int_{\hat{A}}\left(\hat{\tau} d\hat{A}\right)$$

Using $d\hat{A} = \left(\hat{L} * \hat{R}\right) d\theta$, the frictional force is found to be

$$\hat{F}_f = \hat{R} * \hat{L} * \int_0^{2\pi} (\hat{\tau})\, d\theta \tag{16.12}$$

The frictional force (F_f) has been further enhanced by considering the impact of adsorbent thickness (h_{aL}) under thin-film lubrication.

16.4.3 Lubricant flow rate

Equation (16.13) provides the lubricant flow rate inside the short elliptical hydrodynamic bearing:

$$\hat{Q}_z = \left[2\int_0^{\pi} -\left(\frac{\hat{h}^3}{12\hat{\mu}}\right)\left(\frac{d\hat{p}}{d\hat{z}}\right)_{z=\frac{\hat{L}}{2}} \hat{R}\, d\theta \right] \tag{16.13}$$

By taking into account the thickness of the adsorbed layer(h_{aL}), the lubricant flow rate ($\hat{Q}_z$) computation under thin-film lubrication is significantly enhanced. In the thin-film lubrication regime, the lubricant inflow ($\hat{Q}_{Inlet}$) and outflow ($\hat{Q}_{Outlet}$) via the bearing are also obtained.

16.5 RESULTS AND DISCUSSION

A computer algorithm has been created to model the static operating behaviors of a short elliptical bearing $\left(\varepsilon_p = 1.25\right)$ under the thin-film lubrication regime $(\hat{\mu})$. The load-carrying capacity ($\hat{W}$), attitude angle (ϕ), frictional characteristics (F_f), and lubricant flow rate $\left(Q_{Inlet} \;\&\; Q_{Outlet}\right)$ have been carefully examined in relation to these steady-state performance parameters, with an L/D of 0.50 and Gumbel's boundary condition taken into consideration throughout the solution stage. Table 16.1 shows the geometrical and lubrication parameters used in the evaluation of the short elliptical hydrodynamic bearing's performance under thin-film lubrication. In the current simulation study, the two cases are investigated:

(a) $h_{aL} = 0.00$ termed as a short elliptical bearing without contemplating (h_{aL}), and

(b) $h_{aL} = 0.10 \; to \; 0.40$ termed as short elliptical bearing with contemplating(h_{aL}).

The dimensionless changes in the lubricant film thickness (H) with circumferential coordinate (θ) that are obtained both without and with addressing

Table 16.1 Geometrical and lubrication parameters

Parameters	*Range*
Bearing length (L)	50.00 mm
Journal diameter (D)	100.00 mm
Radial clearance (C)	0.025 mm
Aspect ratio (L/D)	0.50
Eccentricity ratio (ε)	0.15 and 0.75 0.10 – 0.90
Dynamic viscosity (μ)	0.05 Pa - s
Adsorbed layer thickness (h_{aL})	0.00 – 0.40
Ellipticity ratio (ε_p)	1.25

the impact of different adsorbent thickness magnitudes (h_{aL}) at $\varepsilon = 0.15$ and $\varepsilon = 0.75$ with$\varepsilon_p = 1.25$, are depicted in Figure 16.2(a, b), respectively. It is noted from Figure 16.2 that when the adsorbent layer thickness (h_{aL}) augments from 0 to selected distinct values, dimensionless lubricant film thickness decreases clearly at particular values of $\varepsilon \,\&\, \varepsilon_p$.

The fluctuation of the dimensionless circumferential pressure distribution (p) with the angular variation (θ) obtained through the suggested thin-film lubrication regime for short elliptical hydrodynamic bearing is depicted in Figure 16.3(a, b). Circumferential pressure variation rises sharply with increasing adsorbent thickness (h_{aL}), especially at higher adsorbent thickness values. As the thickness of the adsorbent layer increases, a dramatic increase in pressure is seen at larger eccentricity ratios and ellipticity ratios.

Figure 16.4(a, b) depicts the fluctuation in the short elliptical hydrodynamic bearing's dimensionless load-carrying capability (W) in relation to the adsorbent layer thickness (h_{aL}). Greater adsorbent layer thickness values are associated with larger load capacities, whereas lower adsorbent layer thickness values result in less variation in load capacity, respectively. It is evident that the phase transformation of the liquid film causes the dimensionless load capacity of the bearing to move toward infinity as the value of $(h_{aL} + \varepsilon)$ approaches unity. In summary, the thickness of the adsorbent layer raises the dimensional load capacity under thin-film lubrication.

Figure 16.5 shows the attitude angle (ϕ) curves under the thin-film lubrication regime for the short elliptical hydrodynamic bearing. For the suggested bearing simulation, the unique selections of $(\varepsilon, \phi)$with adsorbed layer thickness (h_{aL}) are critical to reaching the shaft equilibrium. With an increase in the thickness of the adsorbed layer, the attitude angle decreases;

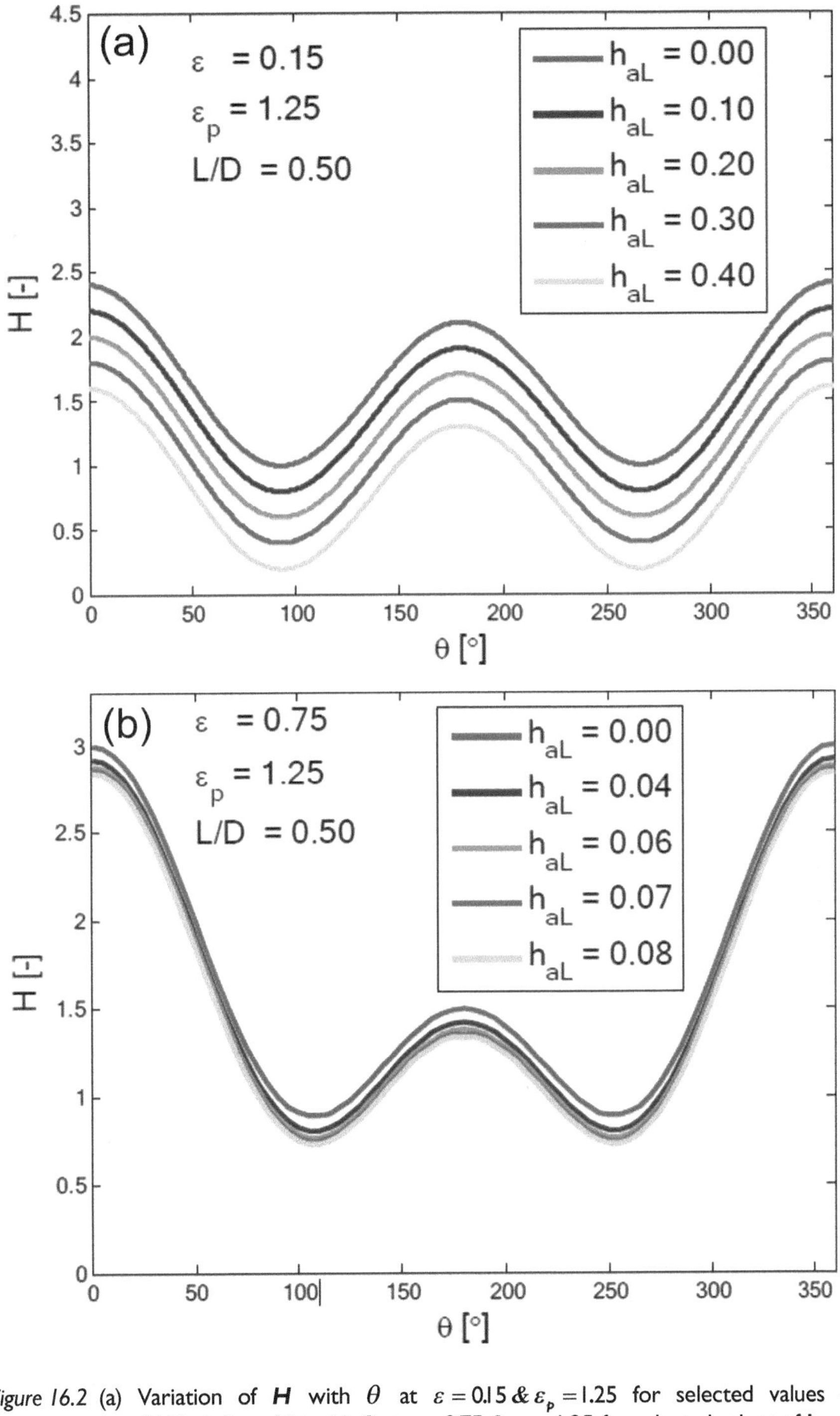

Figure 16.2 (a) Variation of $\boldsymbol{H}$ with θ at $\varepsilon = 0.15$ & $\varepsilon_p = 1.25$ for selected values of $\boldsymbol{h}_{aL}$. (b) Variation of $\boldsymbol{H}$ with θ at $\varepsilon = 0.75$ & $\varepsilon_p = 1.25$ for selected values of $\boldsymbol{h}_{aL}$.

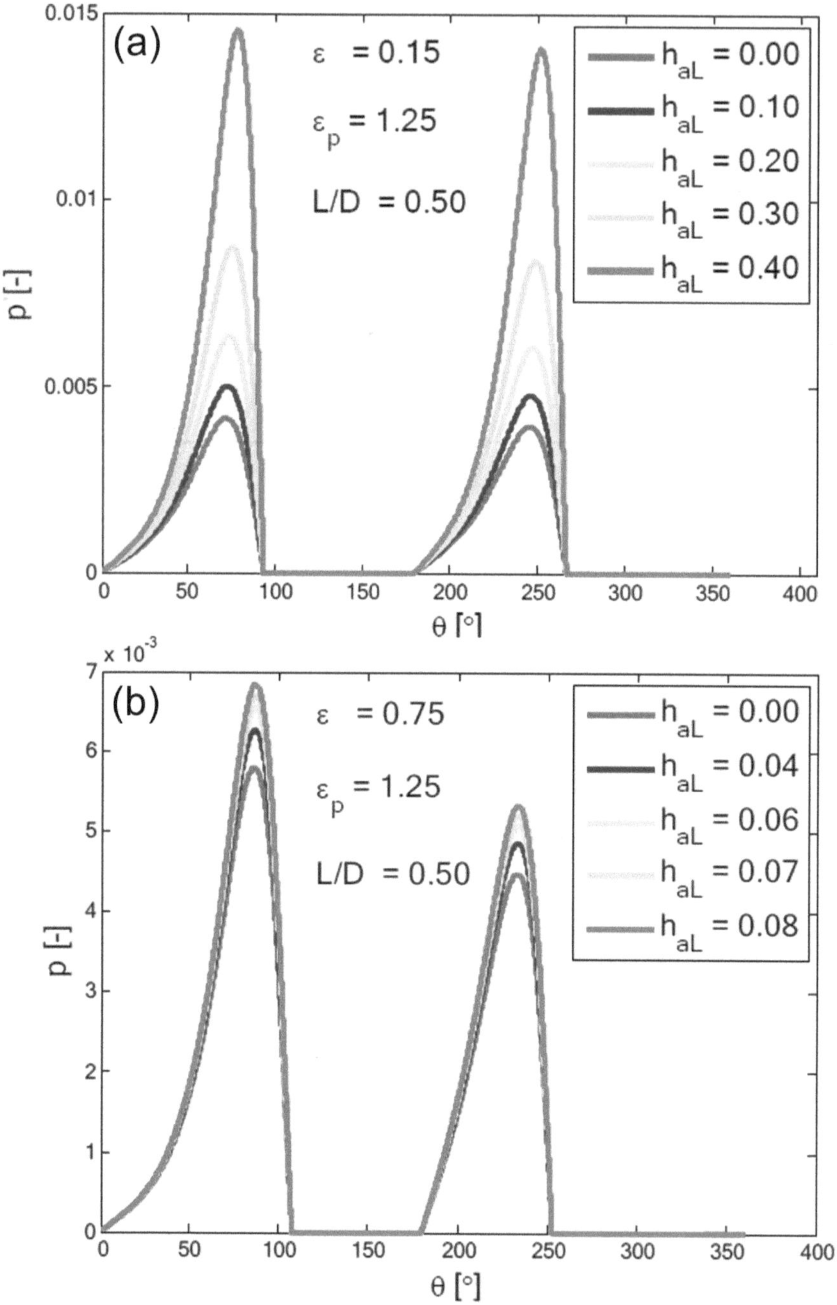

Figure 16.3 (a) Variation of p with θ at $\varepsilon = 0.15$ & $\varepsilon_p = 1.25$ for selected values of $\boldsymbol{h_{aL}}$. (b) Variation of $\boldsymbol{p}$ with θ at $\varepsilon = 0.75$ & $\varepsilon_p = 1.25$ for selected values of $\boldsymbol{h_{aL}}$.

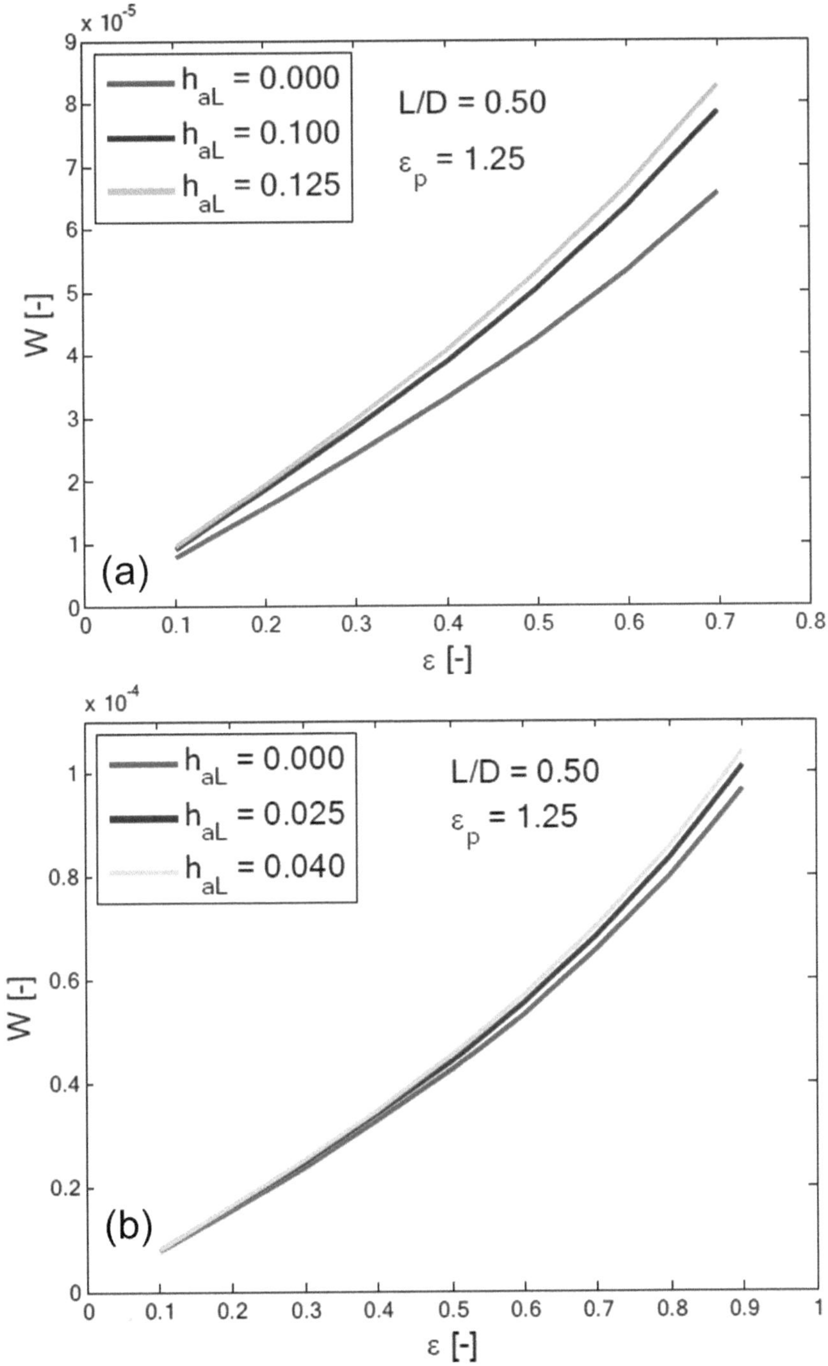

Figure 16.4 (a) Variation of $\boldsymbol{W}$ with ε & ε_p for selected values of $\boldsymbol{h}_{aL}$. (b) Variation of $\boldsymbol{W}$ with ε & ε_p for selected values of $\boldsymbol{h}_{aL}$.

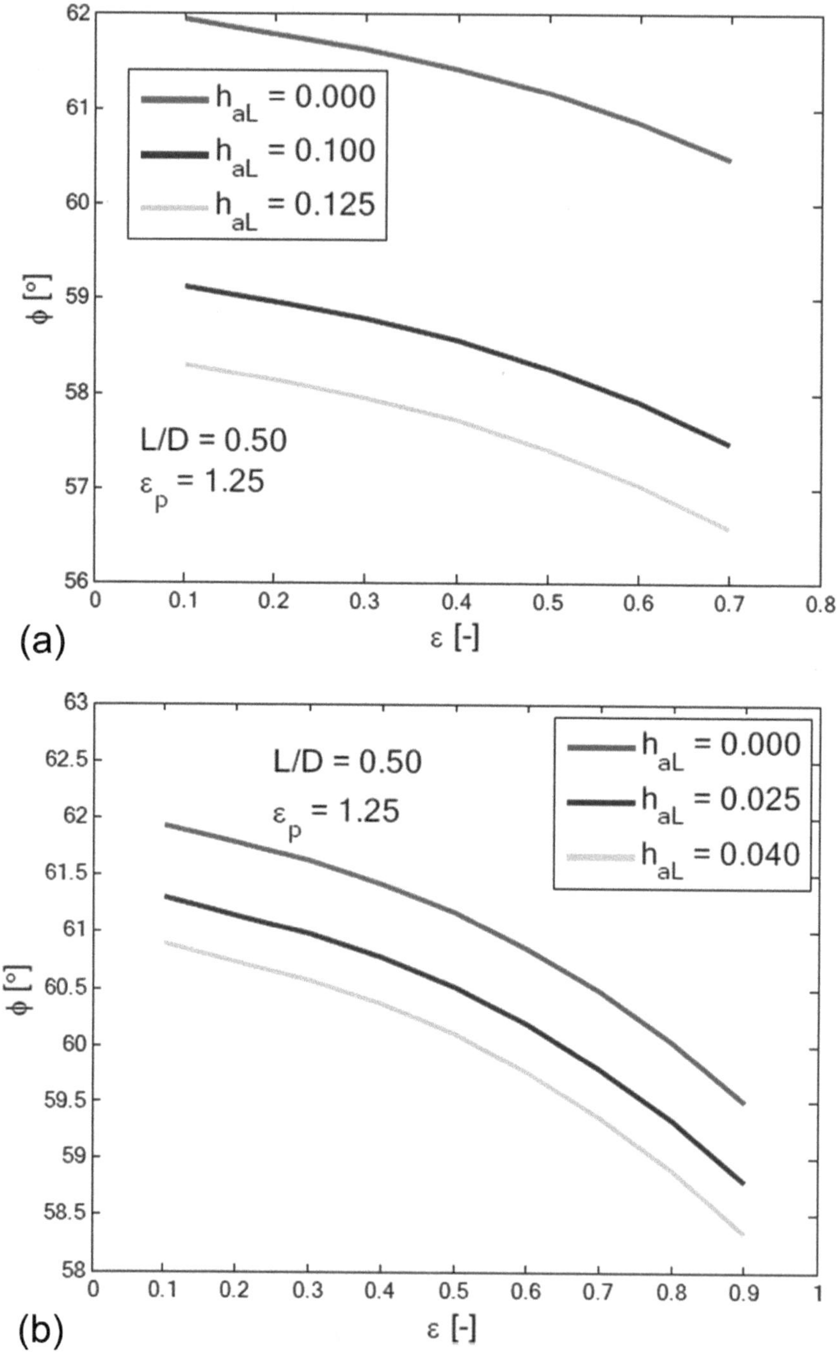

Figure 16.5 Variation of ϕ with ε & ε_p for selected values of h_{aL}.

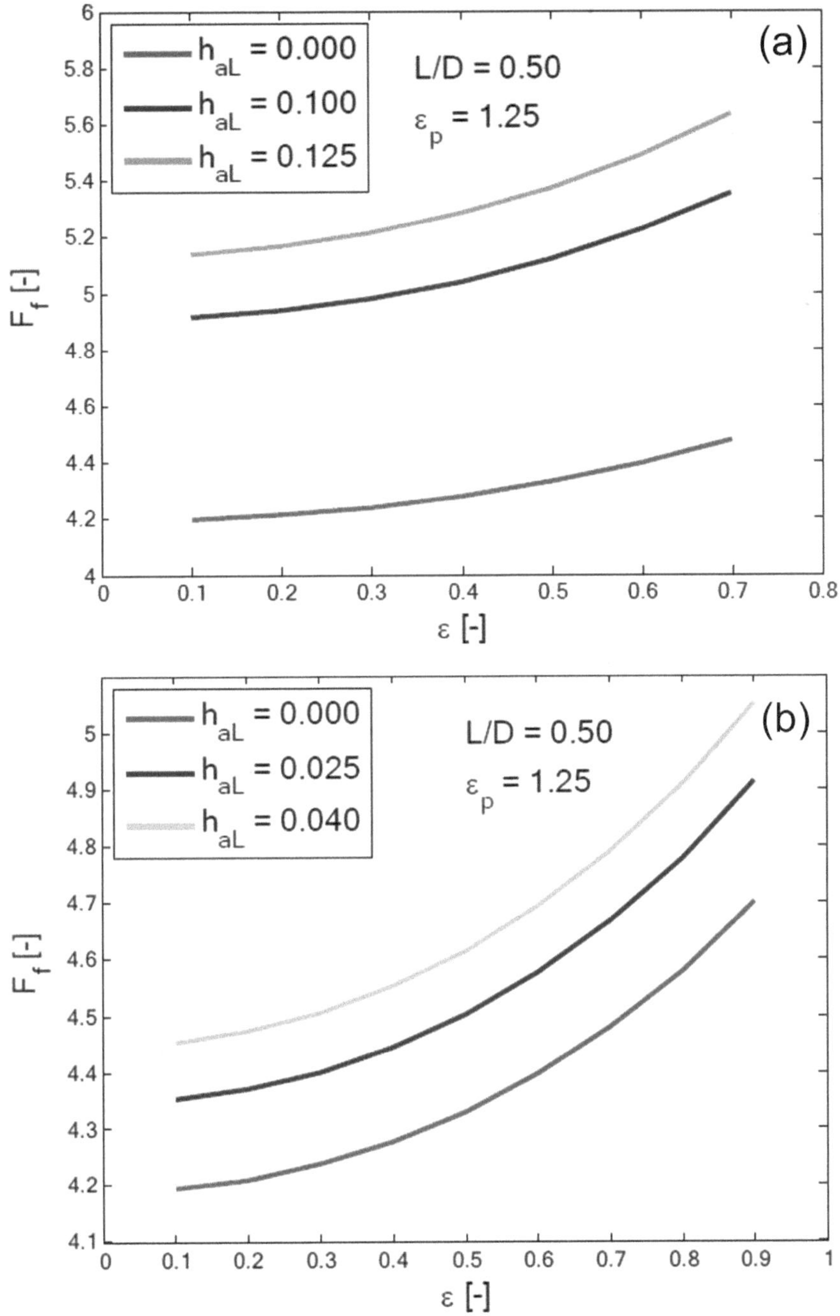

Figure 16.6 (a) Variation of $\boldsymbol{F_f}$ with ε & ε_p for selected values of $\boldsymbol{h_{aL}}$. (b) Variation of $\boldsymbol{F_f}$ with ε & ε_p for selected values of $\boldsymbol{h_{aL}}$.

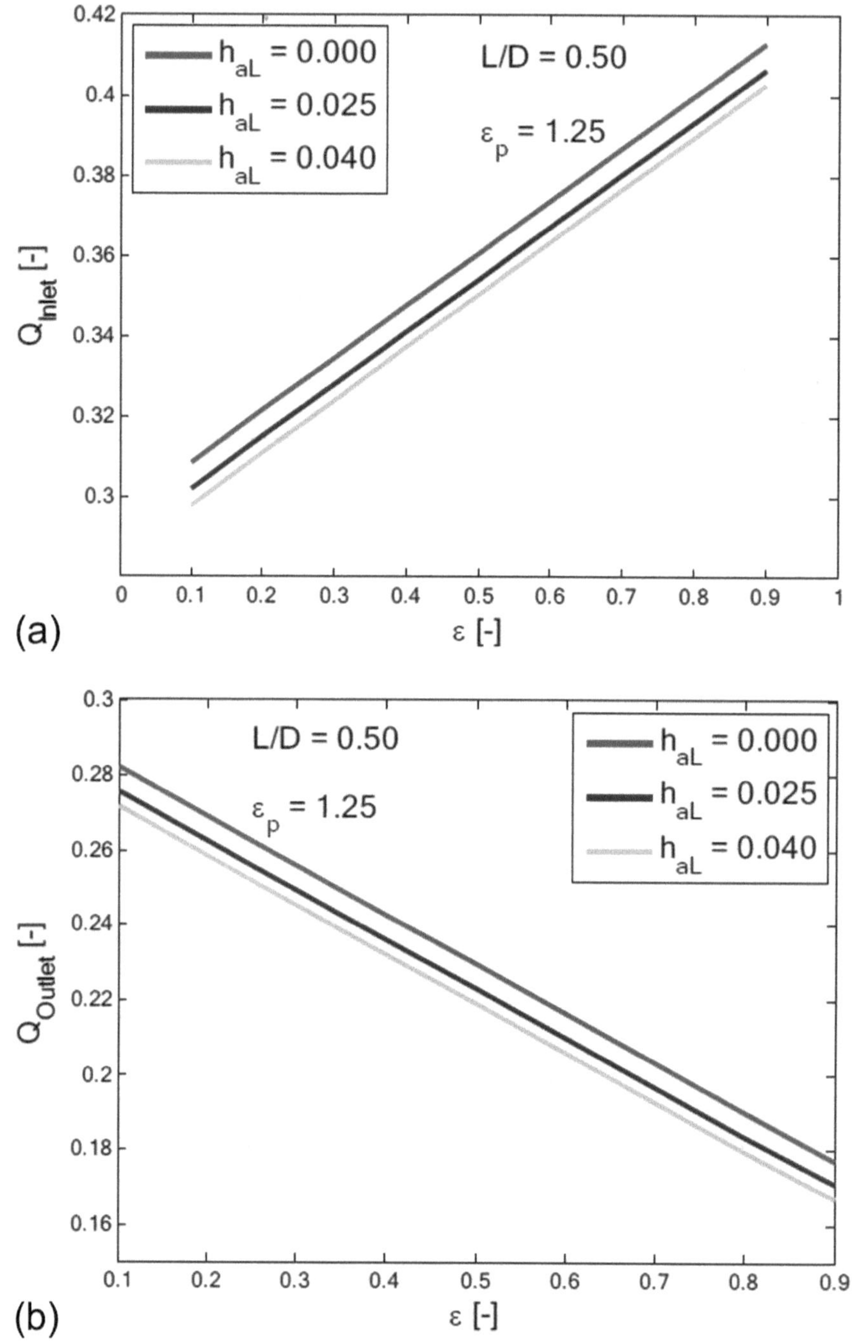

Figure 16.7 Variation of $\mathbf{Q}_{Inlet}$ & $\mathbf{Q}_{Outlet}$ with ε & ε_p for selected values of $\boldsymbol{h}_{aL}$.

these changes become more pronounced at higher values of the adsorbent layer thickness and higher eccentricity ratio.

For the short elliptical bearing, the values of frictional force (F_f) are obtained for different adsorbent thickness magnitudes (h_{aL}) with regard to the eccentricity ratio in Figure 16.6(a, b). The frictional force in the bearing assembly is shown to rise with an increase in adsorbent thickness(h_{aL}). This indicates that, under the suggested lubricant film regime, a more accurate practical estimate of adsorbent thickness for the successful operation of the elliptical bearing assembly is vital.

The relationship between the lubricant flows $(Q_{Inlet}\ \&\ Q_{Outlet})$ and eccentricity ratio (ε) for different adsorbed layer thicknesses (h_{aL}) is depicted in Figure 16.7. Lubricant flow through the elliptical bearing housing is extremely influenced by the thickness of the adsorbent layer.

16.6 CONCLUSIONS

A thin-film lubrication regime of elliptical bearing has been analyzed by simulation. The following conclusions have been made in light of the numerical results:

1. The thickness of the adsorbent layer had a substantial impact on both the load-carrying capacity and pressure fluctuation.
2. In the thin film lubrication regime, the computation of attitude angle and frictional properties is essential to precisely analyze the hydrodynamic bearings.
3. The presence of an adsorbent layer causes a noticeable change in the behavior of the lubricant flow.
4. The lubrication performance of the hydrodynamic bearing is improved by contemplating the adsorbent surface property.
5. The analysis of thin-film lubrication requires the precise selection of the adsorbent layer's magnitude (h_{aL}), the eccentricity ratio (ε), and ellipticity ratio (ε_p) for non-circular bearing.

REFERENCES

1. Kumar, A., Sinhasan, R., Singh, D. V. 1980. Performance characteristics of two-lobe hydrodynamic journal bearings. *Transactions of ASME Journal of Lubrication Technology*, **102**: 425–429.
2. Hashimoto, H., Wada, S., Tsunoda, H. 1984. Performance characteristics of elliptical journal bearings in turbulent flow regime. *Bulletin of JSME*, **27**: 2265–2271.

3. Nair, K. P., Nair, V. P. S., Jayadas, N. H. 2007. Static and dynamic analysis of elastohydrodynamic elliptical journal bearing with micropolar lubricant. *Tribology International,* **40**: 297–305.
4. Chauhan, A., Sehgal, R., Sharma, R. K. 2010. Thermohydrodynamic analysis of elliptical journal bearing with different grade oils. *Tribology International,* **43**: 1970–1977.
5. Israelachivi, J. N. 1986. Measurement of the viscosity of liquids in very thin films. *Science*, **110** (**1**): 263–271.
6. Johnston, G. J., Wayte, R., Spikes, H. A. 1991. The measurement and study of very thin lubrication films in concentrated contact. *Tribology Transactions*, **34**: 187–194.
7. Wu, S. 1994. Empirical determination of effective lubricant rheological parameters. *Tribology Transactions,* 37(**1**): 138–146.
8. Tichy, J. A. 1995. Modeling of thin film lubrication, *Tribology Transactions*, **38** (**1**): 108–118.
9. Tichy, J. A. 1995. A surface layer model for thin film lubrication, *Tribology Transactions*, **38** (**3**): 577–582.
10. Wen, S. 1996. On thin film lubrication, *Lubrication Science*, **8** (**3**): 275–286.
11. Luo, J., Wen, S., Huang, P. 1996. Thin film lubrication Part I: Study on the transition between EHL and thin film lubrication using a relative optical interference intensity technique, *Wear*, **194**: 107–115.
12. Qingwen, Q., Yahong, H., Jun, Z. 1998. An adsorbent layer model for thin film lubrication, *Wear*, **221**: 9–14.
13. Qingwen, Q., Mei, W., Shan, C., Fusheng, Y. 2001. Velocity analysis for layered viscosity model under thin film lubrication, *Tribology International,* **34**: 517–521.
14. Qu, Q. W., Meng, L. X., Chai, S., Niu, S. T. 2001. A new model for thin-film lubrication, *Proceedings of Institution Mechanical Engineers, Part J: Journal of Engineering Tribology*, **215**: 103–107.
15. Qingwen, Q., Mei, W., Lihua, W., Shan, C. 2002. A continuous viscosity model for thin film lubrication, *Tribology International*, **35**: 459–465.
16. Zhang, C. 2005. Research on thin film lubrication: State of the art, *Tribology International*, **38**: 443–448.
17. Liran, M. A., Jianbin, L. 2016. Thin film lubrication in the past 20 years, *Friction*, **4** (**4**): 280–302.
18. Ono, K. 2020. Modified Reynolds equations for thin film lubrication analysis with high viscosity surface layers on both solid surfaces and analysis of micro-tapered bearing, *Tribology International*, **151**: 106515.
19. Harnoy, A. 2003. *Bearing- design in machinery: Engineering tribology and lubrication*. Boca Raton, FL: CRC Press.
20. Wen, S., Huang, P. 2018. *Principles of tribology*, 2nd Edition, Hoboken, NJ: John Wiley & Sons Inc.

Chapter 17

Tribological study to evaluate the performance of viscosity index improver in lubricated contacts

Yatika Gori, Chandra Kishore, Pravin P. Patil, and M. R. Tyagi

17.1 INTRODUCTION

For optimal performance across varying temperatures, most engine and transmission oils incorporate viscosity index improvers (VIIs) or viscosity modifiers (VMs) [1–3]. VIIs are oil-soluble polymer additives that influence the oil's viscosity. The viscosity index, denoted by VI, is a number that indicates the oil's resistance to viscosity change with temperature [4–7]. VI improvers play a crucial role by providing a greater boost to the oil's viscosity at high temperatures compared to low temperatures. This targeted action leads to an enhanced VI, as explored by many researchers [8–12]. Lubricants play a critical role in the smooth operation and longevity of gearboxes. Their performance directly impacts factors like wear, friction, and thermal management, ultimately influencing the overall efficiency and lifespan of the gearbox system [13–18]. Understanding the lubrication mechanisms within a gearbox is crucial. This involves analyzing film thickness, contact pressures, and the interaction between lubricant and gear surfaces [19–23]. Polyisobutylene (PIB) is a common thickener in gear oils. Studying its tribological properties, like friction coefficient and anti-wear behavior, helps assess its effectiveness under varying loads and speeds [24–27]. Temperature plays a significant role in lubricant performance. Analyzing factors like load, rotational speed, and convection heat transfer is essential for understanding thermal behavior within the gearbox [28–30].

VI (Viscosity index) improvers are additives that enhance the oil's viscosity at high temperatures, maintaining a thicker lubricating film. However, these additives can become thinner (shear thinning) under high shear stress, leading to a hydrodynamic and elastohydrodynamic film (EHD film) that's actually less thick than expected based on their low-shear viscosity [31–34]. While VI improvers primarily target the bulk properties of the oil blend, some can also form boundary lubrication films. Studies by Smeeth et al. [35] investigated the film-forming behavior of various VI additives in mineral

DOI: 10.1201/9781003535447-17

oil. Interestingly, some additives generated thicker films than anticipated under low-speed and thin-film conditions. This effect occurred for films below 20 nm and seemed to be caused by highly viscous layers (around 40 times more viscous than the bulk solution) forming on the solid surfaces, likely due to rolling speed. This unique film-forming behavior was observed only with VI improvers containing polar dispersant groups. Researchers believe these polar groups enable the polymer to adhere (adsorb) onto steel surfaces, creating layers with significantly higher viscosity, roughly equivalent to the diameter of a single polymer coil. Notably, the study by Smeeth referenced two polymers (PMA and PMA-O) that exhibited boundary film formation, but it didn't explicitly mention their polar dispersant functionality [35]. Additionally, Smeeth demonstrated that this boundary film formation significantly reduced friction and wear during low-speed, rubbing conditions [35].

A key challenge in utilizing boundary film-forming polymer additives lies in their efficacy on practical engineering components with inherent surface roughness. Real-world contact between components often occurs at the peaks and valleys of these rough surfaces, termed asperity junctions [36–41]. The question remains whether the adsorbed polymer film can effectively form and persist at these asperity contacts. While research has demonstrated the viscosity-enhancing benefits of VI improvers (viscosity index improvers, a type of boundary film-forming polymer additive), their non-Newtonian behavior presents a hurdle. This non-Newtonian nature makes it difficult to accurately predict lubricant film thickness using established theories like elastohydrodynamic lubrication (EHD), further complicating the understanding of friction behavior. Therefore, a crucial area of investigation lies in comprehending the tribological performance (friction, wear, and lubrication) of VI improvers, particularly regarding their impact on lubricant viscosity and its potential for friction reduction and energy conservation [42–48].

This knowledge enables the selection and optimization of lubricants for specific applications, leading to reduced wear and tear, improved fuel efficiency, extended gearbox life, and enhanced overall system performance. Future research directions could involve developing new lubricant formulations with superior tribological properties and exploring advanced modeling techniques that incorporate real-time sensor data for more accurate performance predictions [49–52].

17.2 EXPERIMENTAL TECHNIQUE NON-CONFORMAL CONTACTS

In the present studies, the experimental conditions were designed such that in non-conformal point and line contact, we get boundary /mixed lubrication regimes.

17.2.1 Test lubricants

This experiment investigated the effects of viscosity index (VI) improvers on commercial paraffinic mineral base oil (N150). The base oil's properties are listed in Table 17.1. Polyisobutylene and polymethylacrylate were chosen as VI improvers and added at concentrations of 5% and 6% by weight, respectively. The resulting solutions' viscosities are also presented in Table 17.1. Since data for the polymer solutions' pressure viscosity coefficient wasn't available, the base oil's value from the *CRC Handbook of Lubrication* was assumed. At the applied concentrations, the pressure viscosity coefficient of the solutions likely doesn't deviate significantly from the base oil. Table 17.1 includes a designation system for the used oil samples.

17.2.2 Point contact (four-ball geometry)

The testing apparatus operates under the principle of a lubricated point contact. Three stationary steel balls are immersed in the test oil sample. A fourth steel ball rotates against these stationary balls under controlled conditions of load, speed, and temperature. This rolling contact generates friction between the balls (Figure 17.1). The resulting tangential force acting on the stationary balls is measured by a load cell when the arm of a linear variable displacement transducer (LVDT) makes contact with it. The test allows for the adjustment of input parameters such as normal load, speed,

Table 17.1 Viscometric properties of oil and VII

Sample code	*Description*	*Viscosity in cst*		*Viscosity index*	*α [GPa⁻¹]*
		40°C	*100°C*		
N1	N150	33.3	5.9	100	28.4 E-9
N5	95% N150 + 5% PIB	68.67	10.86	148	
P6	94% N150 + 6% PMA	60.99	10.15	154	

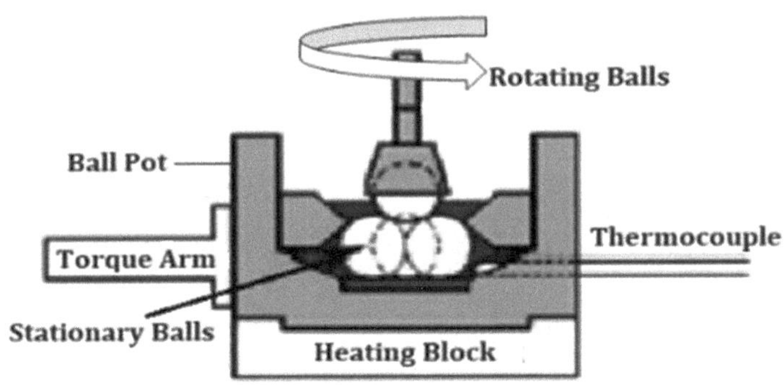

Figure 17.1 Working principle of four-ball tester.

and temperature. Correspondingly, output parameters like friction force, vertical displacement, and metal contact can be continuously monitored and recorded.

17.2.2.1 Test conditions in non-conformal contacts

This investigation utilized EN31 steel balls with a diameter of 12.7 mm. To ensure consistency, fresh balls were employed for each test. Prior to testing, the balls underwent a meticulous cleaning process. This involved oven drying and immersion in toluene for optimal dryness. The tribotests were conducted under a constant load of 50 N. Considering the 8 mm diameter of the balls, this load translates to a maximum Hertzian contact pressure of 2 GPA. The experiment commenced with a rotational speed of 10 rpm, followed by incremental increases to 25, 100, 250, 400, and finally 620 rpm. These varying speeds correspond to a sliding velocity range of 0.004–0.26 m/s. Throughout the tests, the bulk oil temperature within the bath was meticulously maintained at 30°C. A dedicated software suite, Winducom LabVIEW connected to the test rig, facilitated the monitoring and recording of friction data along with metal contact points. To ensure result accuracy, each test was repeated until consistent values were obtained. Upon test completion, the wear scars on the stationary balls were evaluated using a microscope for precise measurement.

17.3 CONFORMAL CONTACT CONDITION (JOURNAL BEARING)

The test apparatus aimed to achieve hydrodynamic lubrication within a specifically designed conformal journal bearing. As shown in Figure 17.2, the system comprised several key components:

- **Bearing assembly:** A highly polished journal (diameter 50 mm, length 1000 mm) made of case-hardened steel (60 HRC) formed the core. This journal was supported on both ends by bearings housed within a spindle housing, which was itself secured to a baseplate. The baseplate incorporated a central cavity to accommodate the test bearing and a lubricant sump.
- **Drive system:** An AC motor drove the journal via a 1:1 pulley and toothed belt arrangement, enabling speeds of up to 3000 rpm.
- **Loading system:** A pneumatic loading system with a maximum capacity of 2000 N applied the desired load on the journal.
- **Lubrication system:** This system delivered a flow of heated oil, consisting of an oil pump, reservoir, temperature controller, filter, tubes, and valves.

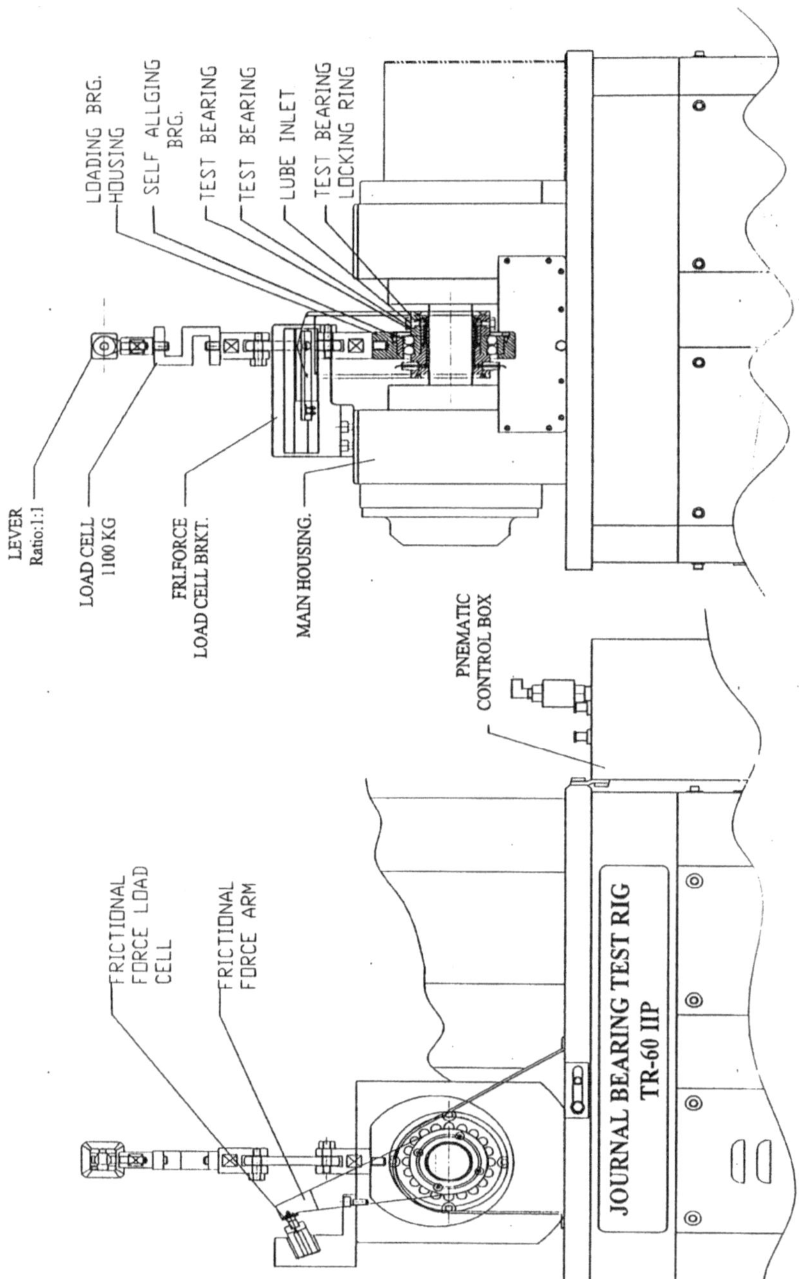

Figure 17.2 Assembly drawing—journal-bearing test rig.

Measurement systems: A load cell measured the friction force, transmitting the signal to a computer for recording and analysis. Film thickness was measured using a linear variable differential transformer (LVDT). Software processed the coefficient of friction (COF), speed, and load data before displaying it on the computer screen.

Lubricants: Lubricants for testing were chosen from Table 17.1, containing either polyisobutylene (PIB) or poly(methylmethacrylate) (PMA) additives.

17.3.1 Test conditions for conformal contact

The tribological performance of the conformal contact configuration was evaluated under controlled conditions. A constant load of 50 N was applied throughout the tests. The sliding speed was progressively increased from 0.004 m/s to 0.3 m/s in increments of 5 rpm every 5 min, corresponding to a speed range of 5 rpm to 115 rpm. A bulk oil temperature of 30°C was maintained in the oil bath for consistency. Prior to testing, the system was meticulously balanced to ensure a zero coefficient of friction when the shaft was stationary. The lubricant supply was then activated for 10 min, allowing the lubricant to fully circulate within the bearing. Any entrapped air bubbles were purged through a dedicated return path. At each designated speed, the coefficient of friction was continuously monitored. Once a stable, periodic oscillation pattern emerged, the average value over a 5-min period was recorded as the coefficient of friction for that specific speed. This timeframe was chosen based on observations, deemed sufficient to capture representative data at each velocity point. Friction and film thickness data were acquired and logged using Winducom LabVIEW software interfaced with the test rig. Each test was repeated until consistent results were achieved. The test materials consisted of a hardened AISI 1020 steel shaft counterposed against an SAE 660 alloy bronze bushing. The bushing offered a high-quality, smooth machine finish, known for its exceptional impact, wear, and corrosion resistance.

17.4 RESULTS AND ANALYSIS

This study investigated the effectiveness of VI improvers as friction modifiers in lubricated contacts. The results offer valuable insights into how VI improvers enhance the anti-wear and friction properties of low-viscosity mineral base oils. The influence of viscosity and speed on friction and wear behavior can be explored by comparing the performance of different VI improver concentrations in the base oil under light load conditions. Non-conformal and conformal contact configurations were studied to simulate real-world scenarios. Non-conformal point and line contacts promote boundary and mixed lubrication regimes, while the conformal journal-bearing contact represents hydrodynamic lubrication.

The results in Figure 17.3 demonstrate an inverse relationship between the percentage of metal contact and sliding speed. As the speed increased from 0 m/s to 0.28 m/s, the percentage of metal contact decreased for all lubricant formulations (from 90% to around 50%). This suggests improved lubrication at higher speeds, potentially due to the formation of a more effective hydrodynamic film. However, the differences in metal contact between lubricant formulations at various speeds were relatively small. Interestingly, at the highest sliding speed, the base oil exhibited the lowest metal contact, while the 5% PIB blend showed the 50% metal contact. This observation, along with the wear scar diameter (WSD) data, suggests a complex interplay between wear behavior and lubrication conditions. Further analysis of these figures may reveal specific mechanisms by which different lubricant formulations influence wear at varying speeds.

Figure 17.4 also reveals an inverse relationship for PMA VI, between percentage metal contact and sliding speed. As the speed increases from 0 m/s to 0.28 m/s, the percentage of metal contact significantly decreases (from 82% to 54%). This trend suggests improved lubrication at higher speeds. It's important to note that the variations in metal contact percentage between different concentrations at each speed are relatively small. At the maximum sliding speed (0.24 m/s), the 6% additive concentration exhibits the lowest metal contact (54%). Wear scar diameter (WSD) measurements will be used

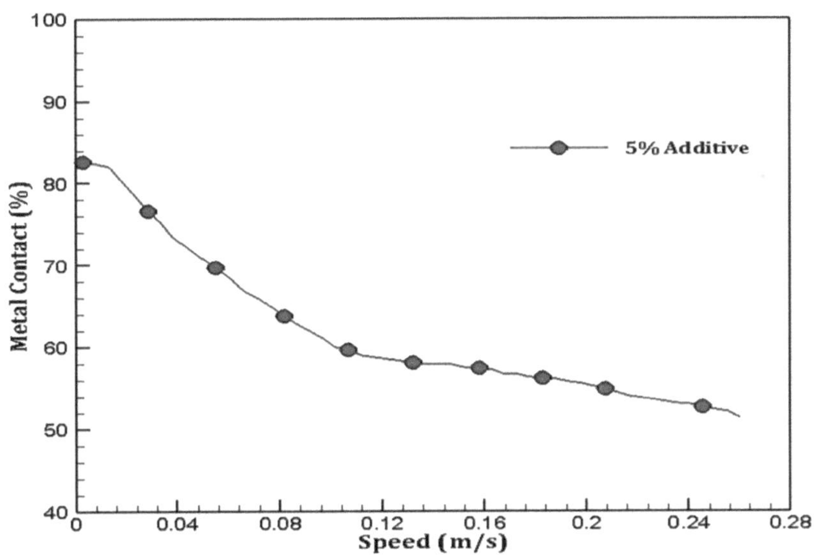

Figure 17.3 Metal contact behavior—PIB.

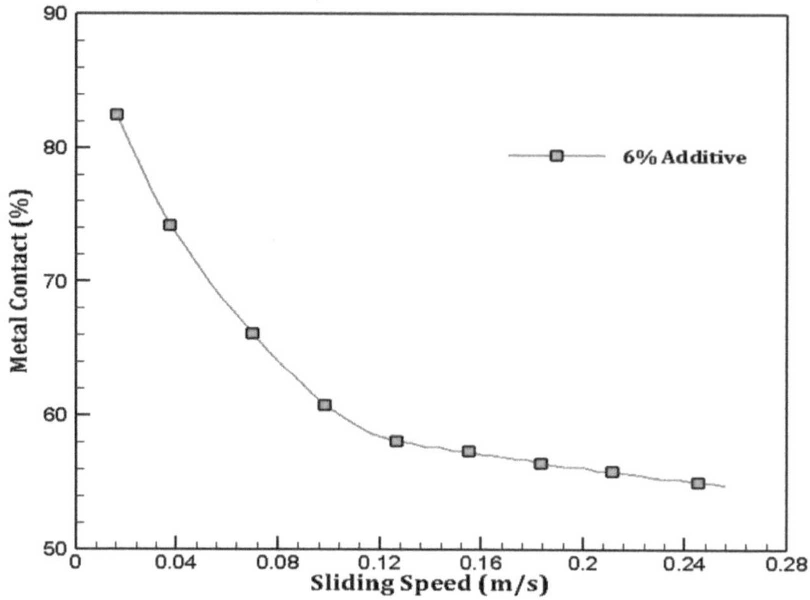

Figure 17.4 Metal contact behavior—PMA.

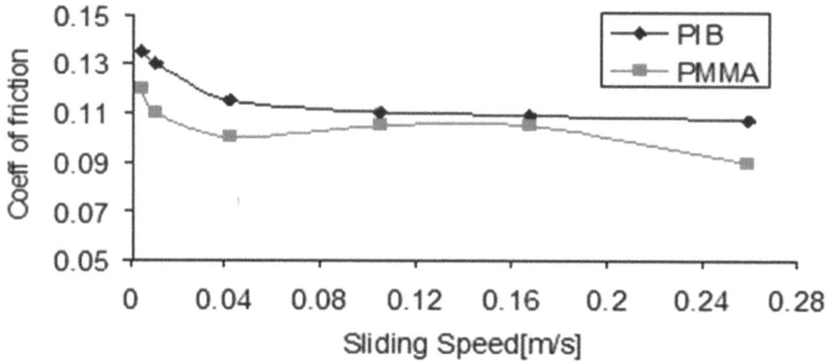

Figure 17.5 Friction behavior of similar viscosity oil of different additives.

to quantify wear reduction and correlate it with the observed metal contact percentages.

Figure 17.5 compares the frictional behavior of PIB and PMA blends with similar viscosities. PMA consistently exhibits lower coefficients of friction (COFs) across all speeds tested. This suggests that PMA offers less resistance to shear forces within the hydrodynamic lubrication regime, likely due to its inherent material properties. Similarly, Figure 17.5 explores the performance of PIB and PMA blends with matching viscosities. Here, PMA demonstrates a lower COF (0.10) and reduced wear throughout the tested speed range (0–0.28 m/s). This improved lubrication performance of PMA is likely attributed to its superior boundary lubrication characteristics compared to PIB. Notably, PIB exhibits the highest COF (0.13) under the same conditions. Interestingly, both figures show a potential trend of minimized COF at specific additive concentrations. This phenomenon could be linked to the non-Newtonian behavior of these materials under applied shear stress. The observed minimum shear resistance might occur at these optimal concentrations, suggesting further investigation to fully understand the underlying mechanisms.

17.5 CONCLUSIONS

This study investigated the effectiveness of VI improvers (VIIs) as friction modifiers in lubricated contacts under light load conditions. The research explored how varying VII percentages in a mineral base oil impacted friction and wear behavior across conformal and counter-formal contact geometries. The results demonstrate the potential of VIIs as lubricant additives, particularly PMA, which exhibited superior performance compared to PIB in reducing friction and wear across both boundary/mixed and hydrodynamic lubrication regimes. This suggests that VIIs, especially PMA, can enhance the tribological properties of low-viscosity mineral base oils. The experimental findings provide valuable insights into the influence of VII type, concentration, and operating conditions (speed) on friction and wear behavior. In future, further studies could explore the effect of VIIs under heavier load conditions and delve deeper into the mechanisms by which VIIs influence lubrication at different concentrations of PIB and PMA by weight.

REFERENCE

1. T. W. Selby (1958), The Non-Newtonian Characteristics of Lubricating Oils.*ASLE Transactions*, Vol. 1, No. 1, pp. 68–81.https://doi.org/10.1080/05698195808972315
2. P. A. Asseff (1971), Some performance characteristics of hydrorefined lubricating oils. *Pet. Hydrocarbons*, Vol. 5, No. 4, pp. 168–170.
3. P. Patil, Y. Gori, A. Kumar, and M. R. Tyagi (2021), "Experimental analysis of tribological properties of polyisobutylene thickened oil in lubricated contacts," *Tribology International*, Vol. 159, No. 106983, pp. 1–7. https://doi.org/10.1016/j.triboint.2021.106983.
4. G. J. Johnston, R. Wayte, and H. A. Spikes (1991), "The measurement and study of very thin lubricant films in concentrated contact," *Tribology Transactions*, Vol. 34, pp. 187–194.

5. A. Dyson and A. R. Wilson (1965–1966), "Film thickness in elastohydrodynamic lubrication by silicone fluids," *The Proceedings of the Institution of Mechanical Engineers, London*, Vol. 180 (3K), pp. 97–112.
6. J. D. Novak and W. O. Winer (1969, July), "The effects of pressure on the non-Newtonian behavior of polymer blended petroleum oils," *Journal of Lubrication Technology*, Vol. 91, pp. 459–463.
7. D. L. Walker, D. M. Sanborn, and W. O. Winer (1975, July), "Molecular degradation of lubricants in sliding elastohydrodynamic contacts," *Journal of Lubrication Technology*, Vol. 97, pp. 390–397.
8. A. Kumar, A. Dwivedi, H. Jaiswal, and P. P. Patil (2015), "Material-based vibration characteristic analysis of heavy vehicle transmission gearbox casing using finite element analysis," Vol. 308, pp. 527–533. https://doi.org/10.1007/978-81-322-2012-1_56.
9. F. Ahmad, V. Tomer, A. Kumar, and P. P. Patil (2016), "FEA simulation based thermo-mechanical analysis of tractor exhaust manifold," pp 173–181. https://doi.org/10.1007/978-81-322-2740-3_18 ISBN: 978-81-322-2740-3.
10. H. Singh and I. B. Gulati (1987), Influence of base oil refining on the performance of viscosity index improvers. *Wear*, Vol. 118, pp. 33–56.
11. P. P. Patil and A. Kumar (2017), "Dynamic structural and thermal characteristics analysis of oil lubricated multi speed transmission gearbox: Variation of load, rotational speed and convection heat transfer," *ISTM*, Vol. 41, pp. 281–291. https://doi.org/10.1007/s40997-016-0063-z.
12. J. M. Palacios and M. L. Bajon (1983), "Effective viscosity of oils containing VI improver and its relation to wear," *Tribology International*, Vol. 16, pp. 97–31.
13. J. Briant, J. Denis, and D. Parc (1989), *Rheological Properties of Lubricants*. Institut Francais du Petrole, Paris.
14. N. Nemes, M. Kovacs and I. Kantor (1986), "Additives for lubricants and operational fluids,"5th International Colloquium Esslingen, West Germany , January 14–16.
15. A. Kumar and P. P. Patil, 2015. "FEA based dynamic vibration response analysis of loose mounted heavy vehicle transmission gearbox casing," *IJAMMC*, Vol. 5, No. 1, pp. 136–141. http://dx.doi.org/10.11127/ijammc.2015.03.02.
16. A. Kumar and P. P Patil (2014), "Dynamic vibration analysis of heavy vehicle truck transmission gearbox housing using FEA," *JESTR*, Vol. 7, No. 4, pp. 66–72.
17. R. G. A. Eckert and D. F. Covey (1986), "Additives for lubricants and operational fluids," 5th International Colloquium Esslingen, West Germany, January 14–16.
18. G. T. Spiess, J. Eo Johnston, and G. Verstrate (1986), "Additives for lubricants and operational fluids," 5th International Colloquium Esslingen, West Germany, January 14–16.
19. V. P. Singh, A. Kumar, C. S. Meena, and G. Dwivedi (eds.) (2024), *Energy Efficient Vehicles: Technologies and Challenges* (1st ed.). CRC Press. https://doi.org/10.1201/9781003464556.
20. V. P. Singh, A. Kumar, C. S. Meena, and G. Dwivedi (2024), "Introduction to energy-efficient vehicles for sustainable transportion," in *Energy Efficient Vehicles: Technologies and Challenges* (pp. 1–18). CRC Press, Boca Raton, FL.

21. N. Nemes, M. Kovacs, and I. Kantor (1986), "Additives for lubricants and operational fluids," 5th International Colloquium Esslingen, West Germany, January 14–16.
22. A. Kumar (2018), "Low to high speed transient structural and thermal temperature measurement of oil-lubricated multi-speed heavy vehicle transmission gearbox system based on FEA," http://doi.org/10.4018/978-1-5225-3722-9.ch001.
23. S. Bhoi, A. Prasad, A. Kumar, R. B. Sarkar, B. Mahto, C. S. Meena, and C. Pandey (2022), "Experimental study to evaluate the wear performance of UHMWPE and XLPE material for orthopedics application," *Bioengineering*, Vol. 9, p. 676. https://doi.org/10.3390/bioengineering9110676.
24. P. Neudorel (1986), "Additives for lubricants and operational fluids," 5th International Colloquium, Esslingen, West Germany, January 14–16.
25. R. K. Upadhyay, V. P. Singh, and A. Kumar (2024), "Sustainable transportion: Policy, planning and implementation," *Review of Economics and Political Science*, Vol. 3, No. 2, pp. 1–22.
26. H. H. Abou El Naga and M. A. Abdel Ghany (1987), Chemical structure bases for oxidation stability of neutral base oils. *ASLE Transactions*, Vol. 30, No. 2, pp. 261–268.
27. V. P. Singh and A. Kumar (2024), "Techno-economic and future aspects of the HEV-EV-FCV," in *Energy Efficient Vehicles: Technologies and Challenges* (pp. 1–18). CRC Press, Boca Raton, FL, pp. 1–20.
28. S. Rana and A. Kumar (2018), "FEA based design and thermal contact conductance analysis of steel and Al rough surfaces," *IJAER*, Vol. 13, No. 16 (2019), pp. 12715–12724.
29. A. Kumar, Y. Gori, and P. P. Patil (2020), "Finite element analysis based thermo-mechanical performance study of heavy vehicle medium duty transmission gearbox," http:// doi.org/10.4018/978-1-7998-4939-1.
30. Institute of Petroleum (1982), *IP Standards for Petroleum and Its Products, Parts I and 2, Institute of Petroleum, Applied Science Publishers UK*. Heyden & Son.
31. A. K. Paul, A. Prasad, and A. Kumar (2022), "Review on artificial neural network and its application in the field of engineering," *Journal of Mechanical Engineering: PRAKASH*, Vol. 1, No. 1, pp. 53–61. https://doi.org/10.56697/JMEP.2022.1107.
32. A. G. Pandey, A. Kumar, P. P. Patil, and V. Kumar (2022), "FEA based contact analysis of railway track and wheel," *AIP Conference Proceedings*, Vol. 2481, pp. 0500101–050010-8. https://doi.org/10.1063/5.0103726.
33. A. Shukla and A. Kumar (2021), "Design and thermal characteristic analysis of gearbox system based on finite element analysis (FEA)," https://doi.org/10.1063/5.0036235.
34. A. Kumar, R. Jain, and P. P. Patil (2016), "Dynamic analysis of heavy vehicle medium duty drive shaft using conventional and composite material," Vol. 149, pp. 1–11. http://doi.org/10.1088/1757-899X/149/1/012156.
35. M. Smeeth et al (1996), "Performance of viscosity index improvers in lubricated contacts," *Langmuir*, Vol. 12, No. 19, pp. 4594–4598.
36. A. Kumar, H. Jaiswal, and P. P. Patil (2017), "Parametric optimization of vibrating heavy vehicle medium duty transmission gearbox housing using response surface method," *International Journal of Vehicle Structures and Systems*, Vol. 9, No. 3, pp. 149–153. http://doi.org/10.4273/ijvss.9.3.04.

37. A. Kumar and P. P. Patil (2016), "FEA simulation based performance study of multi-speed transmission gearbox," *IJMMME*, Vol. 6, No. 1, pp. 57–67. http://doi.org/10.4018/IJMMME.2016010103.
38. G. Brands, "The chemical structure of petroleum fractions: I. Analysis by infrared spectroscopy," *Brennstoff Chem*, Vol. 37, pp. 263–267 (1956); Erdoel Kohle, 11, 700–702 (1958).
39. A. Kumar, H. Jaiswal, R. Jain, P. P. Patil (2014), "Free vibration and material mechanical properties influence based frequency and mode shape analysis of transmission gearbox casing," *Procedia Engineering*, Vol. 97, pp. 1097–1106. http://doi.org/10.1016/j.proeng.2014.12.388.
40. A. Kumar, Y. Gori, C. S. Meena, V. P. Singh, and V. K. Sharma (2022), "Design and analysis of heavy vehicle medium duty transmission gearbox system," pp. 67–86. http://doi.org/10.1201/9781003360001-4.
41. I. V. Kragelsky, M. N. Dobychin, and V. S. Kombalov (1982), *Friction and Wear.* Pergamon Press, Elmsford, p. 458.
42. A. Kumar, S. Rana, Y. Gori, and N. K. Sharma (2021), "Thermal contact conductance prediction using FEM based computational techniques," pp. 183–220. http://doi.org/10.1201/9781003202233-13.
43. A. Kumar and P. P. Patil (2017), "Free vibration and connecting bolt constraint based FEA analysis of heavy vehicle medium duty transmission gearbox housing made from AISI 4130 alloy," pp. 133–146. http://dx.doi.org/10.1016/B978-0-12-810998-4.00007-7.
44. N. S. White (1956), "Small oil-free bearings," *Journal of Research of the National Bureau of Standards*, Vol. 57, pp. 185–189.
45. A. Kumar, Y. Gori, N. Dutt, Y. K. Singla, and A. Maurya (eds.) (2021), "Advanced computational methods in mechanical and materials engineering." https://doi.org/10.1201/9781003202233.
46. K. K. Gajrani, A. Prasad, and A. Kumar (eds.) (2022), "Advances in sustainable machining and manufacturing processes," https://doi.org/10.1201/9781003284574.
47. G. V. Vinogradov, G. M. Bartenev, A. I. Elkin, and V. K. Mikhailov (1970), "Effect of temperature on friction and adhesion of crystalline polymers," *Wear*, Vol. 16, pp. 358–368.
48. K. C. Ludema and D. Tabor (1966), "The friction and viscoelastic properties of polymeric solids," *Wear*, Vol. 9, pp. 329–348.
49. S. Singh, S. K. Behura, A. Kumar, and K. Verma (eds.) (2022), "Nanomanufacturing and nanomaterials design: Principles and applications," https://doi.org/10.1201/9781003220602.
50. V. K. Sharma, A. Kumar, M. Gupta, V. Kumar, D. K. Sharma, and S. Sharma (eds) (2022), "Additive manufacturing in industry 4.0: Methods, techniques, modeling, and nano aspects," https://doi.org/10.1201/9781003360001.
51. R. Kumar, A. Prasad, and A. Kumar (eds.) (2023), "Sustainable smart manufacturing processes in industry 4.0," https://doi.org/10.1201/9781003436072.
52. A. Kumar, V. P. Singh, C. S. Meena, and N. Dutt (eds.) (2023), "Thermal energy systems: Design, computational techniques and applications," https://doi.org/10.1201/9781003395768.

Index

Kolibri 360 GmbH
Ferdinand-Jühlke-Straße 7
99095 Erfurt, Deutschland
produktsicherheit@kolibri360.de